Lecture Notes in Computer Science   8499

Commenced Publication in 1973
Founding and Former Series Editors:
Gerhard Goos, Juris Hartmanis, and Jan van Leeuwen

T0214190

Bruno Sericola   Miklós Telek
Gábor Horváth (Eds.)

Analytical and Stochastic Modelling Techniques and Applications

21st International Conference, ASMTA 2014
Budapest, Hungary, June 30 – July 2, 2014
Proceedings

 Springer

Volume Editors

Bruno Sericola
Inria
Campus de Beaulieu, 35042 Rennes Cedex, France
E-mail: bruno.sericola@inria.fr

Miklós Telek
Gábor Horváth
BME-HIT, Magyar tudósok krt 2, 1117 Budapest, Hungary
E-mail: {telek, ghorvath}@hit.bme.hu

ISSN 0302-9743 e-ISSN 1611-3349
ISBN 978-3-319-08218-9 e-ISBN 978-3-319-08219-6
DOI 10.1007/978-3-319-08219-6
Springer Cham Heidelberg New York Dordrecht London

Library of Congress Control Number: 2014941104

LNCS Sublibrary: SL 2 – Programming and Software Engineering

Typesetting: Camera-ready by author, data conversion by Scientific Publishing Services, Chennai, India

Printed on acid-free paper

Springer is part of Springer Science+Business Media (www.springer.com)

Preface

It is our pleasure to present the proceedings of the 21st International Conference on Analytical and Stochastic Modelling Techniques and Applications (ASMTA), held from June 30 to July 2, 2014, in Budapest, Hungary. The ASMTA conference is a main forum for bringing together researchers of academia and industry to discuss the latest developments in analytical, numerical, and simulation algorithms for stochastic systems, including Markov processes, queueing networks, stochastic Petri nets, process algebras, game theory, etc.

We had submissions from most European contries including Belgium, Denmark, Finland, France, Germany, Hungary, Italy, Latvia, Poland, Spain, The Netherlands, and the UK, but also received contributions from Australia, Azerbaijan, Belarus, China, Japan, Kazakhstan, Korea, and the USA. The international Program Committee reviewed these submissions and decided to accept 18 high-quality papers. The selection procedure was based on at least three, and on average of 3.7 reviews, followed by an active discussion phase. We would like to thank the authors for the high-quality contributions and the Program Committee members for their excellent work and for the time and effort devoted to this conference.

We are grateful for our sponsors, the IEEE UK-RI Computer Chapter and the ECMS - The European Council for Modelling and Simulation. Finally, we would like to thank the EasyChair team and Springer for the editorial support of this conference series.

April 2014

Bruno Sericola
Miklós Telek
Gábor Horváth

Organization

Program Committee

Sergey Andreev	Tampere University of Technology, Finland
Jonatha Anselmi	BCAM - Basque Center for Applied Mathematics, Spain
Christel Baier	Technical University of Dresden, Germany
Simonetta Balsamo	Università Ca' Foscari di Venezia, Italy
Hind Castel-Taleb	Telecom SudPAris, France
Antonis Economou	University of Athens, Greece
Dieter Fiems	Ghent University, Belgium
Jean-Michel Fourneau	Université de Versailles St. Quentin, France
Bo Friis Nielsen	Technical University of Denmark, Denmark
Marco Gribaudo	Politecnico di Milano, Italy
Yezekael Hayel	LIA/University of Avignon, France
Andras Horvath	University of Turin, Italy
Gábor Horváth	Budapest University of Technology and Economics, Hungary
Helen Karatza	Aristotle University of Thessaloniki, Greece
William Knottenbelt	Imperial College London, UK
Lasse Leskelä	Helsinki University of Technology, Finland
Renato Lo Cigno	University of Trento, Italy
Andrea Marin	University of Venice, Italy
Yoni Nazarathy	University of Queensland, Australia
Jose Nino-Mora	Carlos III University of Madrid, Spain
Tuan Phung-Duc	Department of Mathematical and Computing Sciences
Balakrishna J. Prabhu	LAAS-CNRS, France
Marie-Ange Remiche	University of Namur, Belgium
Jacques Resing	Eindhoven University of Technology, The Netherlands
Leonardo Rojas-Nandayapa	The University of Queensland, Australia
Marco Scarpa	University of Messina, Italy
Bruno Sericola	Inria, France
Janos Sztrik	Eszterhazy Karoly College, Debrecen University, Hungary

Miklos Telek	Budapest University of Technology and Economics, Hungary
Nigel Thomas	Newcastle University, UK
Dietmar Tutsch	University of Wuppertal, Germany
Kurt Tutschku	Blekinge Institute of Technology (BTH), Sweden
Benny Van Houdt	University of Antwerp, Belgium
Neil Walton	University of Amsterdam, The Netherlands
Sabine Wittevrongel	Ghent University, Belgium
Verena Wolf	Saarland University, Germany
Katinka Wolter	Freie Universitaet zu Berlin, Germany
Alexander Zeifman	Vologda State Pedagogical University, Russia

Additional Reviewers

Angius, Alessio
Asanjarani, Azam
Gibbons, Daniel
Habachi, Oussama

Krüger, Thilo
Kuhn, Julia
Longo, Francesco
Stojic, Ivan

Table of Contents

Stability Analysis of Some Networks
with Interacting Servers

Rosario Delgado[1,*] and Evsey Morozov[2]

[1] Departament de Matemàtiques. Universitat Autònoma de Barcelona. Edifici C-
Campus de la UAB. 08193 Bellaterra (Cerdanyola del Vallès)- Barcelona, Spain
delgado@mat.uab.cat
Supported by the project MEC ref. MTM2012-33937 and ERDF (European Regional
Development Found) "A way to build Europe"
[2] Institute of Applied Mathematical Research, Russian Academy of Sciences and
Petrozavodsk State University, Russia
emorozov@karelia.ru
Supported by the Program of strategic development of Petrozavodsk State University
for 2012-2016

Abstract. In this work, the fluid limit approach is applied to find stability conditions of two models of queueing networks with interacting
servers. We first consider a two-station queueing model with two customer classes in which customers that are awaiting service at any queue
can move to the other station, whenever it is free, to be served there
immediately. Then we consider a cascade-type three-station system in
which the third station, whenever it is free, can assist the other two stations. In both models, each station is fed by a renewal input with general
i.i.d. inter-arrival times and general i.i.d. service times.

Keywords: cascade networks, fluid limit approach, interacting servers,
stability, X-model.

1 Introduction

In the present paper, we study two variants of the cascade networks considered
in [12]. First, we consider a queueing system consisting of two basic customer
classes, 1 and 2, and two servers. Class-j customers are primarily assigned to
server j, $j = 1, 2$. However, servers are *cross-trained* so that, when become free,
they can serve customers from the other class (that is, from the queue of other
server). Such a model, which is called X-*model* in [19], differs from the two-
station cascade network considered in [12] in which in that paper, the 1st server,
being free, cannot support the 2nd one. Motivation for the study of these models
can be found in [19]. Secondly, we study a *generalized* cascade model consisting
of three stations with three basic customer classes, in which the 3rd station
assists the 2nd station which, in turn, assists the 1st one, as was the case of the
three-station cascade network introduced in [12], but with the novelty that now

---

* Corresponding author.

B. Sericola, M. Telek, and G. Horváth (Eds.): ASMTA 2014, LNCS 8499, pp. 1–15, 2014.

the 3rd station can also assist station 1. For convenience, we call it *tree-cascade system*.

In both models each station has an unlimited-capacity buffer for awaiting customers. Customers of each basic customer class arrive following an independent general renewal input, and have i.i.d. general service times when served at the same station. Interarrival times and service times are possibly different for different customer classes. Service discipline is work-conserving and FCFS within each class. Whenever a station becomes empty while customers are awaiting service in other station, one customer jumps to be served there if it is allowed in the given model. More specifically, a customer can jump from each queue to other station in the X-model, while only jumps from station i to station $j > i$ ($i = 1, 2; j = 2, 3$) are allowed in the tree-cascade system.

There are closely related systems with *flexible* servers, where a server may transfer some service capacity to accommodate workload accumulated in another server, [3,13,16,20,24] and also other with *cross-trained* servers, in which some servers can handle a reduced set of customers types, whereas others accept all types [1,2,21,22]. These models describe a variety of real-life systems, including service centers, production systems, computer networks with rescheduling of jobs, and parallel computing systems where processors have overlapping capabilities [5]. There are also manufacturing applications in which machines may have differing primary functions and some overlapping secondary ones. Such a model is also called *N-model* with static priority rule, see [23] and references therein. The most related papers focus on an optimal server allocation to minimize a cost function, and much less attention is devoted to stability analysis.

Under heavy-traffic regime, the boundary of the stability region is typically defined based on the resource pooling (RP) or complete resource pooling (CRP) assumption when the input rate in a pool of servers must be less than (RP) or equal to (CRP) the maximal service rate of the pool, [4,6,15,17,20]. Sometimes such an assumption is formulated as the existence of a specific solution of a linear programming problem, [4,23]. However concrete service rates are not specified there. At the same time, it is not a trivial problem to delimit the stability region in terms of predefined parameters, see [14,18].

Finally, in [12] stability analysis of a N-station cascade networks in which each station only can support the previous one, is carried out by using the fluid limit approach instead of standard Foster's type arguments. It is worth mentioning that in this model the service discipline is in fact state-dependent, and it can be a source of problems to apply this methodology ([7]).

Our aim is to find stability conditions for the X-model and the tree-cascade system by using *the fluid limit approach* as in [12]. Indeed, for each model we find conditions implying the *positive Harris recurrence* of the underlying Markov process describing the network dynamics. Following [9], we first establish the stability of the fluid limit model associated to the queueing network, which means that the fluid limit of the queue-size process reaches zero in a finite time interval and stays there.

The organization of the paper is as follows. In Section 2 we give main notations. In Section 3 we describe the X-model and the equations that govern the
processes associated to the network. Section 4 is devoted to stability analysis
of such a queueing model. In particular, Section 4.1 introduces the associated
fluid limit model, while in Section 4.2 we prove the stability result (Theorem 1).
Finally, in Section 5 we analyze the tree-cascade system and establish corresponding stability conditions (Theorem 2).

2 Notations

Vector (in)equalities are interpreted component-wise. For any integer $d \geq 1$, let

$$\mathbb{R}_+^d = \{\, v \in \mathbb{R}^d \,:\, v \geq 0 \,\}, \quad \mathbb{Z}_+^d = \{\, v = (v_1, \ldots, v_d) \in \mathbb{R}^d \,:\, v_i \in \mathbb{Z}_+ \,\}.$$

For a vector $v = (v_1, \ldots, v_d)$, let $|v| = \sum_{i=1}^{d} |v_i|$. We say that a sequence of
vectors $\{v^n\}_{n \geq 1}$ converges to a vector v as $n \to \infty$ if $|v^n - v| \to 0$, and denote
it as $\lim_{n \to \infty} v^n = v$. (This convergence is equivalent to the component-wise convergence.) For $n \geq 1$, let $\phi^n \colon [0, \infty) \to \mathbb{R}^d$ be right continuous functions having
limits on the left on $(0, \infty)$, and let function $\phi \colon [0, \infty) \to \mathbb{R}^d$ be continuous. We
say that ϕ^n converges to ϕ as $n \to \infty$ *uniformly on compacts* (u.o.c.) if for any
$T \geq 0$,

$$\|\phi^n - \phi\|_T := \sup_{t \in [0,T]} |\phi^n(t) - \phi(t)| \to 0 \quad \text{as } n \to \infty,$$

and write it as $\lim_{n \to \infty} \phi^n = \phi$. If function ϕ is differentiable at a point $s \in (0, \infty)$
then s is a *regular* point of ϕ, and we denote the derivative by $\dot{\phi}(s)$.

Finally, by convention $\max \emptyset = 0$ and $=_{st}$ denotes stochastic equality.

3 The X-Model

We consider a X-model with two stations, 1 and 2, with infinite-capacity buffers
for the awaiting customers arrived from outside as shown in Figure 1. If a station
becomes free, then an awaiting customer in the buffer of other station (if any)
switches to the free station and starts service there immediately. We call class-j
exogenous customers who arrive at station $j = 1, 2$, and class-(i, j) customers
jumping from station i to station j; $i, j = 1, 2$. (For stability analysis, it does not
matter which customer makes the jump.) In what follows, we use index j (double
index i, j) to denote the quantities related to class-j (class-(i, j)) customers. Let
$\xi_j(\ell), \ell \geq 2$, be the i.i.d. inter-arrival times of class-j arriving after instant 0
$(j = 1, 2)$ and let $\eta_k(\ell), \ell \geq 2$, be the i.i.d. service times of class-k customers
finishing service after instant 0, $k = 1, 2, (1, 2), (2, 1)$. All sequences are assumed
to be mutually independent. We denote the generic elements of these sequences
by ξ_j and η_k, respectively, $j = 1, 2; k = 1, 2, (1, 2), (2, 1)$. The residual arrival

time $\xi_j(1)$ of the first class-j customer entering the system after instant 0 is independent of $\{\xi_j(\ell), \ell \geq 2\}$, $j = 1, 2$. Also the residual service time $\eta_k(1)$ of a class-k customer initially being served, if any, is independent of $\{\eta_k(\ell), \ell \geq 2\}$, and $\eta_k(1) =_{st} \eta_k$ if class k is initially empty, $k = 1, 2, (1, 2), (2, 1)$. Denote the arrival rate $\lambda_j = 1/\mathsf{E}\xi_j$ of class-j customers ($j = 1, 2$) and the service rate $\mu_k = 1/\mathsf{E}\eta_k$ of class-k customers ($k = 1, 2, (1, 2), (2, 1)$), and impose the following standard conditions [9] which in particular imply that λ_j, $\mu_k \in (0, \infty)$:

$$0 < \mathsf{E}\,\eta_k < \infty, \quad k = 1, 2, (1, 2), (2, 1), \tag{1}$$

$$\mathsf{E}\,\xi_j < \infty, \quad j = 1, 2, \tag{2}$$

$$\mathsf{P}(\xi_j \geq x) > 0, \quad j = 1, 2 \quad \text{for any } x \in [0, \infty). \tag{3}$$

Also we assume that inter-arrival times are *spread out*, that is, for some integers $s_j > 1$ and functions $f_j \geq 0$ with $\int_0^\infty f_j(y)\, dy > 0$:

$$\mathsf{P}\left(a \leq \sum_{\ell=2}^{s_j} \xi_j(\ell) \leq b \right) \geq \int_a^b f_j(y)\, dy, \quad \text{for any } 0 \leq a < b, \ j = 1, 2. \tag{4}$$

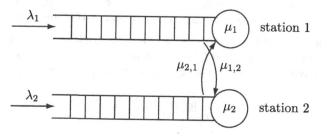

Fig. 1. The X-model

Service discipline is assumed to be *non-preemptive* and also *non-idling*, in which case any server can not be idle if there are waiting customers in *any* buffer.

We introduce the following primitive processes of the X-model: the exogenous arrival process $E = \{E(t) := (E_1(t), E_2(t)), t \geq 0\}$, with $E_j(t) := \max\{n \geq 1 : \sum_{\ell=1}^n \xi_j(\ell) \leq t\}$ the total number of class-j arrivals in interval $[0, t]$, $j = 1, 2$, and the process of served customers $S = \{S(t) := (S_1(t), S_2(t), S_{1,2}(t), S_{2,1}(t)), t \geq 0\}$, where the renewal process $S_k(t) := \max\{n \geq 1 : \sum_{\ell=1}^n \eta_k(\ell) \leq t\}$ is the total number of class-k customers served in $[0, t]$ if the server *devotes all time* to class $k = 1, 2, (1, 2), (2, 1)$. (By definition, $E_j(0) = S_k(0) = 0$.)

Now we also introduce the descriptive processes to measure the performance of the network. For any $t \geq 0$, let $D_k(t)$ be the number of class-k departures (from the network) in interval $[0, t]$, and let $Z_k(t)$ be the number of class-k customers that are being served at time t, so $Z_k(t) \in \{0, 1\}$, $k = 1, 2, (1, 2), (2, 1)$. Also let $T_k(t)$ be the total service time devoted to class-k customers in interval $[0, t]$, that is

$$T_k(t) = \int_0^t 1_{\{Z_k(u)=1\}}\, du.$$

For $j = 1, 2$, let $Y_j(t)$ be the idle time of server j in interval $[0, t]$ and let $Q_j(t)$ be the number of customers in buffer j at time t. Now we introduce the following processes

$$D = \{D(t) := (D_1(t), D_2(t), D_{1,2}(t), D_{2,1}(t)), t \geq 0\},$$
$$Z = \{Z(t) := (Z_1(t), Z_2(t), Z_{1,2}(t), Z_{2,1}(t)), t \geq 0\},$$
$$T = \{T(t) := (T_1(t), T_2(t), T_{1,2}(t), T_{2,1}(t)), t \geq 0\},$$
$$Q = \{Q(t) := (Q_1(t), Q_2(t)), t \geq 0\},$$
$$Y = \{Y(t) := (Y_1(t), Y_2(t)), t \geq 0\}.$$

Processes D, T and Y are nondecreasing and satisfy initial conditions $D(0) = T(0) = Y(0) = 0$, and it is assumed that $Z(0)$ and $Q(0)$ are independent and independent of all above given quantities. We also define the remaining time $U_j(t)$ at time t before next exogenous j-class arrival, $j = 1, 2$, and the remaining service time $V_k(t)$ for the class-k customer being served at time t, $k = 1, 2, (1, 2), (2, 1)$. We assume the processes U, V to be right-continuous, and define $V_k(t) = 0$ if $Z_k(t) = 0$. Note that $U_j(0) = \xi_j(1)$, while $V_k(0) = \eta_k(1)$ if $Z_k(0) = 1$. We introduce the following processes: $U = \{U(t) := (U_1(t), U_2(t)), t \geq 0\}$, $V = \{V(t) := (V_1(t), V_2(t), V_{1,2}(t), V_{2,1}(t)), t \geq 0\}$, and define the process

$$X = (Q, Z, U, V) := \{X(t) = \{(Q(t), Z(t), U(t), V(t)), t \geq 0\}$$

describing the dynamics of the network with the state space $\mathbb{X} = \mathbb{Z}_+^2 \times \{0, 1\}^4 \times \mathbb{R}_+^2 \times \mathbb{R}_+^4$. The process X is a piecewise-deterministic Markov process which satisfies Assumption 3.1 [11] and is a strong Markov process (p. 58, [9]).

Let $x \in \mathbb{X}$ denote the initial state of the network. We denote X as X^x (and analogously for the processes E, S, D, T, Y) if $X(0) = x$, being $X(0) = (Q(0), Z(0), U(0), V(0))$. The following *queueing network equations* hold for $t \geq 0$:

$$D_k(t) = S_k(T_k(t)), \quad k = 1, 2, (1, 2), (2, 1), \tag{5}$$
$$Q_1(t) = Q_1(0) + E_1(t) - (D_1(t) + Z_1(t) + D_{1,2}(t) + Z_{1,2}(t)),$$
$$Q_2(t) = Q_2(0) + E_2(t) - (D_2(t) + Z_2(t) + D_{2,1}(t) + Z_{2,1}(t)),$$
$$(T_1(t) + T_{2,1}(t)) + Y_1(t) = t, \quad (T_2(t) + T_{1,2}(t)) + Y_2(t) = t,$$

$$\int_0^\infty Q_j(t) \, dY_j(t) = 0, \quad j = 1, 2, \tag{6}$$

$$\int_0^\infty Q_1(t) \, dY_2(t) = 0, \quad \int_0^\infty Q_2(t) \, dY_1(t) = 0, \tag{7}$$

$$\int_0^\infty (Q_2(t) + Z_2(t)) \, dT_{1,2}(t) = 0, \quad \int_0^\infty (Q_1(t) + Z_1(t)) \, dT_{2,1}(t) = 0. \tag{8}$$

Equations (5) reflect that $S_k(T_k(t))$ is the total number of class-k customers service completions by time t, while equations (6) correspond to a work-conserving discipline. Relations (7) show that any server cannot be idle if there are customers waiting in the buffer of the other station, while relations (8) mean that server j gives priority to class-j customers over customers waiting in the other station. The rest of the equations are self-explained.

4   Stability Analysis of the X-Model

By definition, the queueing network is *stable* if its associated underlying Markov process X is *positive Harris recurrent*, that is, it has a unique invariant probability measure (see [7,9] for details). By Theorem 4.2 [9], the queueing network is stable whenever the corresponding *fluid limit model* is stable in the sense of Definition 1 below. Thus, to prove stability of the network, it is enough to establish stability of the associated fluid limit model, and this is what we will do, first for the X-model in Theorem 1.

4.1   The Fluid Limit Model

Next proposition is similar to Theorem 4.1 [9] (see also Lemma 3.1 in [10]), and to Proposition 1 [12], so we present the result without proof.

Proposition 1. *For the X-model, for almost all sample paths and any sequence of initial states $\{x_n\}_{n \geq 1} \subset \mathbb{X}$ with $\lim_{n \to \infty} |x_n| = \infty$, there exists a subsequence $\{x_{n_r}\}_{r \geq 1} \subseteq \{x_n\}_{n \geq 1}$ with $\lim_{r \to \infty} |x_{n_r}| = \infty$ such that the following limit exists*

$$\lim_{r \to \infty} \frac{1}{|x_{n_r}|} X^{x_{n_r}}(0) := \bar{X}(0), \tag{9}$$

and the following u.o.c. limit exists

$$\lim_{r \to \infty} \frac{1}{|x_{n_r}|} \left(X^{x_{n_r}}(|x_{n_r}|t), D^{x_{n_r}}(|x_{n_r}|t), T^{x_{n_r}}(|x_{n_r}|t), Y^{x_{n_r}}(|x_{n_r}|t) \right)$$
$$:= (\bar{X}(t), \bar{D}(t), \bar{T}(t), \bar{Y}(t)), \tag{10}$$

where (in an evident notation) $\bar{X}(t) := (\bar{Q}(t), \bar{Z}(t), \bar{U}(t), \bar{V}(t))$. Furthermore, the following equations are satisfied for any $t \geq 0$:

$$\bar{D}_k(t) = \mu_k (\bar{T}_k(t) - \bar{V}_k(0))^+, \quad k = 1, 2, (1,2), (2,1), \tag{11}$$

$$\bar{Z}_k(t) = 0, \quad k = 1, 2, (1,2), (2,1), \tag{12}$$

$$\bar{Q}_1(t) = \bar{Q}_1(0) + \lambda_1 (t - \bar{U}_1(0))^+ - (\bar{D}_1(t) + \bar{D}_{1,2}(t)), \tag{13}$$

$$\bar{Q}_2(t) = \bar{Q}_2(0) + \lambda_2 (t - \bar{U}_2(0))^+ - (\bar{D}_2(t) + \bar{D}_{2,1}(t)), \tag{14}$$

$$(\bar{T}_1(t) + \bar{T}_{2,1}(t)) + \bar{Y}_1(t) = t, \quad (\bar{T}_2(t) + \bar{T}_{1,2}(t)) + \bar{Y}_2(t) = t, \tag{15}$$

$$\int_0^\infty \bar{Q}_j(t)\, d\bar{Y}_j(t) = 0, \quad j = 1, 2, \tag{16}$$

$$\int_0^\infty \bar{Q}_1(t)\, d\bar{Y}_2(t) = 0, \quad \int_0^\infty \bar{Q}_2(t)\, d\bar{Y}_1(t) = 0, \tag{17}$$

$$\int_0^\infty \bar{Q}_2(t)\, d\bar{T}_{1,2}(t) = 0, \quad \int_0^\infty \bar{Q}_1(t)\, d\bar{T}_{2,1}(t) = 0, \tag{18}$$

$$\bar{U}(t) = (\bar{U}(0) - t)^+, \quad \bar{V}(t) = (\bar{V}(0) - t)^+. \tag{19}$$

In addition, for any $0 \leq s \leq t$ and $k = 1, 2, (1, 2), (2, 1)$,

$$0 \leq \bar{T}_k(t + s) - \bar{T}_k(s) \leq t. \tag{20}$$

Remark 1. Any limit $(\bar{X}, \bar{D}, \bar{T}, \bar{Y})$ in (9) and (10) is called a *fluid limit model* associated to the queueing network. Proposition 1 states that any *fluid limit model* satisfies the *fluid model equations* (11)-(19), and also condition (20).

Definition 1. *([7,9,12]) The fluid limit model $(\bar{X}, \bar{D}, \bar{T}, \bar{Y})$ associated to the X-model, with $\bar{X} = (\bar{Q}, \bar{Z}, \bar{U}, \bar{V})$, is stable if there exists $t_1 \geq 0$ (depending on the arrival and service rates only) such that if $|\bar{Q}(0)| + |\bar{U}(0)| + |\bar{V}(0)| = 1$ then*

$$\bar{Q}(t) = 0 \; \text{for all } t \geq t_1. \tag{21}$$

Remark 2. By Lemma 5.3 [9] (see also [8]), hereinafter we will assume without loss of generality that $\bar{U}(0) = \bar{V}(0) = 0$ which, by (19), implies $\bar{U}(t) = \bar{V}(t) = 0$ for all $t > 0$. We denote it $\bar{U} = \bar{V} = 0$.

4.2   The Stability Analysis

In this section, we present stability result of the X-model, whose proof follows the same arguments of Theorem 1 [12]. Denote $r_1 = \mu_1/\mu_{2,1}$ and $r_2 = \mu_2/\mu_{1,2}$, both in $(0, \infty)$.

Theorem 1. *Let conditions (1)-(4) hold and the following assumption take place: either a) $r_1, r_2 \geq 1$; or b) $r_1, r_2 < 1$. Then the X-model is stable if the following two conditions are satisfied:*

$$\begin{cases} (A_1) & \lambda_1 - \mu_1 + \dfrac{\lambda_2 - \mu_2}{r_2} < 0, \\ (A_2) & \dfrac{\lambda_1 - \mu_1}{r_1} + \lambda_2 - \mu_2 < 0. \end{cases} \tag{22}$$

Remark 3. Assumption a) corresponds to the situation in which customers are served at a slower rate when served in the other station of the rightful, while assumption b) reflects just the opposite situation.

Proof. Take a fluid limit model $(\bar{X}, \bar{D}, \bar{T}, \bar{Y})$ with $\bar{X} = (\bar{Q}, \bar{Z}, \bar{U}, \bar{V})$ for the X-model, with $\bar{U} = \bar{V} = 0$ and $|\bar{Q}(0)| = 1$. We introduce the Lyapunov function

$$f(t) = \frac{\bar{Q}_1(t)}{r_1} + \frac{\bar{Q}_2(t)}{r_2}, \quad t \geq 0.$$

Then $\bar{Q}$ and f have the same points of differentiability, and by (13), (14) and (11) we have for any regular point t of $\bar{Q}$,

$$\dot{\bar{Q}}_1(t) = \lambda_1 - \left(\mu_1 \dot{\bar{T}}_1(t) + \mu_{1,2} \dot{\bar{T}}_{1,2}(t)\right),$$
$$\dot{\bar{Q}}_2(t) = \lambda_2 - \left(\mu_2 \dot{\bar{T}}_2(t) + \mu_{2,1} \dot{\bar{T}}_{2,1}(t)\right), \tag{23}$$

implying

$$\dot{f}(t) = \frac{\lambda_1}{r_1} - \left(\frac{\mu_1}{r_1} \dot{\bar{T}}_1(t) + \frac{\mu_{1,2}}{r_1} \dot{\bar{T}}_{1,2}(t)\right) + \frac{\lambda_2}{r_2} - \left(\frac{\mu_2}{r_2} \dot{\bar{T}}_2(t) + \frac{\mu_{2,1}}{r_2} \dot{\bar{T}}_{2,1}(t)\right). \tag{24}$$

Let $t > 0$ be a point of differentiability of f such that $f(t) > 0$, if exist. Then $\bar{Q}_1(t) > 0$ or $\bar{Q}_2(t) > 0$. We will show that in either case, $\dot{f}(t) \leq -C$ for a constant $C > 0$. Then, by Lemma 5.2 [9], f is nonincreasing and $f(t) = 0$ for $t \geq f(0)/C$, where

$$f(0) = \frac{\bar{Q}_1(0)}{r_1} + \frac{\bar{Q}_2(0)}{r_2} \leq \frac{1}{r_1} + \frac{1}{r_2}.$$

(Recall that $\bar{Q}_1(0) + \bar{Q}_2(0) = 1$.) Then (21) will follow with $t_1 = \frac{1}{C}\left(\frac{1}{r_1} + \frac{1}{r_2}\right) > 0$. To finish the proof, we distinguish the following three cases:

a) *Assume that $\bar{Q}_1(t) > 0$ and $\bar{Q}_2(t) = 0$.* As $\bar{Q}_1(t) > 0$, then by (16) $\dot{\bar{Y}}_1(t) = 0$, and it implies by (15)

$$\dot{\bar{T}}_1(t) + \dot{\bar{T}}_{2,1}(t) = 1. \tag{25}$$

Moreover, $\dot{\bar{Y}}_2(t) = 0$ by (17) which, by (15), gives

$$\dot{\bar{T}}_2(t) + \dot{\bar{T}}_{1,2}(t) = 1. \tag{26}$$

Finally, (18) implies $\dot{\bar{T}}_{2,1}(t) = 0$, which in turn by (25) implies $\dot{\bar{T}}_1(t) = 1$. Thus, (24) gives

$$\dot{f}(t) = \frac{\lambda_1 - \mu_1}{r_1} - \frac{\mu_{1,2}}{r_1} \dot{\bar{T}}_{1,2}(t) + \frac{\lambda_2}{r_2} - \frac{\mu_2}{r_2} \dot{\bar{T}}_2(t). \tag{27}$$

Since $\bar{Q}_2 \geq 0$ and $\bar{Q}_2(t) = 0$ is a local minimum, as t is a regular point, then, by Fermat's theorem on stationary points, $\dot{\bar{Q}}_2(t) = 0$. Therefore $\mu_2 \dot{\bar{T}}_2(t) + \mu_{2,1} \dot{\bar{T}}_{2,1}(t) = \lambda_2$ by (23). Since $\dot{\bar{T}}_{2,1}(t) = 0$, we obtain $\dot{\bar{T}}_2(t) = \frac{\lambda_2}{\mu_2}$, and also $\dot{\bar{T}}_{1,2}(t) = 1 - \frac{\lambda_2}{\mu_2}$ by (26). Now, from (27) we can write

$$\dot{f}(t) = \frac{\lambda_1 - \mu_1}{r_1} - \frac{\mu_{1,2}}{r_1}\left(1 - \frac{\lambda_2}{\mu_2}\right) = \frac{1}{r_1}\left(\lambda_1 - \mu_1 - \mu_{1,2} + \frac{\lambda_2}{r_2}\right)$$
$$= \frac{1}{r_1}\left(\lambda_1 - \mu_1 + \frac{\lambda_2 - \mu_2}{r_2}\right) = -C_1,$$

where constant $C_1 := -\frac{1}{r_1}\left(\lambda_1 - \mu_1 + \frac{\lambda_2 - \mu_2}{r_2}\right)$ is positive by condition (A_1) in (22).

b) *Assume that $\bar{Q}_1(t) = 0$ and $\bar{Q}_2(t) > 0$.* By symmetry with the previous case, $\dot{f}(t) = -C_2$, where $C_2 := -\frac{1}{r_2}\left(\frac{\lambda_1 - \mu_1}{r_1} + \lambda_2 - \mu_2\right)$ is positive by condition (A_2) in (22).

c) *Assume that $\bar{Q}_1(t) > 0$ and $\bar{Q}_2(t) > 0$.* As in the previous cases, $\bar{Q}_1(t) > 0$ and $\bar{Q}_2(t) > 0$ imply $\dot{Y}_1(t) = \dot{Y}_2(t) = \dot{T}_{2,1}(t) = \dot{T}_{1,2}(t) = 0$. The latter gives

$$\dot{T}_1(t) = \dot{T}_2(t) = 1$$

by (15). It follows from (24) that

$$\dot{f}(t) = \frac{\lambda_1 - \mu_1}{r_1} + \frac{\lambda_2 - \mu_2}{r_2} := -C_3.$$

Now we show that constant $C_3 > 0$. Indeed, by (22) the three following situations are possible:

  i) $\lambda_1 < \mu_1$ and $\lambda_2 \geq \mu_2$. Under assumption a) condition (A_2) implies $C_3 > 0$, while under assumption b) we arrive to the same conclusion by (A_1). In fact, under a), $-C_3 \leq -r_2 C_2 < 0$, while under b), $-C_3 < -r_1 C_1 < 0$.

 ii) $\lambda_1 \geq \mu_1$ and $\lambda_2 < \mu_2$. In this case, we use (A_1) under assumption a), while use (A_2) under b). Analogously, $-C_3 \leq -r_1 C_1 < 0$ under a), while $-C_3 < -r_2 C_2 < 0$ under b).

iii) $\lambda_1 < \mu_1$ and $\lambda_2 < \mu_2$. In this case it is obvious that $C_3 > 0$.

Therefore, in either case, $\dot{f}(t) \leq -C < 0$, where $C := \min(C_1, C_2, C_3)$.   □

Remark 4. Comparing conditions (22) with stability conditions

$$\begin{cases} \lambda_1 - \mu_1 + \frac{\lambda_2 - \mu_2}{r_2} < 0, \\ \lambda_2 - \mu_2 < 0 \end{cases}$$

for the two-station cascade system (see [12] and references therein), we see that the only difference is that condition (A_2) includes the additional summand $\frac{\lambda_1 - \mu_1}{r_1}$.

5 The Tree-Cascade System

In this section, we consider a three-station queueing system with interacting servers where each server has an infinite-capacity buffer. We call class-j exogenous customers who arrive at station $j = 1, 2, 3$ following independent renewal processes. An awaiting customer at station 1 jumps to station 2 whenever it is free, and an awaiting customer at station 2 jumps to station 3 if it is free. This interaction corresponds exactly to the three-station cascade network studied in [12]. However, in addition, we now allow an awaiting customer at station 1 to jump from the buffer to station 3 whenever it is free. That is, in this model station 3 supports both stations $1, 2$ but gives priority to station 2. We denote by class-(i, j) customers jumping from station i to station $j > i$, $i = 1, 2$. We will

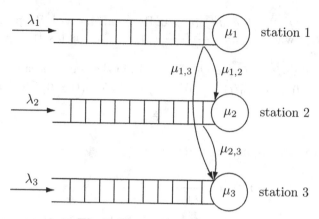

Fig. 2. The tree-cascade system

use the same notations as before, *mutatis mutandis*. In particular, the arrival rates of class-j customers ($j = 1, 2, 3$) are $\lambda_j \in (0, \infty)$, and the service rates of class-k customers ($k = 1, 2, 3, (1, 2), (1, 3), (2, 3)$) are $\mu_k \in (0, \infty)$, see Figure 2.

We also impose standard conditions (1)-(4) on the interarrival and service times, assume *work-conserving* discipline and introduce the following processes:

$$D = \{D(t) := (D_1(t), D_2(t), D_3(t), D_{1,2}(t), D_{1,3}(t), D_{2,3}(t)), t \geq 0\},$$
$$Z = \{Z(t) := (Z_1(t), Z_2(t), Z_3(t), Z_{1,2}(t), Z_{1,3}(t), Z_{2,3}(t)), t \geq 0\},$$
$$T = \{T(t) := (T_1(t), T_2(t), T_3(t), T_{1,2}(t), T_{1,3}(t), T_{2,3}(t)), t \geq 0\},$$
$$Q = \{Q(t) := (Q_1(t), Q_2(t), Q_3(t)), t \geq 0\},$$
$$Y = \{Y(t) := (Y_1(t), Y_2(t), Y_3(t)), t \geq 0\},$$
$$U = \{U(t) := (U_1(t), U_2(t), U_3(t)), t \geq 0\},$$
$$V = \{V(t) := (V_1(t), V_2(t), V_3(t), V_{1,2}(t), V_{1,3}(t), V_{2,3}(t)), t \geq 0\},$$

and define the piecewise-deterministic strong Markov process

$$X = (Q, Z, U, V) := \{X(t) = (Q(t), Z(t), U(t), V(t)), t \geq 0\}$$

describing the dynamics of the network with the state space $\mathbb{X} = \mathbb{Z}_+^3 \times \{0, 1\}^6 \times \mathbb{R}_+^3 \times \mathbb{R}_+^6$. The following *queueing network equations* governing the system hold for all $t \geq 0$:

$$D_k(t) = S_k(T_k(t)), \quad k = 1, 2, 3, (1, 2), (1, 3), (2, 3),$$
$$Q_1(t) = Q_1(0) + E_1(t)$$
$$\quad - (D_1(t) + Z_1(t) + D_{1,2}(t) + Z_{1,2}(t) + D_{1,3}(t) + Z_{1,3}(t)),$$
$$Q_2(t) = Q_2(0) + E_2(t) - (D_2(t) + Z_2(t) + D_{2,3}(t) + Z_{2,3}(t)),$$
$$Q_3(t) = Q_3(0) + E_3(t) - (D_3(t) + Z_3(t)),$$

$$T_1(t) + Y_1(t) = t, \quad (T_2(t) + T_{1,2}(t)) + Y_2(t) = t,$$
$$(T_3(t) + T_{1,3}(t) + T_{2,3}(t)) + Y_3(t) = t,$$

$$\int_0^\infty Q_j(t)\, dY_j(t) = 0, \quad j = 1,2,3,$$

$$\int_0^\infty Q_1(t)\, dY_2(t) = 0, \quad \int_0^\infty (Q_1(t) + Q_2(t))\, dY_3(t) = 0, \tag{28}$$

$$\int_0^\infty (Q_2(t) + Z_2(t))\, dT_{1,2}(t) = 0, \tag{29}$$

$$\int_0^\infty (Q_3(t) + Z_3(t) + Z_{1,3}(t))\, dT_{2,3}(t) = 0, \tag{30}$$

$$\int_0^\infty (Q_2(t) + Q_3(t) + Z_3(t) + Z_{2,3}(t))\, dT_{1,3}(t) = 0. \tag{31}$$

Relations (28) mean that server 2 supports server 1, while server 3 supports both servers 1 and 2. Equations (29) and (30) mean that server j ($j = 2,3$) gives priority to class-j customers, and equation (31) shows that server 3 gives priority to class-2 customers over class-1 ones.

The corresponding fluid limit model $(\bar{X}, \bar{D}, \bar{T}, \bar{Y})$ is defined as above, with $\bar{X} = (\bar{Q}, \bar{Z}, \bar{U}, \bar{V})$ and the assumption $\bar{U} = \bar{V} = 0$. Then the corresponding fluid model equations are the following, for any $t \geq 0$:

$$\bar{D}_k(t) = \mu_k \bar{T}_k(t), \quad k = 1,2,3,(1,2),(1,3),(2,3), \tag{32}$$
$$\bar{Z}_k(t) = 0, \quad k = 1,2,3,(1,2),(1,3),(2,3), \tag{33}$$
$$\bar{Q}_1(t) = \bar{Q}_1(0) + \lambda_1 t - (\bar{D}_1(t) + \bar{D}_{1,2}(t) + \bar{D}_{1,3}(t)), \tag{34}$$
$$\bar{Q}_2(t) = \bar{Q}_2(0) + \lambda_2 t - (\bar{D}_2(t) + \bar{D}_{2,3}(t)), \tag{35}$$
$$\bar{Q}_3(t) = \bar{Q}_3(0) + \lambda_3 t - \bar{D}_3(t), \tag{36}$$
$$\bar{T}_1(t) + \bar{Y}_1(t) = t, \quad (\bar{T}_2(t) + \bar{T}_{1,2}(t)) + \bar{Y}_2(t) = t, \tag{37}$$
$$(\bar{T}_3(t) + \bar{T}_{1,3}(t) + \bar{T}_{2,3}(t)) + \bar{Y}_3(t) = t, \tag{38}$$

$$\int_0^\infty \bar{Q}_j(t)\, d\bar{Y}_j(t) = 0, \quad j = 1,2,3, \tag{39}$$

$$\int_0^\infty \bar{Q}_1(t)\, d\bar{Y}_2(t) = 0, \quad \int_0^\infty (\bar{Q}_1(t) + \bar{Q}_2(t))\, d\bar{Y}_3(t) = 0, \tag{40}$$

$$\int_0^\infty \bar{Q}_2(t)\, d\bar{T}_{1,2}(t) = 0, \quad \int_0^\infty \bar{Q}_3(t)\, d\bar{T}_{2,3}(t) = 0, \tag{41}$$

$$\int_0^\infty (\bar{Q}_2(t) + \bar{Q}_3(t))\, d\bar{T}_{1,3}(t) = 0. \tag{42}$$

We denote $r_{1,2} = \mu_2/\mu_{1,2}$, $r_{1,3} = \mu_3/\mu_{1,3}$ and $r_{2,3} = \mu_3/\mu_{2,3}$, the three quantities in $(0, \infty)$.

Theorem 2. *Let the standard conditions on the inter-arrival and service times hold and also hold:*

$$r_{1,3} \leq r_{1,2}\, r_{2,3}\,. \tag{43}$$

Then, the tree-cascade system is stable if the following conditions hold:

$$\begin{cases} (B_1) & \lambda_1 - \mu_1 + \frac{\lambda_2 - \mu_2}{r_{1,2}} + \frac{\lambda_3 - \mu_3}{r_{1,2}\, r_{2,3}} < 0, \\ (B_2) & \lambda_2 - \mu_2 + \frac{\lambda_3 - \mu_3}{r_{2,3}} < 0, \\ (B_3) & \lambda_3 - \mu_3 < 0. \end{cases} \tag{44}$$

Proof. We consider a fluid limit model $(\bar{X}, \bar{D}, \bar{T}, \bar{Y})$ with $\bar{X} = (\bar{Q}, \bar{Z}, \bar{U}, \bar{V})$ verifying (32)-(42), and assume $|\bar{Q}(0)| = 1$. Let we introduce the following Lyapunov function:

$$f(t) = \bar{Q}_1(t) + \frac{\bar{Q}_2(t)}{r_{1,2}} + \frac{\bar{Q}_3(t)}{r_{1,2}\, r_{2,3}}, \quad t \geq 0\,.$$

Then $\bar{Q}$ and f have the same points of differentiability. It follows from fluid model equations that for any regular point t of $\bar{Q}$,

$$\dot{\bar{Q}}_1(t) = \lambda_1 - \left(\mu_1 \dot{\bar{T}}_1(t) + \mu_{1,2} \dot{\bar{T}}_{1,2}(t) + \mu_{1,3} \dot{\bar{T}}_{1,3}(t) \right), \tag{45}$$

$$\dot{\bar{Q}}_2(t) = \lambda_2 - \left(\mu_2 \dot{\bar{T}}_2(t) + \mu_{2,3} \dot{\bar{T}}_{2,3}(t) \right), \tag{46}$$

$$\dot{\bar{Q}}_3(t) = \lambda_3 - \mu_3 \dot{\bar{T}}_3(t).$$

It implies

$$\dot{f}(t) = \lambda_1 - \left(\mu_1 \dot{\bar{T}}_1(t) + \mu_{1,2} \dot{\bar{T}}_{1,2}(t) + \mu_{1,3} \dot{\bar{T}}_{1,3}(t) \right)$$
$$+ \frac{\lambda_2}{r_{1,2}} - \left(\frac{\mu_2}{r_{1,2}} \dot{\bar{T}}_2(t) + \frac{\mu_{2,3}}{r_{1,2}} \dot{\bar{T}}_{2,3}(t) \right) + \frac{\lambda_3}{r_{1,2}\, r_{2,3}} - \frac{\mu_3}{r_{1,2}\, r_{2,3}} \dot{\bar{T}}_3(t)\,. \tag{47}$$

Let $t > 0$ be a point of differentiability of f such that $\dot{f}(t) > 0$, if exist. Then $\bar{Q}_1(t) > 0$, $\bar{Q}_2(t) > 0$ or $\bar{Q}_3(t) > 0$. We will show that in either case, $\dot{f}(t) \leq -C$ for a constant $C > 0$. Then f is nonincreasing, and $f(t) = 0$ for $t \geq \frac{f(0)}{C}$ (see [9]), where

$$f(0) \leq 1 + \frac{1}{r_{1,2}} + \frac{1}{r_{1,2}\, r_{2,3}}\,.$$

Then (21) is accomplished with $t_1 = \frac{1}{C}\left(1 + \frac{1}{r_{1,2}} + \frac{1}{r_{1,2}\, r_{2,3}}\right)$. We distinguish the following three cases:

a) *Assume that* $\bar{Q}_1(t) > 0$. Then, it follows from (39) and (37) that $\dot{\bar{Y}}_1(t) = 0$ and $\dot{\bar{T}}_1(t) = 1$. Now (40), (37) and (38) give $\dot{\bar{Y}}_2(t) = \dot{\bar{Y}}_3(t) = 0$ and

$$\dot{\bar{T}}_2(t) + \dot{\bar{T}}_{1,2}(t) = 1\,, \tag{48}$$

$$\dot{\bar{T}}_3(t) + \dot{\bar{T}}_{1,3}(t) + \dot{\bar{T}}_{2,3}(t) = 1\,. \tag{49}$$

As a result, (47) becomes

$$\dot{f}(t) = \lambda_1 - \mu_1 + \frac{\lambda_2 - \mu_2}{r_{1,2}} + \frac{\lambda_3 - \mu_3}{r_{1,2}\, r_{2,3}} + \Big(\frac{\mu_3}{r_{1,2}\, r_{2,3}} - \mu_{1,3}\Big)\, \dot{T}_{1,3}(t)$$

$$\leq \lambda_1 - \mu_1 + \frac{\lambda_2 - \mu_2}{r_{1,2}} + \frac{\lambda_3 - \mu_3}{r_{1,2}\, r_{2,3}} = -C_1\,,$$

with $C_1 := -\Big(\lambda_1 - \mu_1 + \frac{\lambda_2-\mu_2}{r_{1,2}} + \frac{\lambda_3-\mu_3}{r_{1,2}\,r_{2,3}}\Big) > 0$ by condition (B_1), where we have used that $\dot{T}_{1,3}(t) \geq 0$ and $\frac{\mu_3}{r_{1,2}\,r_{2,3}} - \mu_{1,3} \leq 0$ by (43).

b) *Assume that* $\bar{Q}_1(t) = 0$ *and* $\bar{Q}_2(t) > 0$. Since $\bar{Q}_1 \geq 0$ and t is a regular point, then we conclude that $\dot{\bar{Q}}_1(t) = 0$. On the other hand, $\bar{Q}_2(t) > 0$ implies by (39) and (40) that $\dot{Y}_2(t) = 0$ and $\dot{Y}_3(t) = 0$, which gives (48) and (49), respectively, by (37) and (38). Moreover, $\dot{T}_{1,2}(t) = \dot{T}_{1,3}(t) = 0$ by (41). As a result, we obtain

$$\dot{T}_2(t) = 1\,, \quad \dot{T}_{1,2}(t) = 0\,,$$
$$\dot{T}_{1,3}(t) = 0\,, \quad \dot{T}_3(t) + \dot{T}_{2,3}(t) = 1\,.$$

If we take into account (45) then (47) becomes

$$\dot{f}(t) = \frac{\lambda_2 - \mu_2}{r_{1,2}} - \frac{\mu_{2,3}}{r_{1,2}}\, \dot{T}_{2,3}(t) + \frac{\lambda_3 - \mu_3\, \dot{T}_3(t)}{r_{1,2}\, r_{2,3}}$$

$$= \frac{\lambda_2 - \mu_2}{r_{1,2}} - \frac{\mu_3}{r_{1,2}\, r_{2,3}}\big(1 - \dot{T}_3(t)\big) + \frac{\lambda_3 - \mu_3\, \dot{T}_3(t)}{r_{1,2}\, r_{2,3}}$$

$$= \frac{\lambda_2 - \mu_2}{r_{1,2}} + \frac{\lambda_3 - \mu_3}{r_{1,2}\, r_{2,3}} := -C_2\,,$$

where $C_2 > 0$ by condition (B_2).

c) *Assume that* $\bar{Q}_1(t) = \bar{Q}_2(t) = 0$ *and* $\bar{Q}_3(t) > 0$. As above, equalities $\bar{Q}_1(t) = \bar{Q}_2(t) = 0$ imply $\dot{\bar{Q}}_1(t) = \dot{\bar{Q}}_2(t) = 0$, while $\bar{Q}_3(t) > 0$ implies

$$\dot{T}_{1,3}(t) = \dot{T}_{2,3} = 0\,, \quad \dot{T}_3(t) = 1\,.$$

Now it is seen from (45) and (46) that (47) becomes

$$\dot{f}(t) = \frac{\lambda_3 - \mu_3\, \dot{T}_3(t)}{r_{1,2}\, r_{2,3}} = \frac{\lambda_3 - \mu_3}{r_{1,2}\, r_{2,3}} := -C_3\,,$$

where $C_3 > 0$ by condition (B_3).

Thus, in either case $\dot{f}(t) \leq -C < 0$, where $C := \min(C_1, C_2, C_3)$. $\qquad \square$

Remark 5. The following stability conditions have been obtained in [12] for the three-station cascade system:

$$\begin{cases} \lambda_1 - \mu_1 + \frac{\lambda_2 - \mu_2}{r_{1,2}} < 0, \\ \lambda_2 - \mu_2 + \frac{\lambda_3 - \mu_3}{r_{2,3}} < 0, \\ \lambda_3 - \mu_3 < 0. \end{cases}$$

The only difference with (44) is that, in condition (B_1), we have the additional summand $\frac{\lambda_3 - \mu_3}{r_{1,2} r_{2,3}}$, which reflects the fact that in the tree-cascade system server 3 also accepts awaiting customers from the 1st station.

6   Conclusion

In this paper, we apply the fluid limit approach to find stability conditions of two models of queueing networks with interacting servers. In both models, each station has a renewal input with general i.i.d. inter-arrival times and general i.i.d. service times. The first, so-called X-model, contains two stations with two classes of customers. Each station, when is free, helps to serve customers awaiting in queue of other station. Then a cascade-type three-station system is considered, in which the third station, whenever it is free, assists the other two stations. Following conventional methodology of the fluid analysis [9], we first construct the network equations describing stochastic dynamics of the system and then obtain deterministic fluid limit model. Finally, for each model, we find some conditions and relevant Lyapunov functions to prove that the fluid limit of the queue-size process reaches zero in a finite time interval and stays there. It means that under these conditions the corresponding underlying Markov process describing the network dynamics is positive Harris recurrent.

Acknowledgments. The authors wish to thank the anonymous referees for careful reading and helpful comments on the paper.

References

1. Agnihothri, S.R., Mishra, A.K., Simmons, D.E.: Workforce cross-training decisions in field service systems with two job types. Journal of the Operational Research Society 54(4), 410–418 (2003)
2. Ahghari, M., Balcioglu, B.: Benefits of cross-training in a skill-based routing contact center with priority queues and impatient customers. IIE Transactions 41, 524–536 (2009)
3. Ahn, H.-S., Duenyas, I., Zhang, Q.R.: Optimal control of a flexible server. Adv. Appl. Probab. 36, 139–170 (2004)
4. Andradottir, S., Ayhan, H., Down, G.D.: Dynamic server allocation for queueing networks with flexible servers. Operations Reserach 51(6), 952–968 (2003)
5. Bell, S.L., Williams, R.J.: Dynamic scheduling of a server system with two parallel servers: Asymptotic optimality of a continuous review threshold policy in heavy traffic. In: Proceedings of the 38 Conference on Decision and Control, Phoenix, Arizona, pp. 2255–2260 (December 1999)

6. Bell, S.L., Williams, R.J.: Dynamic scheduling of a system with two parallel servers in heavy traffic with complete resource pooling: Asymptotic optimality of a continuous review threshold policy. Ann. Probab. 11, 608–649 (2001)
7. Bramson, M.: Stability of queueing networks, École d'Été de Probabilités de Saint-Flour XXXVI-2006. LNM, vol. 1950. Springer (2008)
8. Chen, H.: Fluid approximations and stability of multiclass queueing networks: Work-conserving disciplines. Ann. Appl. Prob. 5(3), 637–665 (1995)
9. Dai, J.G.: On positive Harris recurrence of multiclass queueing networks: A unified approach via fluid limit models. Ann. Appl. Prob. 5(1), 49–77 (1995)
10. Dai, J.G., Meyn, S.P.: Stability and convergence of moments for multiclass queueing networks via fluid limit models. IEEE Trans. Aut. Cont. 40(11), 1889–1904 (1995)
11. Davis, M.H.A.: Piecewise deterministic Markov processes: a general class of non-diffusion stochastic models. J. Roy. Statist. Soc. Ser. B 46, 353–388 (1984)
12. Delgado, R., Morozov, E.: Stability analysis of cascade networks via fluid models. Preprint (2014),
    http://gent.uab.cat/rosario_delgado/sites/gent.uab.cat.rosario_delgado/files/_cascade_networks.pdf
13. Down, D.G., Lewis, M.E.: Dynamic Load Balancing in Parallel Queueing Systems: Stability and Optimal Control. European Journal of Operational Research 168, 509–519 (2006)
14. Foley, R.D., McDonald, D.R.: Large deviations of a modified Jackson network: Stability and rough asymptotics. Annals of Applied Probability 15(1B), 519–541 (2005)
15. Harrison, M., Lopez, M.J.: Heavy traffic resource pooling in parallel-server systems. Queueing Systems 33, 339–368 (1999)
16. Kirkizlar, E., Andradottir, S., Ayhan, H.: Robustness of efficient server assignment policies to service time distributions in finite-buffered lines. Naval Research Logistics 57(6), 563–582 (2010)
17. Mandelbaum, A., Stolyar, A.L.: Scheduling flexible servers with convex delay costs: Heavy-traffic optimality of the generalized $c\mu$-rule. Operations Research 52(6), 836–855 (2004)
18. Morozov, E., Steyaert, B.: Stability analysis of a two-station cascade queueing network. Ann. Oper. Res. 13, 135–160 (2013)
19. Perry, O., Whitt, W.: Achieving Rapid Recovery in an Overload Control for Large-Scale Service Systems. Preprint (2013),
    http://www.columbia.edu/~ww2040/Recover_Submit.pdf
20. Stolyar, A.L., Tezcan, T.: Control of systems with flexible multi-server pools: a shadow routing approach. Queueing Systems 66, 1–51 (2010)
21. Tekin, E., Hopp, W.J., Van Oyen, M.P.: Pooling strategies for call center agent cross-training. IIE Transactions 41(6), 546–561 (2009)
22. Terekhov, D., Beck, J.C.: An extended queueing control model for facilities with front room and back room operations and mixed-skilled workers. European Journal of Operational Research 198(1), 223–231 (2009)
23. Tezcan, T.: Stability analysis of N-model systems under a static priority rule. Queueing Syst. 73, 235–259 (2013)
24. Tsai, Y.C., Argon, N.T.: Dynamic server assignment policies for assembly-type queues with flexible servers. Naval Research Logistics 55(3), 234–251 (2008)

Markovian Agents Population Models to Study Cancer Evolution

Francesca Cordero[1], Chiara Fornari[1], Marco Gribaudo[2], and Daniele Manini[1]

[1] Dip. di Informatica, Università di Torino
Corso Svizzera 185, 10149 Torino, Italy
{cordero,manini,fornari}@di.unito.it
[2] Dip. di Elettronica e Informazione, Politecnico di Milano
via Ponzio 34/5, 20133 Milano, Italy
marco.gribaudo@polimi.it

Abstract. We introduce a new Markovian Agents formalism, called Population Markovian Agent Models, a technique able to describe systems characterized by large populations of entities whose properties and interactions may depend on their position. We apply this approach to the analysis of the cancer evolution, and to comprehend the mechanisms underlying the Cancer Stem Cells hierarchy whose characterization is crucial in the study of tumor progression. We exploit the model to consider movement of cell populations in a bi-dimensional space, and use it to derive the system evolution.

1 Introduction

In this work we present the *Population Markovian Agent Model* (PMAM), a technique able to describe systems characterized by large populations of entities whose properties and interactions may depend on their position. We spatially extend some of the concepts presented in [9] were the e-coli bacteria was studied with an object oriented technique, and in [3] where we developed a framework based on the Markovian agent formalism to model next generation cellular networks. In this formalism, agents can move to other locations, can increase in number or decrease (either spontaneously or induced by other agents), or they can multiply during the transitions. We apply this approach to the analysis of the cancer evolution. Current evidence indicates that many cancers arise from a small population of cells named Cancer Stem Cells (CSCs) [20], which have undergone malignant transformation driven by frequent genetic mutations [17]. The application of the PMAM to describe the behaviour of such cell populations allowed us to observe their proliferation through the host tissue. The model output could be exploited to comprehend the mechanisms underlying the CSC hierarchy, whose characterization is crucial in the study of tumor progression and in predicting treatment response, as shown in [10].

The rest of the paper is organized as follows. In Section 2, we introduce the PMAM formalism. The CSC biological model is presented in Section 3. In Section 4, we describe the PMAM model of the cancer subpopulations together

B. Sericola, M. Telek, and G. Horváth (Eds.): ASMTA 2014, LNCS 8499, pp. 16–32, 2014.
© Springer International Publishing Switzerland 2014

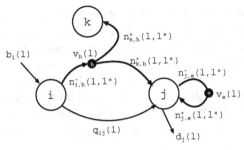

Fig. 1. A Population Markovian Agent (PMA) showing birth, death, spontaneous transitions and population reactions

with the results obtained by applying it to single location and bi-dimensional scenario.

2 Population Markovian Agents

Population Markovian Agents are an extension of Markovian Agents [14], a formalism used to describe spatial system, where elements interact by inducing a transition in neighboring agents. Such models are solved using Mean Field Analysis [13,1]. In this case, agents do not communicate using messages as ordinary Markovian Agents, but they can influence each other via *induction*: the rate of jumping from one state to another can be influenced by the number of agents in a given state at a given location. Moreover, in PMAM, agents can move to other locations, can increase in number or decrease (either spontaneously or induced by other agents), or they can multiply during the transitions.

2.1 Population Markovian Agents Models

A *Population Markovian Agent Model* (PMAM) is a tuple $(L, M, Init)$. L is a set of locations, that can be either discrete or continuous: to simplify the presentation, in this work we will focus only on discrete locations $L = \{l_1, \ldots, l_{|L|}\}$. M is a set of *Population Markovian Agents* (PMA), $Init(l)$ is the initialization function, that describes the initial state for all the agents in each location: they will be explained in more detail in the following. In this work we will only present results concerning a single class of agents: this limitation however can be easily removed following the ideas presented in [7].

A PMA describes the evolution of a single agent, and it is defined by a tuple $(\Omega, Q, b, d, R, v, N)$. The first four components define the common Markovian Agent behavior, while the last three account for the population evolution of the system. A graphical representation of a PMA is given in Figure 1. We have that: $\Omega = \{\sigma_1, \ldots \sigma_{|\Omega|}\}$ is the set of states in which the agent can be. States are visually represented as circles.

$Q = (\tilde{Q}(l), \hat{Q}^{[k]}(l, l'))$, where $\tilde{Q}(l) = |\tilde{q}_{ij}(l)|$ and $\hat{Q}^{[k]}(l', l) = |\hat{q}_{ij}^{[k]}(l', l)|$, with $\sigma_i, \sigma_j, \sigma_k \in \Omega$ and $l, l' \in L$, are the *transition rate matrices*. $\tilde{q}_{ij}(l)$ is the rate at which an agent in location l jumps from state σ_i to σ_j due to local activities. $\hat{Q}_{ij}^{[k]}(l',l)$ accounts for transitions caused by inductions, and represents the rate at which an agent in state σ_k in position l' induces jumps from state σ_i to state σ_j in position l. Standard transitions are drawn as normal arcs that go from the source to the destination state, that are labeled with the corresponding transition rate. Both matrices $\tilde{Q}(l)$ and $\hat{Q}^{[k]}(l',l)$ are infinitesimal generators: their diagonal term is equal to the sum of the other transitions out of the state (that is, $\tilde{q}_{ii}(l) = -\sum_{j \neq i} \tilde{q}_{ij}(l)$ and $\hat{q}_{ii}^{[k]}(l',l) = -\sum_{j \neq i} \hat{q}_{ij}^{[k]}(l',l)$), and their row sum is equal to zero.

$b = (\tilde{b}(l), \hat{b}^{[k]}(l,l'))$, with $\tilde{b}(l) = |\tilde{b}_i(l)|$ and $\hat{b}^{[k]}(l) = |\hat{b}_i^{[k]}(l',l)|$, are respectively the spontaneous and induced *births vectors*. In particular, $\tilde{b}_i(l)$ represents the rate at which agents are created in state σ_i at location l, and $\hat{b}_i^{[k]}(l',l)$ is the rate at which an agent in state σ_k in position l' induces birth of agents in state i at location l. Birth is graphically represented with an arrow that enters one state, labeled with the birth rate.

In a similar way, $d = (\tilde{d}(l), \hat{d}^{[k]}(l,l'))$, where $\tilde{d}(l) = |\tilde{d}_i(l)|$ and $\hat{d}^{[k]}(l',l) = |\hat{d}_i^{[k]}(l',l)|$ are respectively the spontaneous and induced *death vectors*, where $\tilde{d}_i(l)$ represents the rate at which agents are destroyed in state σ_i at location l, and $\hat{d}_i^{[k]}(l',l)$ is the rate at which an agent in state σ_k and position l' induces decrease in the number of agents in state i at location l. Graphically, death is represented by arrows exiting a state, labeled with the death rate.

$R = \{r_1, \ldots, r_{|R|}\}$ is a set of *reactions*, that allow agents to move, duplicate or merge. Such events can occur either locally or in neighbor locations. Graphically they are represented as small filled circles.

$v(l) = |v_h(l)|$ is the *reaction vector* with $r_h \in R$ and $l \in L$. In particular, $v_h(l)$ represents the speed at which reaction r_h occurs in location l. Since reaction rates can have very complex expressions, they can be function of the complete state of the model. Reaction rates are graphically represented as labels associated to the corresponding reactions.

Finally, $N = (N^+(l,l''), N^-(l,l''))$ where $N^+(l,l'') = |n_{i,k}^+(l,l'')|$ and $N^-(l,l'') = |n_{i,k}^-(l,l'')|$ describe the effects that a reaction $r_k \in R$, happening in location l, has on the number of agents in state $\sigma_i \in \Omega$ at location l''. The value of $n_{i,k}^+(l,l'')$ represents the number of agents that are added to state σ_i in location l'' when reaction r_k takes place, while $n_{i,k}^-(l,l'')$ accounts for the agents that are removed. Graphically, additive effects are represented as arcs that go from one reaction to the state, and subtractive effects are drawn as arrows going from the state to the reaction. In both cases, arcs are labeled with the replication factor ($n_{i,k}^+(l,l'')$ or $n_{i,k}^-(l,l'')$): if there is an arc with no label, then $n_{i,k}^+(l,l'') = 1$ or $n_{i,k}^-(l,l'') = 1$. The formalization of terms v and N is quite similar to one used in computational system biology (see [11]). Self-loop arcs are also allowed to represent reactions that multiply or divide the number of agents in a given location. In this case both $n_{j,e}^-(l,l'') > 0$ and $n_{j,e}^+(l,l'') > 0$ for the same reaction r_e and state σ_j (see for example reaction r_e for state σ_j in the right part of Figure 1).

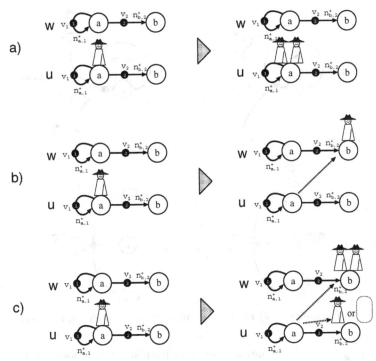

Fig. 2. Examples of different effects achievable with r, v and N: a) duplication, b) movement, c) combination of duplication, halving and movement

The definitions of v and N are powerful enough to model several different features: duplication or decimation of agents in a state, movement of agents (deterministic or probabilistic) to other locations, duplication or decimation combined with state jump and location movement. In the following, we will describe how such features can be obtained. To simplify the description, we focus on an agent with two states ($\Omega = \{\sigma_a, \sigma_b\}$), and two reactions ($R = \{r_1, r_2\}$) available in two locations ($L = \{u, w\}$). In particular we consider how N can be defined for an agent in position u: for all the σ_i, r_k, l and l' non directly described, we suppose $n^{\pm}_{i,k}(l, l'') = 0$.

Duplication or decimation of agents. We want to model duplication of an agent in state a (Figure 2.a) at position u. We first add a self-loop arc on state σ_a involving reaction r_1. We then set $n^-_{a,1}(u, u) = 1$ and $n^+_{a,1}(u, u) = 2$, meaning that whenever the agent chooses the self-loop transition, it will remain in state σ_a, and in the same location u, but it will be multiplied by a factor of 2. In general a reaction r_k with $n^-_{i,k}(l, l) = 1$ and $n^+_{i,k}(l, l) = a$, either multiplies (if $a > 1$) or decimates ($a < 1$) an agent in state σ_i at location l.

Movement of agents. We want to model the movement of an agent in state σ_a at location u, to state σ_b of location w. (Figure 2.b). We add an arc from state σ_a to reaction r_2, and one from r_2 to σ_b. We set $n^-_{a,2}(u, u) = 1$ and $n^+_{b,2}(u, w) = 1$.

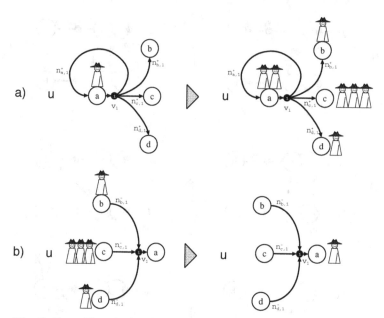

Fig. 3. Examples of: a) multi-state duplication, b) multi-state fusion

In general $n_{i,k}^+(l, l') = 1$, defines a movement from state σ_i of location l to state σ_j of location l' due to a reaction r_k.

Combined movement, duplication and decimation of agents. Figure 2.c) shows a very general case, where the agent in state σ_a at location u jumps to state σ_b due to reaction r_2: during the transition, two copies of the agent are generated in location w, and the original agent might remain in location u with probability $1/2$, this is achieved by setting $n_{b,2}^+(u, w) = 2$ and $n_{b,2}^+(u, u) = 0.5$.

Multi-state duplication accounts for cases in which one agent during its transition is duplicated into several states. For example, in Figure 3.a, the agent in state σ_a gives birth to six other agents due to reaction r_1: one remains in state σ_a together with the original agent, one goes to state σ_b, three to state σ_c and one to state σ_d. This effect can be achieved by defining $n_{a,1}^+(u, u) = 1$, $n_{a,1}^+(u, u) = 2$, $n_{b,1}^+(u, u) = 1$, $n_{c,2}^+(u, u) = 3$ and $n_{d,1}^+(u, u) = 1$. This methodology can be easily extended to support multi-state duplication in different locations.

Multi-state fusion is the exact opposite of multi-state duplication: agents in several states are merged together. In Figure 3.c, for instance, one agent in state σ_b, three from σ_c and one from σ_d are merged together to generate one agent in state σ_a. This is achieved by setting: $n_{b,1}^-(u, u) = 1$, $n_{c,2}^-(u, u) = 3$, $n_{d,1}^-(u, u) = 1$ and $n_{a,1}^+(u, u) = 1$.

2.2   Analysis

As for conventional Markovian Agents, we describe a state of the system as a vector $x(l) = |x_i(l)|$, with one component for each state $\sigma_i \in \Omega$, per each location $l \in L$. $x_i(l)$ represents the number of agents in state σ_i at location l, and the evolution of the model is approximated using a mean field technique. The transient evolution of the PMAM model can be evaluated by solving the following system of non-linear differential equations:

$$\begin{cases} x(l,0) = Init(l) \\ \dfrac{dx(l,t)}{dt} = \mathcal{B}(l,x) + x(l,t) \cdot (\mathcal{D}(l,x) + \mathcal{K}(l,x)) + \mathcal{M}(l,x) \end{cases} \quad (1)$$

In Equation (1), $Init(l) = |init_i(l)|$ is a vector whose elements $init_i(l)$ determine the initial number of agents in state σ_i for each location l. Vector $\mathcal{B}(l,x)$ represents the birth term and takes into account both the spontaneous generation of agents, and the birth induced by neighbor locations. It can be computed as:

$$\mathcal{B}(l,x) = \tilde{b}(l) + \sum_{l' \in L} \sum_{\sigma_k \in \Omega} \hat{b}^{[k]}(l',l) \cdot x_k(l') \quad (2)$$

The induced birth rate (second term on the r.h.s. of Equation (2)) can be affected by the number of agents in any state σ_k and in any location l' of the model. In particular, the induced birth rate is proportional to both the number of agents $x_k(l')$, and the induction factor $\hat{b}^{[k]}(l',l)$. In a similar way, the death term (D) can be defined:

$$\mathcal{D}(l,x) = diag\left(\tilde{d}(l) + \sum_{l' \in L} \sum_{\sigma_k \in \Omega} \hat{d}^{[k]}(l',l) \cdot x_k(l') \right) \quad (3)$$

where $diag(y)$ is a diagonal matrix composed by the elements of vector y. Note that in Equation (1), $\mathcal{D}(l,t)$ is left multiplied by the count vector $x(l,t)$: in this way the actual death rate is proportional both to the number of agents in the inducing location $x_k(l')$, and to the number of agents in the considered position $x(l)$. Matrix $\mathcal{K}(l,x)$ represents the main transition kernel, and is defined as:

$$\mathcal{K}(l,x) = \tilde{Q}(l) + \sum_{l' \in L} \sum_{\sigma_k \in \Omega} \hat{Q}^{[k]}(l',l) \cdot x_k(l') \quad (4)$$

Note that since both matrices $\tilde{Q}$ and $\hat{Q}$ are infinitesimal generators, also $\mathcal{K}(l,x)$ is an infinitesimal generator. Finally, vector $\mathcal{M}(l,x)$ accounts for the change of agents in a location l due to the *reactions* used to model duplication, decimation or movement of agents. It can be defined as:

$$\mathcal{M}(l,x) = -\sum_{l'' \in L} v(l,x) \cdot \left[N^-(l,l'') \right]^T + \sum_{l'' \in L} v(l'',x) \cdot \left[N^+(l'',l) \right]^T \quad (5)$$

where $[N]^T$ denotes the transpose of matrix N. The first term on the r.h.s. of Equation (5) accounts for the change in the number of agents occurring in one state due to elements removed by the reactions. In particular, it accounts for all the agents that are leaving location l directed to a reaction happening in location l''. The second term on the r.h.s. considers agents that are entering a location l, coming from a reaction that takes place in a location l''. The definition of $v(l, x)$ must depend on the total state of the system x, and it must be correctly defined to prevent reactions to happen when there are not enough agents in the involved states: an improperly defined function $v(l, x)$ can lead to negative counts in the number of agents. Determining the condition for which a function $v(l, x)$ does not lead to negative counts, it is an important topic that will be investigated in future woks: here we will limit ourselves to consider functions coming from system biology, that are known to correctly behave as reaction rates. In particular we refer to the Law of Mass Action (Waage and Guldberg, 1864), which tells that the reaction rate is proportional to the probability of a collision of the reactants, that in turn is proportional to the concentration of reactants (agents in our case), elevated to the multiplicity required to start the reaction:

$$v_h(l, x) = \gamma_h \prod_{j:\exists l'' \in L, n_{j,h}^-(l,l'')>0} \left(\sum_{l' \in L} x_j(l') \right)^{\sum_{l'' \in L} n_{j,h}^-(l,l'')} \tag{6}$$

Here γ_v represents the speed at which the reaction occurs. Note that since the *soruce* agents of reaction r_h in location l can arrive from any location l'' such that $n_{j,h}^-(l, l'') > 0$, the total count of agents required to engage a reaction must be computed with the sum $\sum_{l'' \in L} n_{j,h}^-(l, l'')$. For the same reason, the number of agents involved in the reaction, must be computed with the sum over all the possible input locations, i.e., $\sum_{l' \in L} x_j(l')$.

3   The Biological Model

Recent studies in cancer biology have lead to a new perspective in tumor progression, known as the CSC theory. It states that the growth and evolution of many cancers are driven by a small population of cells named CSC [20], and that CSC-based tumors are hierarchically structured, and characterized by different subpopulations of cells: CSCs, Progenitor Cells (PCs) and Totally differentiated Cells (TCs). Moreover, such heterogeneity is considered the cause of the failure of many conventional therapies. Indeed, although several treatments induce death on the differentiated tumor cell subpopulations (i.e. non-CSCs), they are ineffective on CSCs, which resist to most of the common drugs and vaccinations, causing tumor relapse [19]. It is hence fundamental to fully comprehend the mechanisms underlying the CSC hierarchy, whose characterization is crucial in the study of tumor progression and in predicting treatment response.

In this paper we propose a model describing the CSC-based tumor growth and able to reproduce the overall dynamics among cell subpopulations during tumor progression. Specifically, starting from the CSC theory we have defined relationships (7), which express how subpopulations change during tumor evolution. Notice that, over the subpopulations defined by the CSC theory (i.e. CSCs, PCs and TCs) we have introduced a new set of cells, called Dead/quiescent Cells (DCs), since they represent a significant contribution to the tumor volume.

a) $CSC \xrightarrow{P_{sy}\omega_{CSC}} CSC + CSC$ g) $PC_1 \xrightarrow{\eta_2} PC_2$

b) $CSC \xrightarrow{(1-P_{sy})\omega_{CSC}} CSC + PC_1$ h) $PC_1 \xrightarrow{\delta_2} DC$

c) $CSC \xrightarrow{\eta_1} PC_1$ i) $PC_2 \xrightarrow{\eta_3} TC$

d) $CSC \xrightarrow{\delta_1} DC$ j) $PC_2 \xrightarrow{\delta_2} DC$ (7)

e) $PC_1 \xrightarrow{\gamma_{PC}} CSC$ k) $TC \xrightarrow{\delta_3} DC$

f) $PC_1 \xrightarrow{\omega_{PC}} PC_2 + PC_2$ l) $DC \xrightarrow{\delta_4} 0$

Equations (7) describe both dynamics among cell subpopulations, and those cellular events which are responsible for changes in each subpopulation. More precisely, in accordance with the CSC theory, several factors are responsible for the CSC variation [20]. Firstly, they can self-renew either symmetrically (P_{sy} ω_{CSC}), or asymmetrically (($1 - P_{sy})\omega_{CSC}$). Then, a furhter progression mode - called differentiation (η_1) or CSC commitment - can also occur contributing to give rise to the multipotent progenitor cell subpopulation. PCs, on turn, evolve through two stages: the first one, PC_1, able to proliferate (ω_{CSC}) and differentiate (η_2) giving rise to the second stage, PC_2. On the other side, PCs$_2$ develop into non-proliferative terminally differentiated cells, TCs, only through differentiation (η_3). We considered also the de-differentiation (γ_{PC}) of PCs$_1$ into CSCs, as described in [5]. De-differentiation is the process in which specialized cells take on a more primitive stage, probably in response to environment signals [15]. Lastly, CSCs, PCs and TCs are also affected by a death rate, (δ_i), which is specific for each cell type ($i = 1, 2, 3$ for CSCs, PCs, TCs, respectively). Therefore, the DC subpopulation increases as a result of the death of the viable cell species, but it decreases due to cell lysis (δ_4).

A schematic representation of the described subpopulation dynamics is shown in Figure 4, where the hierarchical organization of CSC-based tumors is highlighted too. Specifically, Figure 4 shows how subpopulations differ considering their differentiation degree and their proliferative potential. Differentiation is highest in TCs - which are completely differentiated - and, conversely, proliferation potential is highest in CSCs, which can divide either symmetrically or asymmetrically. Using a derivation similar to the Generalized Mass Action law [8], relations (7) were translated into the following system of Ordinary Differential Equations (ODEs):

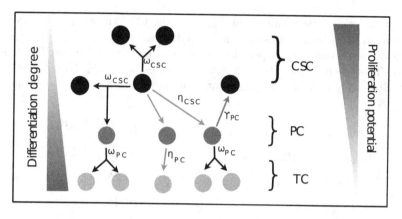

Fig. 4. CSC hierarchical model. Simplified schematic representation of the cell sub-population dynamics and interactions. The proliferative potential and differentiation degree are highlighted for each subpopulation: moving from the CSC compartment to the TC one, cells became more differentiated and lose their proliferative ability.

$$\frac{dS_{CSC}}{dt} = P_{sy}\omega_{CSC}S_{CSC} + \gamma_{PC}S_{PC_1} - \eta_1 S_{CSC} - \delta_1 S_{CSC}$$

$$\frac{dS_{PC_1}}{dt} = (1 - P_{sy})\omega_{CSC}S_{CSC} - \omega_{PC}S_{PC_1} - \gamma_{PC}S_{PC_1} +$$
$$+ \eta_1 S_{CSC} - \eta_2 S_{PC_1} - \delta_2 S_{PC_1}$$

$$\frac{dS_{PC_2}}{dt} = 2\omega_{PC}S_{PC_1} + \eta_2 S_{PC_1} - \eta_3 S_{PC_2} - \delta_2 S_{PC_2}$$

$$\frac{dS_{TC}}{dt} = \eta_3 S_{PC_2} - \delta_3 S_{TC}$$

$$\frac{dS_{DC}}{dt} = \delta_1 S_{CSC} + \delta_2(S_{PC_1} + S_{PC_2}) + \delta_3 S_{TC} - \delta_4 S_{DC} \qquad (8)$$

where S_{CSC}, S_{PC}, S_{TC} and S_{DC} are the total numbers of CSCs, PCs, TCs and DCs, respectively.

Although in cancer many mechanisms of cell control are disrupted, the feedback control - typical of lineage progression - still plays an important role in tumor progression. More precisely, the lineage progression of the original tissue - as well as many other intrinsic feedback controls - continue to operate, even if altered by cell mutations. As described by the CSCs theory, indeed, tumor cells appear to progress through a lineage stages-like evolution, as in normal tissues. Therefore, auto growth limitation, expressed by feedback regulatory controls, were introduced in model (8). Specifically, such mechanisms are defined to control cell proliferation of both CSCs and PCs, i.e. to model an unbounded cell division. The following two functions represent the feedback regulating proliferation:

$$\omega_{CSC}(t) = \frac{\omega_{CSC}}{1 + h_{CSC}S_{TC}(t)}, \tag{9}$$

$$\omega_{PC}(t) = \frac{\omega_{PC}}{1 + h_{PC}S_{TC}(t)}, \tag{10}$$

where h_{CSC} and h_{PC} define the strength of the feedback. Notice that, in equations (9),(10) the number of TCs acts as a brake on cell growth, being CSC and PC proliferation rates defined as decreasing functions of this quantity. Indeed, a high number of cells in the system induces a decrease in the cell proliferation rates.

Model Parameters All rates and initial cellular concentrations are tuned starting from those found in the literature [21,18,6] and by experimental evidences. Specifically, considering the tumor-initiation ability of CSCs, population dynamics were investigated using only few CSCs, with respect to the total, as initial condition. Parameters were then retrieved by tuning the system to reproduce the tumor mass growth trend observed in BALB/c mice, after a subcutaneous injection of 10^5 cancer cells. All the values used in the numerical experiments are reported in Table 1.

Table 1. Parameter values of cell population model (8)

Symbols	Values	Biological meaning
ω_{CSC}	6	CSC proliferation
ω_{PC}	60	PC proliferation
P_{sy}	0.5	CSC symmetrical proliferation probability
η_1	1	CSC differentiation
η_2	4	PC_1 differentiation
η_3	6	PC_2 differentiation
γ_{PC}	0.0001	PC_1 de-differentiation
δ_1	0.0001	CSC death
δ_2	0.001	PC death
δ_3	0.001	TC death
δ_4	0.001	DC death
h_{CSC}	10^{-12}	CSC feedback intensity
h_{PC}	10^{-12}	PC feedback intensity
k_{CSC}	0.016	initial CSC concentration
k_{PC}	0.109	initial PC concentration
k_{TC}	$1 - (k_{CSC} + k_{PC})$	initial TC concentration
S_0	10^5	initial number of cell
S_{CSC0}	$S_0 k_{CSC}$	initial number of CSCs
S_{PC0}	$S_0 k_{PC}$	initial number of PCs
S_{TC0}	$S_0 k_{TC}$	initial number of TCs
S_{DC0}	0	initial number of DCs

4   Experiments

Here we study the biological case presented in the previous section: first we describe the corresponding PMAM model and analyze the case with a single location, also showing its equivalence to the conventional biological models. Then, we extend the model to consider movement of agents in a bi-dimensional space, and exploit it to derive new results. The initial phase of the increment of tumor volume, that in our case corresponds to an increment of the number of TC (see Figure 6), is comparable with respect to experimental data published in [12].

Figure 5 shows the PMAM model of the system shown in Equation 7. For each subpopulation in the model, there is a corresponding state in the PMA, and for each subpopulation dynamic (expressed as evolution equation), there is a corresponding arc in the PMA. The evolution of the populations is modeled by changes of state: for example, a CSC cell that evolves in a PC_1 cell is modeled by a state transition from CSC to PC_1 in the agent. In particular, reactions that transform a subpopulation into another (e.g. Equations (7.c-e,g-k)) are modeled with a normal state transition at the rate corresponding to the speed of the reaction: for example this is done with the arc from state CSC to state PC_1 at rate η_1 for Equation (7.c), with the arc from state PC_1 to state PC_2 at rate η_2 for Equation (7.g) and so on. Removal of subpopulation elements (Equation (7.l)) are instead modeled with death-arc, as the one that exits state CSC at rate δ_4. Equations that models an increase of subpopulation are modeled with reactions. Equation (7.a), that models a duplication of a CSC cell, is modeled with reaction r_1 that performs a self-loop on state CSC. In this case, the speed $v(l)_1 = P_{sy}\omega_{CSC}$ corresponds to the rate of the evolution equation and we have $n^+_{CSC,1}(l, l'') = 2$. Equation (7.f) represents the combination of a subpopulation change (from PC_1 to PC_2) with a duplication. This is done by connecting an arc from state PC_1 to reaction r_3, and one from reaction r_3 to state PC_2. The speed matches the one of Equation (7.f) (that is $v(l)_3 = \omega_{PC}$), and we have $n^+_{PC_2,3}(l, l'') = 2$ to define the duplication. Finally, Equation (7.b) adds a CSC and a PC_1 starting from a CSC. This is modeled by reaction r_2 at speed $v(l)_2 = (1 - P_{sy})\omega_{CSC}$. Since two subpopulations are generated, two output arcs exiting from reaction r_2 connect it respectively with states CSC and PC_1. Both subpopulations are generated with multiplicity one, so we have $n^+_{PC_1,2}(l, l'') = 1$ and $n^+_{CSC,2}(l, l'') = 1$. Note that in all the three reactions, we have implemented the feedback mechanism defined in Equations (9) and (10) for both ω_{CSC} and ω_{PC}.

4.1   Single Location Model

We first consider a model with a single location $L = \{l\}$. In this case, if we apply the technique proposed in Section 2.2 to the model presented in Figure 5, we obtain exactly the same equations as the ones presented in Equation (8). We have solved the model using a custom experimental Markovian Agents tool on a standard MacBook Air, that was able to compute the solution in about one second. The tool analyzes the PMAM models using a mean field approximation.

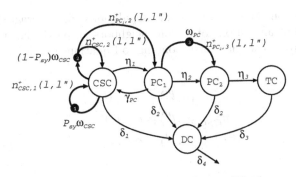

Fig. 5. Model of the CSC population diffusion

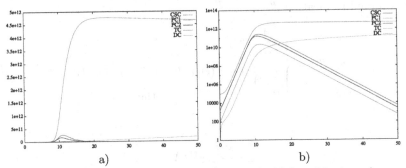

a)   b)

Fig. 6. Single location model: a) linear scale, b) logarithmic scale

For each cell, and for each state, a non-linear differential equation is solved to capture the evolution of the populations in the model. Figure 6 shows the results both in linear and logarithmic scale. As we can see from the linear scale representation in Figure 6a, at around time $T = 8$ the subpopulation of the TC cells starts to explode, and at around $T = 12$, CSC, PC_1 and PC_2 reach their maximum count and start to decrease. The TC subpopulation is limited at around 10^{12} due to the feedback mechanism defined in Equations (9) and (10). If we examine the logarithmic version of Figure 6b, we can appreciate the fact that there is a constant growth of DC cells as $T \to \infty$ and that in the beginning (around $T = 1$) there is first a decrease in both PC_1 and PC_2 cells that start to grows immediately after.

4.2 Spatial Model

We then consider the same system on a bi-dimensional grid of 21x21 locations[1] $L = \{l_{ij} : 1 \le i, j \le 21\}$. Thanks to the proposed PMAM formalism, we can express the possibility of reactions to expand in neighbor cells. To simplify the

---

[1] The choice of using a two-dimensional space has been made to simplify the presentation of the results. The proposed technique could be easily extended to consider tridimensional models.

model, we focus on the case where new populations can be created not only in the same cell, but also in one of its four neighbors: in particular we modify the arcs that exit from reactions corresponding to Equations (7.a,b,f). In particular, we set:

$$n^+_{CSC,1}(l, l'') = 1 + \phi(l, l'')$$ (11)

$$n^+_{CSC,2}(l, l'') = \phi(l, l'')$$ (12)

$$n^+_{PC_2,3}(l, l'') = \phi(l, l'')$$ (13)

(14)

where $\phi(l, l'')$ represents the probability that a subpopulation generated in location l is routed to location l''.

$$\phi(l, l'') = \begin{cases} \dfrac{1}{G(l)} \psi(l) & \text{if } l'' = l \\ \dfrac{1}{G(l)} [1 - \psi(l)] \, \psi(l'') & \text{if } l'' \in neigh(l) \\ 0 & \text{otherwise} \end{cases}$$ (15)

with:

$$\psi(l) = e^{-\left(\dfrac{\sum_k x_k(l)}{\chi_{max}}\right)^{\alpha}}$$ (16)

$$G(l) : \sum_{l' \in L} \phi(l, l') = 1$$ (17)

$$neigh(l_{ij}) = \{l'_{i'j'} : (i - i')^2 + (j - j')^2 = 1\}$$ (18)

$\psi(l)$ is the probability that a location l accepts a cell: it is controlled by a maximum value χ_{max} and it tends to zero if the total number of cells in location l is greater than χ_{max}, or to one if the total population in l is less than χ_{max}. Parameter α controls the speed at which the transition from zero to one occurs: higher values of α models a more deterministic behavior, while smaller values causes smoother transitions. In our experiments we set $\alpha = 2$ and $\chi_{max} = 10^{11}$. $G(l)$ is a normalization constant, to make $\phi(l, l')$ a proper probability distribution in l'. Finally $neigh(l)$ represents the set of the cells that share one of the borders with l (i.e. its closest neighbors).

We evaluate the model starting in two different configurations. First we consider that all the subpopulations start in the center of the space, using the same proportion used for the single location case presented in Section 4.1. As it can be seen in Figure 7, the evolution of the model remains concentrated in the center until the total population of cells reaches χ_{max}. Then neighbor cells start to be filled and expansion starts to grow at around $T = 8$. As for the single location case, at $T = 12$ the concentration of PC cells starts to decrease. At time $T = 20$, we only have a high concentration of TC cells and all the other subpopulations are almost negligible.

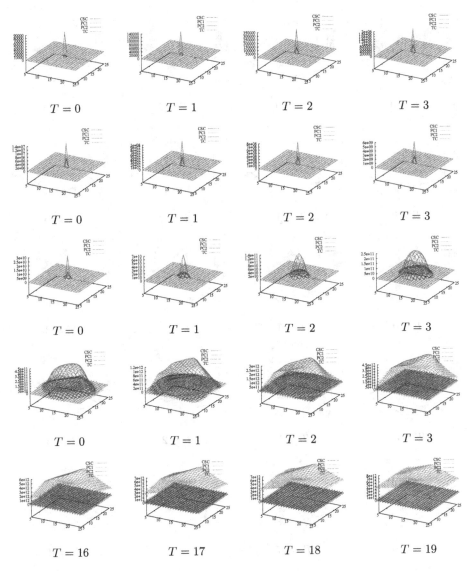

Fig. 7. Central diffusion

We then consider a different initial distribution that resemble the one of a *mammosphere* [4]. In particular we imagine that there is a nucleus of CSC cells, followed by a ring of PC cells, and finally an outer membrane of TC cells. The evolution of this model can be followed in Figure 8. As it can be seen, without considering other evolution rules, the outer TC and CSC cells tend to implode toward the center, and the initial ring topology is almost canceled out already at $T = 3$. From that point on, the model evolves in a way similar to the one presented in Figure 7: in this case however the concentration of cells is more

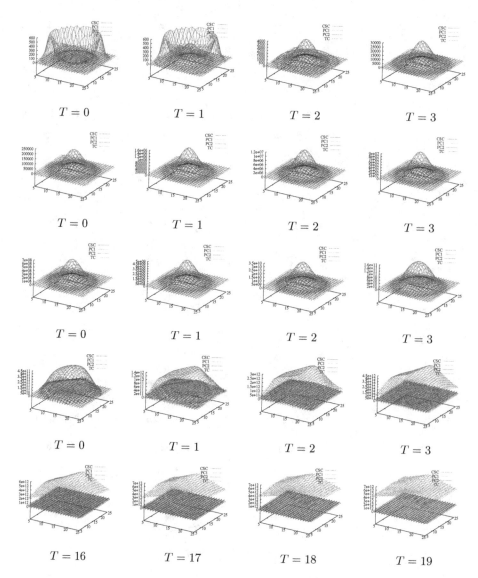

Fig. 8. Mammosphere simulation

evenly spread across the space, and there is a more abrupt reduction in the concentration of non-TC cells after $T = 12$.

5   Conclusions

In this paper we have presented a new Markovian Agents based formalism that is able to consider spatial evolution of populations of entities. We have applied to a simple biological example to test its suitability to capture spatial phenomena that can happen in reaction occurring in an area. The validation of our results in term of spatial growth will be given by a in vivo and in vitro experiments. In detail, mammospheres generated from TUBO cell line will be frozen then sliced into sections. Immunofluorescence of these frozen sections was performed according to the antibodies - selected by previous analysis - associated to each cell subpopulation, e.g. CSC, PC and TC. These fluorescence imaging was performed using a confocal laser scanning microscope will give to us the possibility to obtain the spatial distribution of each populationa of cells. In future works, this technique could be applied to study the effect of localized therapies, trying to determine optimal prescription strategies (such as diffusing a medicine in the center of tumor, or in some points on its boundaries). The same technique can also be used to model other spatial based system, such as wireless networks [2], economical evolution of firms [16] and many more.

References

1. Bobbio, A., Gribaudo, M., Telek, M.: Analysis of large scale interacting systems by mean field method. In: 5th International Conference on Quantitative Evaluation of Systems - QEST 2008, St. Malo (2008)
2. Gribaudo, M., Manini, D., Chiasserini, C.: Studying mobile internet technologies with agent based mean-field models. In: Dudin, A., De Turck, K. (eds.) ASMTA 2013. LNCS, vol. 7984, pp. 112–126. Springer, Heidelberg (2013)
3. Chiasserini, C., Gribaudo, M., Manini, D.: Traffic offloading/onloading in multi-rat cellular networks. In: Proc. of IFIP Wireless Days Conference (WD 2013). IEEE Press (2013)
4. Grange, C., Lanzardo, S., Cavallo, F., Camussi, G., Bussolati, B.: Sca-1 identifies the tumor-initiating cells in mammary tumors of balb-neut transgenic mice. Neoplasia 10 (2008)
5. Chaffer, C., Brueckmann, I., Scheel, C., Kaestli, A.J., Wiggins, P.A., Rodrigues, L.O., Brooks, M., Reinhardt, F., Su, Y., Polyak, K., Arendt, L.M., Kuperwasser, C., Bierie, B., Weinberg, R.A.: Normal and neo-plastic nonstem cells can spontaneously convert to a stem-like state. PNAS 108, 7950–7955 (2011)
6. Turner, C., Kohandel, M.: Investigating the link between epithelial-mesenchymal transition and the cancer stem cell phenotype: A mathematical approach. Journal of Theoretical Biology 265, 329–335 (2010)
7. Bruneo, D., Scarpa, M., Bobbio, A., Cerotti, D., Gribaudo, M.: Markovian agent modeling swarm intelligence algorithms in wireless sensor networks. Perform. Eval. 69(3-4), 135–149 (2012)
8. Voit, E.O.: Computational Analysis of Biochemical Systems. Cambridge University Press (2000)
9. Cordero, F., Manini, D., Gribaudo, M.: Modeling biological pathways: an object-oriented like methodology based on mean field analysis. In: The Third International Conference on Advanced Engineering Computing and Applications in Sciences (ADVCOM), pp. 193–211. IEEE Computer Society Press (2009)
10. Youssefpour, H., Li, X., Lander, A.D., Lowengrub, J.S.: Multispecies model of cell lineages and feedback control in solid tumors. Journal of Theoretical Biology 304, 39 (2012)

11. Edwards, J.S., Palsson, B.O.: How will bioinformatics influence metabolic engineering? Biotechnology and Bioengineering 58(2-3), 162–169 (1998)
12. Conti, L., Lanzardo, S., Arigoni, M., Antonazzo, R., Radaelli, E., et al.: The noninflammatory role of high mobility group box 1/toll-like receptor 2 axis in the self-renewal of mammary cancer stem cells. FASEB Journal (2010)
13. Benaim, M., LeBoudec, J.Y.: A class of mean field interaction models for computer and communication systems. Performance Evaluation 65(11-12), 823–838 (2008)
14. Gribaudo, M., Cerotti, D., Bobbio, A.: Analysis of on-off policies in sensor networks using interacting markovian agents. In: 4th International Workshop on Sensor Networks and Systems for Pervasive Computing - PerSens 2008, Hong Kong (2008)
15. Gupta, P.B., Chaffer, C.L., Weinberg, R.A.: Cancer stem cells: mirage or reality? Nature Medicine 15(9), 1010–1012 (2009)
16. Pisano, P., Manini, D., Gribaudo, M., Pironti, M.: Strategic focus and business model organization: The main field analysis approach. Journal of Modern Accounting and Auditing 7(7) (2012)
17. Bjerkvig, R., Tysnes, B.B., Aboody, K.S., Najbauer, J., Terzis, A.J.A.: The origin of the cancer stem cells: Current controversis and new insights. Nature Review Cancer 5 (2005)
18. Molina-Pena, R., Álvarez, M.: A simple mathematical model based on the cancer stem cell hypothesis suggests kinetic commonalities in solid tumor growth. Plos One 7 (2012)
19. Pardal, R., Clarke, M.F., Morrison, S.J.: Applying the principles of stem-cell biology to cancer. Nature Review Cancer 3 (2003)
20. Tang, T.G.: Understanding cancer stem cell heterogeneity and plasticity. Cell Research 22 (2012)
21. Zhu, X., Zhou, X., Lewis, M., Xia, L., Wong, S.: Cancer stem cell, niche and egfr decide tumor development and treatment response: A bio-computational simulation study. Journal of Theoretical Biology 269, 138–149 (2011)

Modelling Interacting Epidemics in Overlapping Populations

Marily Nika[1], Dieter Fiems[2], Koen De Turck[2], and William J. Knottenbelt[1]

[1] Department of Computing, Imperial College London,
London SW7 2AZ, United Kingdom
{marily,wjk}@doc.ic.ac.uk
[2] Ghent University, Department TELIN, St-Pietersnieuwstraat 41, 9000 Gent,
Belgium
{dieter.fiems,kdeturck}@telin.ugent.be

Abstract. Epidemic modelling is fundamental to our understanding of biological, social and technological spreading phenomena. As conceptual frameworks for epidemiology advance, it is important they are able to elucidate empirically-observed dynamic feedback phenomena involving interactions amongst pathogenic agents in the form of syndemic and counter-syndemic effects. In this paper we model the dynamics of two types of epidemics with syndemic and counter-syndemic interaction effects in multiple possibly-overlapping populations. We derive a Markov model whose fluid limit reduces to a set of coupled SIR-type ODEs. Its numerical solution reveals some interesting multimodal behaviours, as shown in our case studies.

Keywords: Epidemics, Social Networks, Syndemic, Counter-syndemic.

1 Introduction

You think because you understand 'one' you must also understand 'two', because one and one make two. But you must also understand 'and'...

Rumi (13th century Persian Poet)

Epidemics of various kinds have been an important focus of study throughout human history. As health care standards have risen and information technology has advanced over the past half century, our preoccupation with epidemics of a biological nature has lessened while our obsession with epidemics of a social and technological nature has dramatically increased. This has been accompanied by a growing realisation that many of the epidemiological techniques used in the modelling of biological diseases can be readily transplanted into social and technological domains such as content and information diffusion, rumour spreading, gossiping protocols and viral marketing.

There is one recent but crucial respect in which our conceptual understanding of biological epidemics has advanced dramatically. In particular, it has become increasingly realised that it is important to study the interplay between

B. Sericola, M. Telek, and G. Horváth (Eds.): ASMTA 2014, LNCS 8499, pp. 33–45, 2014.

pathogenic agents and between pathogenic agents and their environment. The corresponding field of study is known as *synepidemiology* in which the subjects of study are *syndemics* and *counter-syndemics* [35]. A syndemic is a set of mutually reinforcing health problems whose combined impact is more devastating than sum of the health problems in isolation (e.g. the risk developing tuberculosis is estimated to be between 12–20 times high for people with HIV [21]), while a counter-syndemic concerns a set of mutually inhibiting health problems whose combined impact is not as high as the sum of the health problems in isolation (e.g. studies suggest that a measles infection can temporarily inhibit the replication of the HIV virus [27]). Very lately, there has been a growing awareness that syndemics may also exist in a technological context: e.g. the purchase of a smartphone may make the purchase of the corresponding accessories and applications more likely [30].

In this paper, we extend the well-known Susceptible-Infected-Recovered (SIR) compartmental epidemiological model to support the interplay of multiple interacting epidemics. Our focus is on a scenario of two potentially-interacting epidemics spreading across a set of overlapping subpopulations. In this context, we derive a Markov model which describes the state changes of an individual with respect to each epidemic and whose transition rates incorporate syndemic and counter-syndemic interactions. The fluid limit of this Markov model reduces to a set of coupled SIR-type ODEs, the solution of which describes the evolution of the number of individuals infected by each epidemic.

The remainder of this paper is organised as follows. Section 2 presents an historical perspective on conceptual frameworks and modelling efforts pertinent to the field of epidemic modelling in the biological, social and technological domains. Section 3 presents our approach in extending the SIR model to support interacting epidemics, while Section 4 presents case studies of two interacting SIR epidemics propagating through two intersecting populations with various degrees of overlap. Section 5 concludes and considers avenues for future work.

2   Background

Human societies have been ravaged by biological epidemics throughout history with recurrent deadly outbreaks of bubonic plague, smallpox, yellow fever, cholera and influenza [38]. As shown in Fig. 1, the predominant early theories of disease causation were mostly supernatural, astronomical or religious, with causal agents including evil spirits, planetary motion and divine retribution. From the Middle Ages until Victorian times, it was also believed that if one inhaled *miasmas* – toxic vapors that emanated from swamps or decaying organic matter – disease would result [32]. Progress towards a more scientific and data–based approach began to be made from 1600 onwards with the collection of the first public health statistics, by John Graunt (1620–1674) [11] and others. One of the most famous studies was by John Snow of the 1854 London Cholera epidemic [36] in which he identified a particular water pump as the likely source of the outbreak.

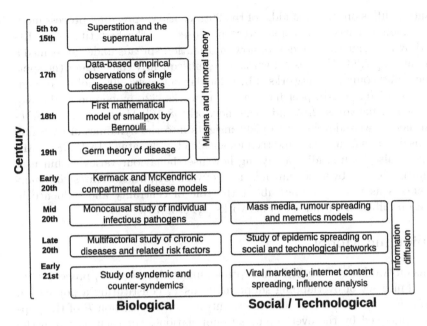

Fig. 1. The historical development of conceptual frameworks for epidemics

Predictive mathematical models for epidemics were relatively slow to develop, despite their utility in understanding, managing and forecasting epidemics. One of the earliest was by Daniel Bernoulli who carried out a study of the effects of smallpox vaccination in 1766 [9]. But arguably the most significant breakthrough came with the compartmental disease models proposed by Kermack and McKendrick in 1927 [19]. These elegantly express disease dynamics as coupled ODEs. The most well-known model is the Susceptible-Infected-Recovered (SIR) model. SIR features a closed population of individuals divided into three evolving sub-populations: $S(t)$ tracks those susceptible to become infected by the disease at time t, $I(t)$ tracks those infected by the disease with rate β and $R(t)$ tracks those who have recovered from the disease at rate γ.

The science of epidemiology has made rapid advances in recent times and has moved from monocausal studies of infectious diseases to multifactor studies of chronic diseases (e.g. obesity). It has become increasingly realised that many diseases feature a complex web of interconnected risk factors (the so-called *web of causation*), which may include relationships with other diseases and relationships between diseases and the environment. The latter point of view is central to the science of *synepidemiology* [12]. Related mathematical models have been concurrently evolving, with some studies of the dynamics of two possibly-dependent co-infections in single populations [4, 25].

Of course it is not only disease which spreads in an epidemic fashion and researchers have proved adept at progressively transplanting the corresponding theory into sociological and technological domains, especially those related to information diffusion. In the middle of the 20th century, spreading-process models for rumours, ideas and memes were proposed for the first time [5, 7, 13], followed by mathematical models of how information spreads under mass media dissemination [17, 23]. Various networks have subsequently come under the spotlight including computer networks [14], vehicular networks [39], mobile and ad-hoc networks [20], peer-to-peer file-sharing networks [22], mobile networks [34], wireless sensor networks [2, 6] and social networks [8, 16]. More recently, mathematical models were developed to yield insights into the dynamics of emerging infectious diseases from social and technological network data [3, 15, 18, 28, 29, 31]. There have also been studies analysing how user behaviour varies within user communities defined by a recommendation network [24], which creates *viral marketing* effects as well as studies about the role of centrality and influence in information diffusion within social networks [1, 26, 33, 37].

3　Epidemic Model

We focus on two interacting SIR (susceptible, infected, recovered) processes living on a finite set of overlapping subpopulations P_i constituting a population $P = \cup_i P_i$. For notational convenience, we introduce the partition $\mathcal{P}$ of the population P induced by the overlapping sub-populations. For each part p in the partition, let its neighbourhood $\mathcal{N}(p)$ be a set of parts which includes p. Moreover, the size of the population of part p is denoted by $n(p)$.

Remark 1. The neighbourhood of any part will be used to relate an individual's view-of-the-world to its infection rate. To make this concrete, consider a simple example where there are two subpopulations with a non-empty intersection. These overlapping subpopulations induce a partition with 3 parts: the two parts of individuals that belong to one subpopulation and not to the other, and the part corresponding to the intersection. As individuals in the intersection belong to both sub-populations, their neighbourhood includes all parts. The individuals that only belong to a single sub-population only see their own sub-population. Their neighbourhood therefore consists of their own part and the intersection.

Any individual of the population is susceptible to, infected by or recovered from any of two epidemics. The state of an individual is described by a pair (k, ℓ), with $k, \ell \in \{s, i, r\}$, where s, i and r stand for susceptible, infected and recovered, respectively and where k and ℓ refer to the first and second epidemic, respectively. We consider a Markovian epidemic model and its fluid limit. At any point in time, the state of the Markov chain is described by the number of individuals in the different states and in the different parts.

Prior to introducing the Markov chain, some additional notation is required. Let $x^p_{(k,\ell)}$ be the number of individuals of part p that are in state (k, ℓ), and let $\mathbf{x}$ be the vector with elements $x^p_{(k,\ell)}$, for $p \in \mathcal{P}$ and $k, l \in \{s, i, r\}$. The state space $\mathcal{X}$ of the Markov chain is defined as the set of vectors $\mathbf{x}$ such that,

$$x^p_{(k,\ell)} \in \mathbb{N} = \{0, 1, 2, \ldots\}, \quad \sum_{k,\ell \in \{s,i,r\}} x^p_{(k,\ell)} = n(p) \quad \text{for all } p \in \mathcal{P}.$$

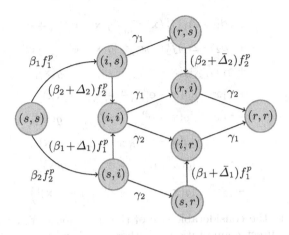

Fig. 2. Transition rates for an individual in part p

Moreover, let $\mathbf{e}^p_{(k,\ell)}$, for $p \in \mathcal{P}$ and $k, l \in \{s, i, r\}$ be the obvious unit vectors of the state space $\mathcal{X}$. The following parameters describe the transition rates for changing states. An (s, s)-individual in part p gets infected by the first and second epidemics with rates $\beta_1 f_1^p(\mathbf{x})$ and $\beta_2 f_2^p(\mathbf{x})$, respectively. Here, $f_1^p(\mathbf{x})$ and $f_2^p(\mathbf{x})$ are the fractions of individuals that are infected by epidemic 1 and 2 in the neighbourhood of $p \in \mathcal{P}$,

$$f_1^p(\mathbf{x}) = \frac{\sum_{q \in \mathcal{N}(p)} \left(x^q_{(i,r)} + x^q_{(i,s)} + x^q_{(i,i)} \right)}{\sum_{q \in \mathcal{N}(p)} n(q)}, \tag{1}$$

$$f_2^p(\mathbf{x}) = \frac{\sum_{q \in \mathcal{N}(p)} \left(x^q_{(r,i)} + x^q_{(s,i)} + x^q_{(i,i)} \right)}{\sum_{q \in \mathcal{N}(p)} n(q)}. \tag{2}$$

If such an individual is already infected by or has already recovered from the other epidemic, the infection rate is modified. An (s, i) individual in part p gets infected by the first epidemic with rate $(\beta_1 + \Delta_1) f_1^p(\mathbf{x})$, while an (s, r) individual gets infected by the first epidemic with rate $(\beta_1 + \bar{\Delta}_1) f_1^p(\mathbf{x})$. Modified infection rates are defined likewise for the second epidemic. Finally, the recovery rates of an individual from epidemic 1 and 2 are constant and equal to γ_1 and γ_2, respectively. For clarity, the transition rates for an individual in part p are depicted in Figure 2. The infinitesimal generator $\mathcal{A}$ of this Markov chain is:

$$\mathcal{A}g(\mathbf{x}) = \sum_{p \in \mathcal{P}} \left(\beta_1 f_1^p(\mathbf{x}) x^p_{(s,s)} [g(\mathbf{x} - \mathbf{e}^p_{(s,s)} + \mathbf{e}^p_{(i,s)}) - g(\mathbf{x})] \right.$$

$$+ \beta_2 f_2^p(\mathbf{x}) x^p_{(s,s)} [g(\mathbf{x} - \mathbf{e}^p_{(s,s)} + \mathbf{e}^p_{(s,i)}) - g(\mathbf{x})]$$

$$+ (\beta_1 + \Delta_1) f_1^p(\mathbf{x}) x^p_{(s,i)} [g(\mathbf{x} - \mathbf{e}^p_{(s,i)} + \mathbf{e}^p_{(i,i)}) - g(\mathbf{x})]$$

$$+ (\beta_2 + \Delta_2) f_2^p(\mathbf{x}) x^p_{(i,s)} [g(\mathbf{x} - \mathbf{e}^p_{(i,s)} + \mathbf{e}^p_{(i,i)}) - g(\mathbf{x})]$$

$$+ (\beta_1 + \bar{\Delta}_1) f_1^p(\mathbf{x}) x_{(s,r)}^p [g(\mathbf{x} - \mathbf{e}_{(s,r)}^p + \mathbf{e}_{(i,r)}^p) - g(\mathbf{x})]$$

$$+ (\beta_2 + \bar{\Delta}_2) f_2^p(\mathbf{x}) x_{(r,s)}^p [g(\mathbf{x} - \mathbf{e}_{(r,s)}^p + \mathbf{e}_{(r,i)}^p) - g(\mathbf{x})]$$

$$+ \gamma_1 x_{(i,s)}^p [g(\mathbf{x} - \mathbf{e}_{(i,s)}^p + \mathbf{e}_{(r,s)}^p) - g(\mathbf{x})]$$

$$+ \gamma_1 x_{(i,i)}^p [g(\mathbf{x} - \mathbf{e}_{(i,i)}^p + \mathbf{e}_{(r,i)}^p) - g(\mathbf{x})]$$

$$+ \gamma_1 x_{(i,r)}^p [g(\mathbf{x} - \mathbf{e}_{(i,r)}^p + \mathbf{e}_{(r,r)}^p) - g(\mathbf{x})]$$

$$+ \gamma_2 x_{(s,i)}^p [g(\mathbf{x} - \mathbf{e}_{(s,i)}^p + \mathbf{e}_{(s,r)}^p) - g(\mathbf{x})]$$

$$+ \gamma_2 x_{(i,i)}^p [g(\mathbf{x} - \mathbf{e}_{(i,i)}^p + \mathbf{e}_{(i,r)}^p) - g(\mathbf{x})]$$

$$+ \gamma_2 x_{(r,i)}^p [g(\mathbf{x} - \mathbf{e}_{(r,i)}^p + \mathbf{e}_{(r,r)}^p) - g(\mathbf{x})] \Big), \tag{3}$$

for $\mathbf{x} \in \mathcal{X}$. Due to the considerable size of the state space $\mathcal{X}$, even for modest population sizes, direct computation of either transient or stationary distributions is quite forbidding. As we are mainly interested in the dynamics when the population is large, we focus on the fluid limit of the process. However, the original Markov chain will also be simulated and compared with the fluid limits.

More specifically, we consider a sequence of Markov chains with generators $\mathcal{A}_N$ such that the population size is N for the Nth Markov chain and we keep track of the fractions of populations, such that components of the state space $\mathcal{X}_N$ of the Nth Markov chain live on a lattice with step size $1/N$, and the unit vectors have size $1/N$ as well. By contrast, the transition rates increase by N as we need to translate from population fractions to population sizes. Setting $\epsilon := 1/N$, we get the following generator:

$$\mathcal{A}_{\epsilon^{-1}} g(\mathbf{x}) = \epsilon^{-1} \sum_{p \in \mathcal{P}} \Big(\beta_1 f_1^p(\mathbf{x}) x_{(s,s)}^p [g(\mathbf{x} - \epsilon \mathbf{e}_{(s,s)}^p + \epsilon \mathbf{e}_{(i,s)}^p) - g(\mathbf{x})]$$

$$+ \beta_2 f_2^p(\mathbf{x}) x_{(s,s)}^p [g(\mathbf{x} - \epsilon \mathbf{e}_{(s,s)}^p + \epsilon \mathbf{e}_{(s,i)}^p) - g(\mathbf{x})]$$

$$+ (\beta_1 + \Delta_1) f_1^p(\mathbf{x}) x_{(s,i)}^p [g(\mathbf{x} - \epsilon \mathbf{e}_{(s,i)}^p + \epsilon \mathbf{e}_{(i,i)}^p) - g(\mathbf{x})]$$

$$+ (\beta_2 + \Delta_2) f_2^p(\mathbf{x}) x_{(i,s)}^p [g(\mathbf{x} - \epsilon \mathbf{e}_{(i,s)}^p + \epsilon \mathbf{e}_{(i,i)}^p) - g(\mathbf{x})]$$

$$+ (\beta_1 + \bar{\Delta}_1) f_1^p(\mathbf{x}) x_{(s,r)}^p [g(\mathbf{x} - \epsilon \mathbf{e}_{(s,r)}^p + \epsilon \mathbf{e}_{(i,r)}^p) - g(\mathbf{x})]$$

$$+ (\beta_2 + \bar{\Delta}_2) f_2^p(\mathbf{x}) x_{(r,s)}^p [g(\mathbf{x} - \epsilon \mathbf{e}_{(r,s)}^p + \epsilon \mathbf{e}_{(r,i)}^p) - g(\mathbf{x})]$$

$$+ \gamma_1 x_{(i,s)}^p [g(\mathbf{x} - \epsilon \mathbf{e}_{(i,s)}^p + \epsilon \mathbf{e}_{(r,s)}^p) - g(\mathbf{x})]$$

$$+ \gamma_1 x_{(i,i)}^p [g(\mathbf{x} - \epsilon \mathbf{e}_{(i,i)}^p + \epsilon \mathbf{e}_{(r,i)}^p) - g(\mathbf{x})]$$

$$+ \gamma_1 x_{(i,r)}^p [g(\mathbf{x} - \epsilon \mathbf{e}_{(i,r)}^p + \epsilon \mathbf{e}_{(r,r)}^p) - g(\mathbf{x})]$$

$$+ \gamma_2 x_{(s,i)}^p [g(\mathbf{x} - \epsilon \mathbf{e}_{(s,i)}^p + \epsilon \mathbf{e}_{(s,r)}^p) - g(\mathbf{x})]$$

$$+ \gamma_2 x_{(i,i)}^p [g(\mathbf{x} - \epsilon \mathbf{e}_{(i,i)}^p + \epsilon \mathbf{e}_{(i,r)}^p) - g(\mathbf{x})]$$

$$+ \gamma_2 x_{(r,i)}^p [g(\mathbf{x} - \epsilon \mathbf{e}_{(r,i)}^p + \epsilon \mathbf{e}_{(r,r)}^p) - g(\mathbf{x})] \Big). \tag{4}$$

We can deduce the (candidate) fluid limit by Taylor expansion of this generator around $\epsilon = 0$, from which we find a limiting generator of the form

$\hat{A}g = \mathbf{h}(\mathbf{x}) \cdot \nabla g$, for a certain $9|\mathcal{P}|$-dimensional vector function $\mathbf{h}$. Note that a generator of this form corresponds to a deterministic process satisfying the system of differential equations $\dot{\mathbf{x}}(t) = \mathbf{h}(\mathbf{x}(t))$.

In order to prove this limit rigourously, it needs to be checked that both the pre-limit processes and the limit process are Feller processes [10], which basically boils down to checking the so-called Hille-Yosida conditions. We believe that a careful proof of this statement falls outside the scope of this paper, but remark that due to the compactness of the state space (in the prelimit as well as in the limit), the proof is not as involved as is sometimes the case. Below we detail the set of differential equations, where we have dropped the dependence of t for notational convenience.

After some manipulations we find the following fluid limit which not only generalises syndemics in a single population but also epidemics on a stratified population:

$$\dot{x}^p_{(s,s)} = -\beta_1 y^p_1 x^p_{(s,s)} - \beta_2 y^p_2 x^p_{(s,s)}$$
$$\dot{x}^p_{(i,s)} = \beta_1 y^p_1 x^p_{(s,s)} - (\beta_2 + \Delta_2) y^p_2 x^p_{(i,s)} - \gamma_1 x^p_{(i,s)}$$
$$\dot{x}^p_{(s,i)} = \beta_2 y^p_2 x^p_{(s,s)} - (\beta_1 + \Delta_1) y^p_1 x^p_{(s,i)} - \gamma_2 x^p_{(s,i)}$$
$$\dot{x}^p_{(i,i)} = (\beta_2 + \Delta_2) y^p_2 x^p_{(i,s)} + (\beta_1 + \Delta_1) y^p_1 x^p_{(s,i)} - (\gamma_1 + \gamma_2) x^p_{(i,i)}$$
$$\dot{x}^p_{(r,s)} = \gamma_1 x^p_{(i,s)} - (\beta_2 + \bar{\Delta}_2) y^p_2 x^p_{(r,s)}$$
$$\dot{x}^p_{(r,i)} = (\beta_2 + \bar{\Delta}_2) y^p_2 x^p_{(r,s)} + \gamma_1 x^p_{(i,i)} - \gamma_2 x^p_{(r,i)}$$
$$\dot{x}^p_{(i,r)} = (\beta_1 + \bar{\Delta}_1) y^p_1 x^p_{(s,r)} + \gamma_2 x^p_{(i,i)} - \gamma_1 x^p_{(i,r)}$$
$$\dot{x}^p_{(s,r)} = \gamma_2 x^p_{(s,i)} - (\beta_1 + \bar{\Delta}_1) y^p_1 x^p_{(s,r)}$$
$$\dot{x}^p_{(r,r)} = \gamma_1 x^p_{(i,r)} + \gamma_2 x^p_{(r,i)}$$
$$y^p_1 = \frac{\sum_{q \in \mathcal{N}(p)} \left(x^q_{(i,s)} + x^q_{(i,i)} + x^q_{(i,r)} \right)}{\sum_{q \in \mathcal{N}(p)} \nu(p)}$$
$$y^p_2 = \frac{\sum_{q \in \mathcal{N}(p)} \left(x^q_{(s,i)} + x^q_{(i,i)} + x^q_{(r,i)} \right)}{\sum_{q \in \mathcal{N}(p)} \nu(q)},$$

for $p \in \mathcal{P}$. The fractions y^p_1 and y^p_2 were introduced in the set of ODEs to simplify notation: $y^p_i(t)$ is the fraction of individuals that are infected by epidemic i in the neighbourhood of p.

4   Case Studies

With the ODEs established we now focus on some numerical examples. To limit the number of parameters, we investigate the spread of two epidemics, say e_1 and e_2, on two intersecting populations. For both epidemics, the spreading and recovery parameters are set to $\beta_i = 0.4$ and $\gamma_i = 0.1$ ($i = 1, 2$), respectively.

There are two populations. Population $P1$ constitutes 30% of the total population. The population $P2$ constitutes 70% of the total population. The fraction of the individuals in the intersection of both populations – referred to as the degree of overlap – is denoted by ν and assumed to be 0.01% unless indicated otherwise. For a fixed ν, $30\% - \nu/2$ and $70\% - \nu/2$ of the individuals are in $P1$ and not in $P2$ and in $P2$ and not in $P1$, respectively.

For all case studies $\bar{\Delta}_1 = \Delta_1$ and $\bar{\Delta}_2 = \Delta_2$. Epidemic e_1 begins in the non-intersecting population $P1$ at time 0, and epidemic e_2 begins in the non-intersecting population $P2$ at time 0. The initial number of infected individuals is 1% for each epidemic, and no individuals are infected by both epidemics at the start. With the parameters fixed, we now investigate how spreading of the epidemics is affected by (i) the size of the intersection, (ii) syndemic effects and (iii) counter-syndemic effects.

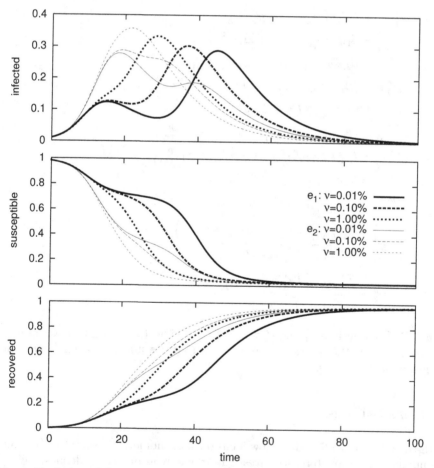

Fig. 3. Evolution of the fractions of infected, susceptible and recovered individuals for epidemics e_1 and e_2 and for different sizes of the intersection ν as indicated

Case Study 1: Influence of Degree of Overlap. Fig. 3 shows the influence of the degree of overlap ν between the populations on the spread of e_1 and e_2. We see that the smaller the intersection, the more significant the delay of the propagation of the epidemics between the populations. With values of ν above 1%, the results are increasingly indistinguishable from epidemics spreading in a single population. The multimodality of the spread over time is quite apparent. The epidemics first reach their peak in the population in which they originated. Only after sufficiently many individuals in the intersection are affected, spreading in the other population starts, reaching its peak considerably later, even though the spreading mechanism is exactly the same in both populations and for both epidemics. Finally note that the first peak of e_2 is considerably higher than the first peak of e_1 while the opposite is observed for the second peak which is in line with the sizes of the populations the epidemics originate from.

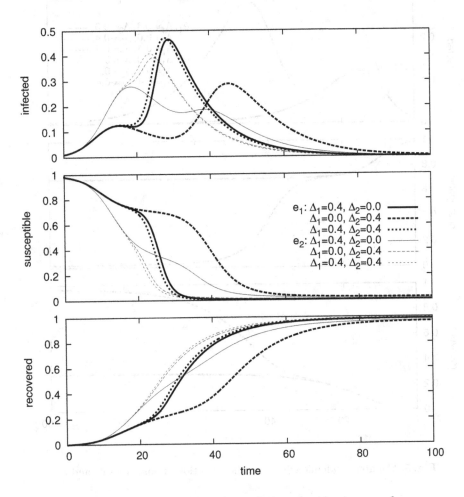

Fig. 4. Syndemic effects on the evolution of epidemics e_1 and e_2

Case Study 2: The Impact of Syndemic Effects. Fig. 4 shows how syndemic effects affect the evolution of the epidemics. We consider three cases. For $\Delta_1 = \beta = 0.4$ and $\Delta_2 = 0$, the second epidemic reinforces spreading of the first. Specifically, if an individual is infected by the second epidemic, its infection rate for the first epidemic is doubled. For $\Delta_2 = \beta = 0.4$ and $\Delta_1 = 0$, the first epidemic reinforces spreading of the second in a similar manner. Finally, for $\Delta_1 = \Delta_2 = \beta = 0.4$, both epidemics reinforce each other. For $\Delta_1 = 0, \Delta_2 = 0.4$ for e_1 corresponds to the case where there are no syndemic effects on e_1. Comparison with the other e_1 curves clearly reveals the syndemic effects. Particularly note that when both epidemics reinforce each other, the peak of e_1 is sooner and a little higher.

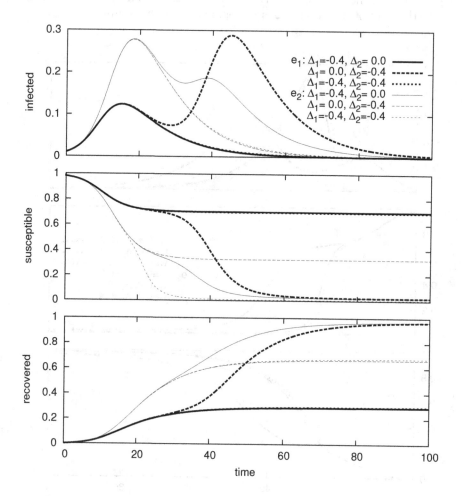

Fig. 5. Counter-syndemic effects on the evolution of epidemics e_1 and e_2

Case study 3: The Impact of Counter-Syndemic Effects. Fig. 5 shows the impact of counter-syndemic effects. We consider three cases. For $\Delta_1 = -\beta = -0.4$ and $\Delta_2 = 0$, an individual infected by the second epidemic is immune to the first epidemic. For $\Delta_2 = -\beta = -0.4$ and $\Delta_1 = 0$, an individual infected by the first epidemic is immune to the second epidemic. Finally, for $\Delta_1 = \Delta_2 = -\beta = -0.4$, immunity works both ways. As similar effects apply for both epidemics, we focus on e_1. Clearly, for $\Delta_2 = -\beta = -0.4$ and $\Delta_1 = 0$, the first epidemic is not affected by syndemic effects. Hence, the e_1 curve for $\Delta_2 = -\beta = -0.4$ and $\Delta_1 = 0$ can be used as reference. Comparing this curve with the other e_1 curves clearly illustrates counter-syndemic effects. In fact, the second peak of the epidemic is no longer present. This is explained by noting that this peak was reached in the population where the second epidemic originates. By the time the first epidemic reaches this population, most of its individuals are already immune. Finally, note that a large proportion of the population remains susceptible to the first epidemic.

5   Conclusion

It is important that the sophistication of mathematical modelling techniques keeps pace with our evolving understanding of the dynamics of epidemic processes, especially as they become applied in myriad domains beyond the biological. Our present paper has made some progress in this direction by considering models of syndemic and counter-syndemic interactions between two SIR epidemics in multiple overlapping populations. The results from this kind of analysis can give insights into epidemic forecasting and optimal strategies for managing the response to outbreaks.

Much more remains to be done. For example, while the present work targets fluid limits, other scalings leading to diffusion limits may shed light on the variance of outcomes. In addition, our populations are assumed to be static when a more realistic model might assume some dynamic movement of individuals between populations (practically realised as facilities to join and leave populations). Practical case studies could also be carried out in application areas ranging from computer viruses to extreme ideologies.

References

1. Avrachenkov, K., De Turck, K., Fiems, D., Prabhu, B.J.: Information dissemination processes in directed social networks. CoRR, abs/1311.2023 (2013)
2. Boukerche, A.: Epidemic Models, Algorithms, and Protocols in Wireless Sensor and Ad Hoc Networks. In: De, P., Das, S.K. (eds.) Algorithms and Protocols for Wireless Sensor Networks, ch. 3. Wiley-IEEE Press (2008)
3. Cha, M., Kwak, H., Rodriguez, P., Ahn, Y.-Y., Moon, S.: I tube, you tube, everybody tubes: Analyzing the world's largest user generated content video system. In: IMC 2007: Proc. 7th ACM SIGCOMM, pp. 1–14. ACM, NY (2007)
4. Chen, L., Ghanbarnejad, F., Cai, W., Grassberger, P.: Outbreaks of coinfections: The critical role of cooperativity. Europhysics Letters 104 (December 2013)

5. Daley, D.J., Kendall, D.G.: Epidemics and rumours. Nature 204, 1118 (1964)
6. Datta, A., Quarteroni, S., Aberer, K.: Autonomous gossiping: A self-organizing epidemic algorithm for selective information dissemination in wireless mobile ad-hoc networks. In: Bouzeghoub, M., Goble, C., Kashyap, V., Spaccapietra, S. (eds.) ICSNW 2004. LNCS, vol. 3226, pp. 126–143. Springer, Heidelberg (2004)
7. Dawkins, R.: The Selfish Gene. Oxford University Press, Oxford (1976)
8. De Cuypere, E., De Turck, K., Wittevrongel, S., Fiems, D.: Markovian sir model for opinion propagation. In: Proceedings of the 2013 25th International Teletraffic Congress (ITC), pp. 1–7. IEEE (2013)
9. Dietz, K., Heesterbeek, J.A.P.: Daniel Bernoulli's epidemiological model revisited. Mathematical Biosciences 180, 1–21 (2002)
10. Ethier, S.N., Kurtz, T.G.: Markov processes – characterization and convergence. Wiley Series in Probability and Mathematical Statistics: Probability and Mathematical Statistics. John Wiley & Sons Inc., New York (1986)
11. Gaunt, J.: Natural and Political Observations Mentioned in a following index, and made upon the Bills of Mortality (1662), http://www.neonatology.org/pdf/graunt.pdf
12. Getz, L.: Sustainable and responsible preventive medicine: Conceptualising ethical dilemmas arising from clinical implementation of advancing medical technology. PhD thesis. Norweigan University of Science and Technology, Trondheim (2006)
13. Goffman, W., Newill, V.A.: Generalization of epidemic theory: An application to the transmission of ideas. Nature 204, 225–228 (1964)
14. Hedetniemi, S.M., Hedetniemi, S.T., Liestman, A.L.: A survey of gossiping and broadcasting in communication networks. Networks 18(4), 319–349 (1988)
15. Hu, H.-W., Lee, S.-Y.: Study on influence diffusion in social network. International Journal of Computer Science and Electronics Engineering (IJCSEE) 1 (2013)
16. Iribarren, J.L., Moro, E.: Information diffusion epidemics in social networks. Physical Review (2009)
17. Jaewon, Y., Leskovec, J.: Modeling information diffusion in implicit networks. In: 2010 IEEE 10th International Conference on Data Mining (ICDM), pp. 599–608 (December 2010)
18. Karnik, A., Saroop, A., Borkar, V.: On the diffusion of messages in online social networks. Performance Evaluation 70(4), 271–285 (2013)
19. Kermack, W.O., McKendrick, A.G.: Contributions to the mathematical theory of epidemics–I. Bull. Math. Biol. 53(1-2), 33–55 (1927, 1991)
20. Khelil, A., Becker, C., Tian, J., Rothermel, K.: An epidemic model for information diffusion in MANETs. In: Proc. 5th ACM International Workshop on Modeling Analysis and Simulation of Wireless and Mobile Systems, pp. 54–60 (2002)
21. Kwan, C., Ernst, J.: HIV and tuberculosis: a deadly human syndemic. Clinical Microbiology 24(2), 351–376 (2011)
22. Leibnitz, K., Hoßfeld, T., Wakamiya, N., Murata, M.: Modeling of epidemic diffusion in peer-to-peer file-sharing networks. In: Ijspeert, A.J., Masuzawa, T., Kusumoto, S. (eds.) BioADIT 2006. LNCS, vol. 3853, pp. 322–329. Springer, Heidelberg (2006)
23. Lerman, K., Ghosh, R.: Information contagion: An empirical study of the spread of news on Digg and Twitter social networks. In: Proc. 4th International AAAI Conference on Weblogs and Social Media, ICWSM 2010 (2010)
24. Leskovec, J., Adamic, L.A., Huberman, B.A.: The dynamics of viral marketing. ACM Trans. Web 1 (May 2007)
25. Martcheva, M., Pilyugin, S.S.: The role of coinfection in multi-disease dynamics. SIAM Journal of Applied Mathematics 66, 843–872 (2006)

26. Mochalova, A., Nanopoulos, A.: On the role of centrality in information diffusion in social networks. In: Proc. ECIS 2013, p. 101 (2013)
27. Moss, W.J., Scott, S., Ndhlovu, Z., Monze, M., Cutts, F.T., Quinn, T.C., Griffin, D.E.: Suppression of human immunodeficiency virus type 1 viral load during acute measles. Pediatric Infectious Disease Journal 28(1), 63–65 (2009)
28. Myers, S., Leskovec, J.: Clash of the Contagions: Cooperation and Competition in Information Diffusion. In: Proc. IEEE International Conference on Data Mining, ICDM 2012 (2012)
29. Netrapalli, P., Sanghavi, S.: Learning the graph of epidemic cascades. In: Proc. 12th ACM SIGMETRICS/PERFORMANCE Joint International Conference on Measurement and Modeling of Computer Systems (SIGMETRICS 2012), pp. 211–222. ACM, New York (2012)
30. Newman, M.E.J., Ferrario, C.R.: Interacting epidemics and coinfection on contact networks. PLoS ONE 8(8), e71321(2013)
31. Nika, M., Ivanova, G., Knottenbelt, W.J.: On celebrity, epidemiology and the internet. In: Proc. 7th International Conference on Performance Evaluation Methodologies and Tools (VALUETOOLS 2013), Turin, Italy (December 2013)
32. Parker, R.: Miasma: Pollution and Purification in Early Greek Religion. Clarendon paperbacks. Clarendon Press (1990)
33. Pi, S.-M., Liu, Y.-C., Chen, T.-Y., Li, S.-H.: The influence of instant messaging usage behavior on organizational communication satisfaction. In: Proc. 41st Annual Hawaii International Conference on System Sciences, HICSS 2008, p. 449. IEEE Computer Society, Washington, DC (2008)
34. Rhodes, C.J., Nekovee, M.: The opportunistic transmission of wireless worms between mobile devices. CoRR, abs/0802.2685 (2008)
35. Singer, M.: Introduction to Syndemics. Wiley (2009)
36. Snow, J.: On the Mode of Communication of Cholera. John Churchill (1855)
37. Weng, L., Flammini, A., Vespignani, A., Menczer, F.: Competition among memes in a world with limited attention. Scientific Reports 2(335) (2013)
38. Wikipedia. List of epidemics, http://en.wikipedia.org/wiki/List_of_epidemics
39. Wischhof, L., Ebner, A., Rohling, H.: Information dissemination in self-organizing intervehicle networks. IEEE Transactions on Intelligent Transportation Systems 6(1), 90–101 (2005)

A Dynamic Page-Refresh Index Policy for Web Crawlers

José Niño-Mora

Statistics Department, Carlos III University of Madrid, Spain
jnimora@alum.mit.edu
http://alum.mit.edu/www/jnimora

Abstract. This paper consider a Markovian model for the optimal dynamic scheduling of page refreshes in a local repository of copies of randomly evolving remote web pages. A limited number of refresh agents, e.g., crawlers for web search engines, are used to visit the remote pages for refreshing their copies, which raises the need for effective scheduling policies. Maintaining the copies results in utilities and costs, which are incorporated into a performance objective to be optimized. The paper develops a low-complexity closed-form heuristic dynamic index policy, and an upper bound on the optimal performance, by adapting a general approach of Whittle. The existence and evaluation of the index are resolved by methods introduced earlier by the author. A numerical study provides evidence showing that the proposed policy is consistently near optimal and may substantially outperform a myopic baseline policy.

Keywords: web page refresh policies, web crawlers, dynamic scheduling, index policies, Whittle index, Markov decision processes.

1   Introduction and Model Description

We consider a Markovian model for the optimal dynamic scheduling of page refreshes in a local repository of copies of randomly evolving remote content sources, e.g., web pages. In order to prevent the latter from becoming obsolete, a collection of refresh agents, e.g., web *crawlers*, is used to visit the remote sources for refreshing their local copies, bringing them up-to-date. Typically, the number of such agents is much smaller than the number of content sources, which raises the need for effective scheduling policies for prioritizing which local copies to refresh at each time. Further, there might be refresh costs, due, e.g., to contract agreements or to the energy expended, which creates tradeoffs between the utility and the cost of maintaining the freshness of the local copies.

Such issues have attracted extensive research attention, motivated by applications such as web crawler scheduling in search engines for maintaining local copies of remote web pages (see, e.g., [1–5]), and refresh scheduling for apps data in smartphones (see [6]). Most of such work considers static (state-independent) refresh scheduling policies. Work on dynamic refresh policies is limited, and has mostly focused on the case of a single content source, as in [3, 6].

B. Sericola, M. Telek, and G. Horváth (Eds.): ASMTA 2014, LNCS 8499, pp. 46–60, 2014.
© Springer International Publishing Switzerland 2014

In contrast, this paper aims to develop an effective, low-complexity dynamic refresh scheduling policy, which comes close to optimizing a performance objective involving both freshness and cost metrics, in the setting of a *Markov decision process* (MDP) model.

We thus consider an MDP model for the dynamic refresh scheduling of n local copies of remote content sources using $m \leqslant n$ refresh agents, which we will call crawlers. We assume that time is slotted into discrete periods $t \geqslant 0$, and that a crawler spends one time slot to visit a content source and refresh its local copy. Thus, at each time slot, no more than m local copies can be refreshed.

The system controller keeps track over time of the *state* of each local copy $k = 1, \ldots, n$, which is given by its *age* $X_k(t) \in \mathbb{N} \triangleq \{1, 2, \ldots\}$, the number of time slots since its last refresh. We consider that when local copy k is x_k-slots old ($X_k(t) = x_k$), it yields a *utility* $U_k(x_k)$ in the current slot, which satisfies the following natural assumption.

Assumption 1. *For each k, $U_k(x_k)$ is nonincreasing in the age x_k.*

Such utilities can be used as powerful modeling devices to encapsulate information on both the stochastic evolution of updates and the relative value of each content source. Thus, suppose that content source k is updated at random time slots according to a discrete-time renewal process $N_k(t)$ with inter-renewal times distributed as T_k. For example, we might define, for a given reward $r_k > 0$,

$$U_k(x_k) \triangleq r_k \mathsf{P}\{x_k\text{-old copy } k \text{ is fresh}\} = r_k \mathsf{P}\{T_k > x_k\}. \qquad (1)$$

We could also take, for a given cost-per-missed-update $h_k > 0$,

$$U_k(x_k) \triangleq -h_k \mathsf{E}[\text{number of missed updates in } x_k\text{-old copy } k] = -h_k \mathsf{E}[N_k(x_k)]. \qquad (2)$$

More generally, we could consider a nondecreasing function $H_k(n_k) \geqslant 0$ measuring the cost of missing k content source updates in copy k, and take

$$U_k(x_k) \triangleq -\mathsf{E}[H_k(N_k(x_k))]. \qquad (3)$$

If the controller refreshes copy k at the start of slot t, then the its age is reset back to $X_k(t+1) = 1$ and a *refresh cost* $c_k \geqslant 0$ is incurred. If the copy is not refreshed, then $X_k(t+1) = X_k(t) + 1$ and no cost is incurred.

We use a binary action process $A_k(t) \in \{0, 1\}$ for each copy k, where $A_k(t) = 1$ if the copy is refreshed in slot t and $A_k(t) = 0$ otherwise. At each time t, the controller observes the system state $\mathbf{X}(t) = \big(X_k(t)\big)_{k=1}^n \in \mathbb{N}^n$, and then chooses a joint control action $\mathbf{A}(t) = \big(A_k(t)\big)_{k=1}^n$, subject to the *refresh capacity constraints*

$$\sum_{k=1}^n A_k(t) \leqslant m, \quad t = 0, 1, 2, \ldots \qquad (4)$$

We denote in the sequel the state and action dependent one-slot *net rewards* for copy k by $R_k(x_k, a_k) \triangleq U_k(x_k) - c_k a_k$.

Actions are selected according to a *scheduling policy* π, which is chosen from the class $\Pi(m)$ of control policies that at each time slot t select an $\mathbf{A}(t)$ satisfying (4) based on the state-action history $\mathscr{H}(t) \triangleq ((\mathbf{X}(s), \mathbf{A}(s))_{s=0}^{t-1}; \mathbf{X}(t))$. We will denote henceforth by $X_k^{\pi, \mathbf{x}^0}(t)$ and $A_k^{\pi, \mathbf{x}^0}(t)$ the state and action processes for copy k under policy π starting from the system state $\mathbf{x}^0 = (x_k^0)_{k=1}^n \in \mathbb{N}^n$.

We consider the following infinite-horizon performance optimization problems: (1) find a discount-optimal policy for a given discount factor $\beta \in (0, 1)$,

$$\underset{\pi \in \Pi}{\text{maximize}} \sum_{t=0}^{\infty} \sum_{k=1}^{n} \beta^t R_k(X_k^{\pi, \mathbf{x}^0}(t) A_k^{\pi, \mathbf{x}^0}(t)); \tag{5}$$

and (2) find a long-run average-optimal policy,

$$\underset{\pi \in \Pi}{\text{maximize}} \liminf_{T \to \infty} \frac{1}{T} \sum_{t=0}^{T} \sum_{k=1}^{n} R_k(X_k^{\pi, \mathbf{x}^0}(t) A_k^{\pi, \mathbf{x}^0}(t)). \tag{6}$$

Problems (5) and (6) belong to the class of *multiarmed restless bandit problems* (MARBPs), which are in general computationally intractable. See [7]. Our main goal is to obtain new heuristic dynamic scheduling policies for such problems, which have low complexity and attain a nearly optimal performance.

Given their intuitive appeal and ease of implementation, we consider policies of *priority-index* type. These are based on attaching an *index* $\varphi_k(x_k)$ to each copy k, which is nondecreasing in its age x_k. At each time t, copies with higher index values $\varphi_k(X_k(t))$ get higher priority for being refreshed. Only copies whose current index value is positive are considered. Ties are broken arbitrarily.

In particular, we will consider the index introduced by Whittle in [8] for restless bandits, adapted to the present model. To deploy such an index certain technical issues need to be resolved, such as establishing its existence (*indexability*) and devising an efficient index-computing scheme. For such purposes, we will deploy the methodology introduced and developed by the author in [9–12].

The main contributions of this paper are as follows. The paper carries out an analysis for the model of concern, establishing its indexability and deriving closed formulae for the Whittle index. Further, the paper reports the results of a numerical study on randomly generated instances, which shows that the policy consistently attains a near-optimal performance, and substantially outperforms the performance of a baseline myopic policy.

The remainder of the paper is organized as follows. §2 shows how to adapt Whittle's relaxation and index to the present model. §3 applies a general methodology to establish the model's indexability (existence of the Whittle index) and to evaluate the Whittle index. §4 reports the results of the numerical study. Finally, §5 ends the paper with some concluding remarks.

Detailed analyses and proofs of the results outlined herein will be presented in a full version of this paper.

2   Whittle's Relaxation and Index Policy

The following discussion focuses on the discounted problem (5), although the approach and results apply also to the long-run average problem (6).

2.1   Relaxed Problem and Performance Bound

Along the lines in [8], one can obtain an upper bound $v^R(\mathbf{x}^0)$ on the optimal value $v^*(\mathbf{x}^0)$ of (5) by solving the following *relaxed problem*, where it is assumed that each copy is separately controlled, as if in isolation, under its own policy $\pi_k \in \Pi_k$, and there is a coupling constraint which relaxes (4):

$$\text{maximize} \sum_{k=1}^{n} \sum_{t=0}^{\infty} \beta^t R_k\left(X_k^{\pi_k,x_k^0}(t), A_k^{\pi_k,x_k^0}(t)\right)$$

$$\text{subject to: } \pi_k \in \Pi_k, \quad k = 1,\ldots,n \tag{7}$$

$$\sum_{k=1}^{n} \sum_{t=0}^{\infty} \beta^t A_k^{\pi_k,x_k^0}(t) \leqslant \frac{m}{1-\beta}.$$

Note that, in (7), $X_k^{\pi_k,x_k^0}(t)$ and $A_k^{\pi_k,x_k^0}(t)$ denote the state and action processes for copy k under the single-copy policy π_k starting from state $x_k^0 \in \mathbb{N}$, and Π_k is the class of admissible refresh policies for copy k taken in isolation.

2.2   Lagrangian Relaxation and Decomposition

To address (7) we pursue a Lagrangian approach (see [13, Ch. 5]), attaching a multiplier $\lambda \geqslant 0$ to the coupling constraint, which yields the *Lagrangian relaxation*

$$\text{maximize} \frac{m\lambda}{1-\beta} + \sum_{k=1}^{n} \sum_{t=0}^{\infty} \beta^t \left\{ R_k\left(X_k^{\pi_k,x_k^0}(t), A_k^{\pi_k,x_k^0}(t)\right) - \lambda A_k^{\pi_k,x_k^0}(t) \right\} \tag{8}$$

$$\text{subject to: } \pi_k \in \Pi_k, \quad k = 1,\ldots,n.$$

For any multiplier value $\lambda \geqslant 0$, (8) is a relaxation of (7) in that its optimal value $v^L(\mathbf{x}^0;\lambda)$ is an upper bound for $v^R(\mathbf{x}^0)$. Further, (8) is much easier to solve that (7), as it decouples into the n separate single-copy subproblems

$$P_k(\lambda): \quad \underset{\pi_k \in \Pi_k}{\text{maximize}} \sum_{t=0}^{\infty} \beta^t \left\{ R_k\left(X_k^{\pi_k,x_k^0}(t), A_k^{\pi_k,x_k^0}(t)\right) - \lambda A_k^{\pi_k,x_k^0}(t) \right\}, \tag{9}$$

for $k = 1, \ldots, n$. Now, Lagrangian theory ensures the existence of an optimal multiplier value $\lambda^* \geq 0$ (which depends on $\mathbf{x}^0$) such that the Lagrangian relaxation of (7) is exact, in that $v^L(\mathbf{x}^0; \lambda^*) = v^R(\mathbf{x}^0)$. An optimal multiplier λ^* can be obtained by solving the scalar convex optimization problem

$$\underset{\lambda \geq 0}{\text{minimize}} \; \frac{m\lambda}{1 - \beta} + \sum_{k=1}^{n} v_k^*(x_k^0; \lambda), \tag{10}$$

where $v_k^*(x_k^0; \lambda)$ is the optimal value of (9).

2.3 Indexability and Index Policy

Consider the parametric problem collection $\{P_k(\lambda) : \lambda \in \mathbb{R}\}$ in (9) for copy k. Note that λ represents an extra refresh charge (added to the regular cost c_k).

Let us say that copy k's subproblem collection $\{P_k(\lambda) : \lambda \in \mathbb{R}\}$ is *indexable* if there is a function $\varphi_k^W \colon \mathbb{N} \to \mathbb{R}$ such that, for any given $\lambda \in \mathbb{R}$, it is optimal in problem $P_k(\lambda)$ —regardless of the initial state x_k^0— to refresh the copy when $X_k(t) = x_k$ if and only if $\varphi_k^W(x_k) \geq \lambda$, and it is optimal not to refresh it if and only if $\varphi_k^W(x_k) \leq \lambda$. We will then refer to φ_k^W as the *Whittle index* for copy k.

If each copy is indexable, we readily obtain a low-complexity approach to solve (10) and thus compute $v^R(\mathbf{x}^0)$. Further, the $\varphi_k^W(x_k)$ can be used as priority-indices to obtain a dynamic index policy for problem (5).

3 Indexability Analysis

This section reports the results of an analysis for establishing that the single-copy refresh model of concern is indexable and for evaluating its Whittle index.

3.1 Sufficient Indexability Conditions

We next outline the sufficient indexability conditions developed in [9–12], formulating them as they apply to the present setting for ease of exposition. Since the focus is on a single-copy dynamic refresh subproblem (9), we drop henceforth the subscript k and write the copy's state and action at time t, under policy π and starting from x, as $X^{\pi,x}(t)$ and $A^{\pi,x}(t)$.

For any given copy refresh policy $\pi \in \Pi$ and initial state $x \in \mathbb{N}$, we evaluate the policy's performance using two *performance metrics*: the *refresh metric*

$$G(x, \pi) \triangleq \sum_{t=0}^{\infty} \beta^t A^{\pi,x}(t),$$

and the *reward metric*

$$F(x, \pi) \triangleq \sum_{t=0}^{\infty} \beta^t R\big(X^{\pi,x}(t), A^{\pi,x}(t)\big).$$

We will be concerned with the class of *threshold policies*. For any given threshold $z \in \overline{\mathbb{N}} \triangleq \mathbb{N} \cup \{0, \infty\}$, consider the *z-policy*, defined as the stationary deterministic policy that refreshes the copy in state x if and only if $x > z$. We will denote its performance metrics by $G(x, z)$ and $F(x, z)$.

For a given a and z, let $\langle a, z \rangle$ be the policy that takes action a at time $t = 0$ and then follows the z-policy aftewards. Define the *marginal refresh metric*

$$g(x, z) \triangleq G(x, \langle 1, z \rangle) - G(x, \langle 0, z \rangle), \tag{11}$$

and the *marginal reward metric*

$$f(x, z) \triangleq F(x, \langle 1, z \rangle) - F(x, \langle 0, z \rangle). \tag{12}$$

Further, if $g(x, z) \neq 0$, define the *marginal productivity* (MP) metric

$$\varphi^{\mathrm{MP}}(x, z) \triangleq \frac{f(x, z)}{g(x, z)}. \tag{13}$$

Finally, if $g(x, x) \neq 0$ for every state $x \in \mathbb{N}$, define the copy's *MP index* φ^{MP} by (with a slight abuse of notation)

$$\varphi^{\mathrm{MP}}(x) \triangleq \varphi^{\mathrm{MP}}(x, x), \quad x \in \mathbb{N}. \tag{14}$$

Adapting the author's results in [9–12] to the present setting, we next define the concept of *PCL-indexability* —after *partial conservation laws* (PCLs).

Definition 1. We say that the copy refresh model is *PCL-indexable* (with respect to threshold policies) if the following conditions hold:

(i) positive marginal refresh metric: $g(x, z) > 0$ for $x \in \mathbb{N}, z \in \overline{\mathbb{N}}$;
(ii) monotone MP index: φ^{MP} is monotone nondecreasing.

For the countably infinite state case, the following result is proven in [11].

Theorem 2. *A PCL-indexable model is indexable, with its Whittle index being given by its MP index:* $\varphi^{\mathrm{W}} = \varphi^{\mathrm{MP}}$.

3.2 Evaluation of Performance Metrics

For any $z \in \overline{\mathbb{N}}$, the refresh metric $G(\cdot, z)$ is characterized by the equations

$$G(x, z) = \begin{cases} \beta G(x + 1, z), & 1 \leqslant x \leqslant z \\ 1 + \beta G(1, z), & x > z. \end{cases}$$

We can solve such equations in closed form. For a threshold $z \in \mathbb{N}$, we obtain

$$G(x, z) = \begin{cases} \beta^{z-x} G(z, z), & 1 \leqslant x < z \\ \beta/(1 - \beta^{z+1}), & x = z \\ 1 + \beta^z G(z, z), & x > z. \end{cases}$$

Further, for $z = 0$ (always refresh) we have $G(x, 0) \equiv 1/(1 - \beta)$ and, for $z = \infty$ (never refresh), $G(x, \infty) \equiv 0$.

As for the reward metric $F(\cdot, z)$, it is characterized by

$$F(x, z) = \begin{cases} U(x) + \beta F(x + 1, z), & 1 \leqslant x \leqslant z \\ U(0) - c + \beta F(1, z), & x > z. \end{cases}$$

We can also solve it in closed form. For a threshold $z \in \mathbb{N}$, we obtain

$$F(x, z) = \begin{cases} \displaystyle\sum_{t=0}^{z-x-1} \beta^t U(x + t) + \beta^{z-x} F(z, z), & 1 \leqslant x < z \\[2em] \dfrac{U(z) - \beta c + \beta \displaystyle\sum_{t=0}^{z-1} \beta^t U(t)}{1 - \beta^{z+1}}, & x = z \\[2em] U(0) - c + \beta F(1, z), & x > z. \end{cases}$$

For $z = 0$ (always refresh) we have $F(x, 0) \equiv \{U(0) - c\}/(1 - \beta)$, whereas for $z = \infty$ (never refresh) we have $F(x, \infty) = \sum_{t=0}^{\infty} \beta^t U(x + t)$.

3.3   Evaluation of Marginal Metrics and MP Index

Concerning the marginal refresh metric, we have

$$g(x, z) = 1 + \beta G(1, z) - \beta G(x + 1, z) = \begin{cases} 1 - \beta^{z-x}(1 - \beta^x)G(z, z), & 1 \leqslant x < z \\ (1 - \beta)\{1 + \beta G(1, z)\}, & x \geqslant z. \end{cases} \tag{15}$$

In particular, $g(x, 0) \equiv 1$ and $g(x, \infty) \equiv 1$. Further,

$$g(x, x) = G(x + 1, x) - G(x, x) = \frac{1 - \beta}{1 - \beta^{x+1}}.$$

As for the marginal reward metric, we have

$$f(x, z) = U(0) - c + \beta F(1, z) - U(x) - \beta F(x + 1, z), \tag{16}$$

from which we obtain

$$f(x, x) = F(x + 1, x) - F(x, x) = \frac{(1 - \beta)\left\{\displaystyle\sum_{t=0}^{x-1} \beta^t U(t) - c\right\} - (1 - \beta^x)U(x)}{1 - \beta^{x+1}}.$$

We thus obtain the *MP index* in closed form:

$$\varphi^{\mathrm{MP}}(x) = \frac{f(x, x)}{g(x, x)} \triangleq \sum_{t=0}^{x-1} \beta^t U(t) - \frac{1 - \beta^x}{1 - \beta} U(x) - c, \quad x \in \mathbb{N}. \tag{17}$$

Note that φ^{MP} is generated by the recursion (with $\Delta U(x) \triangleq U(x) - U(x-1)$)

$$\varphi^{\text{MP}}(x) = \begin{cases} -c - \Delta U(1), & \text{if } x = 1 \\ \varphi^{\text{MP}}(x-1) - \dfrac{1 - \beta^x}{1 - \beta} \Delta U(x), & \text{if } x \geqslant 2. \end{cases} \tag{18}$$

It is of interest to consider the case $\beta = 0$, which yields the *myopic index*

$$\varphi^{\text{M}}(x) = U(0) - U(x) - c. \tag{19}$$

3.4 PCL-Indexability and Whittle Index

From the above we readily obtain the following result under Assumption 1.

Proposition 1. *The discounted copy refresh model is PCL-indexable.*

Hence, by Theorem 2, the present copy refresh model's Whittle index is given by its MP index: $\varphi^{\text{W}} = \varphi^{\text{MP}}$.

Note that, as the age $x \to \infty$, the index $\varphi^{\text{W}}(x)$ converges to the limit

$$\varphi^{\text{W}}(\infty) = \sum_{t=0}^{\infty} \beta^t U(t) - c. \tag{20}$$

3.5 Long-Run Average Criterion

We next draw on the above results for the discounted criterion to obtain corresponding results for the long-run average criterion.

Under the long-run average criterion we consider the refresh and reward performance metrics given by

$$\bar{G}(x, \pi) \triangleq \limsup_{T \to \infty} \frac{1}{T} \sum_{t=0}^{T} A^{\pi, x}(t)$$

and

$$\bar{F}(x, \pi) \triangleq \liminf_{T \to \infty} \frac{1}{T} \sum_{t=0}^{T} R\big(X^{\pi, x}(t), A^{\pi, x}(t)\big).$$

We define the long-run average marginal metrics $\bar{g}(x, z)$, $\bar{f}(x, z)$ and $\bar{\varphi}^{\text{MP}}(x, z)$ as for the discounted criterion, as well as the long-run average MP index $\bar{\varphi}^{\text{MP}}(x)$.

We can evaluate such metrics from their discounted counterparts using that

$$\bar{G}(x, z) = \lim_{\beta \nearrow 1} (1 - \beta)G(x, z), \quad \bar{F}(x, z) = \lim_{\beta \nearrow 1} (1 - \beta)F(x, z).$$

Such metrics do not depend on the initial state x, being given by

$$\bar{G}(x, z) = \begin{cases} \dfrac{1}{z + 1}, & z \in \mathbb{N} \cup \{0\} \\ 0, & z = \infty \end{cases}$$

and

$$\bar{F}(x,z) = \begin{cases} \dfrac{\displaystyle\sum_{t=0}^{z} U(t) - c}{z+1}, & z \in \mathbb{N} \cup \{0\} \\ 0, & z = \infty. \end{cases}$$

As for the long-run average marginal metrics, we have

$$\bar{g}(x,z) = \lim_{\beta \nearrow 1} g(x,z) = \begin{cases} 1 - \dfrac{x}{z+1}, & 1 \leqslant x < z \\ \dfrac{1}{z+1}, & x \geqslant z. \end{cases}$$

for $z \in \mathbb{N}$, whereas $\bar{g}(x,0) = \bar{g}(x,\infty) \equiv 1$.

On the other hand,

$$\bar{f}(x,x) = \lim_{\beta \nearrow 1} f(x,x) = \frac{\displaystyle\sum_{t=0}^{x-1} U(t) - xU(x) - c}{x+1}.$$

We thus obtain the long-run *MP index* in closed form:

$$\bar{\varphi}^{\mathrm{MP}}(x) = \frac{\bar{f}(x,x)}{\bar{g}(x,x)} = \sum_{t=0}^{x-1} U(t) - xU(x) - c, \quad x \in \mathbb{N}. \tag{21}$$

Note that $\bar{\varphi}^{\mathrm{MP}}(x)$ is generated by the recursion

$$\bar{\varphi}^{\mathrm{MP}}(x) = \begin{cases} -c - \Delta U(1), & \text{if } x = 1 \\ \bar{\varphi}^{\mathrm{MP}}(x-1) - x\Delta U(x), & \text{if } x \geqslant 2. \end{cases} \tag{22}$$

From the above we readily obtain the following result under Assumption 1.

Proposition 2. *The long-run average copy refresh model is PCL-indexable.*

As for the discounted case, Theorem 2 yields that the present copy refresh model's long-run Whittle index is given by its long-run MP index: $\bar{\varphi}^{\mathrm{W}} = \bar{\varphi}^{\mathrm{MP}}$. Note that, as the age $x \to \infty$, the index $\bar{\varphi}^{\mathrm{W}}(x)$ tends to

$$\bar{\varphi}^{\mathrm{W}}(\infty) = \sum_{t=0}^{\infty} U(t) - c. \tag{23}$$

3.6 Geometrically Decreasing Utility

A relevant special case is that where the utility function decreases geometrically with the copy's age, with $U(x) = r\alpha^x$ for some constants $r > 0$ and $\alpha \in (0,1)$. Note that this corresponds to the case of (1) when the remote source's inter-update time T is geometrically distributed with $P\{T = x\} = (1 - \alpha)\alpha^{x-1}$.

Then, the discounted single-copy Whittle index has the evaluation

$$\varphi^{\mathrm{W}}(x) = r\left\{\frac{1 - (\alpha\beta)^x}{1 - \alpha\beta} - \frac{\alpha^x - (\alpha\beta)^x}{1 - \beta}\right\} - c, \tag{24}$$

while the long-run Whittle index is

$$\bar{\varphi}^{\mathrm{W}}(x) = r\left\{\frac{1 - \alpha^x}{1 - \alpha} - x\alpha^x\right\} - c. \tag{25}$$

3.7 Linearly Decreasing Utility

Another relevant case is that where the utility decreases linearly with the copy's age until a minimum of zero for age y, so $U(x) = \max(0, y - x)r$ for some $r > 0$ and $y \in \mathbb{N}$, so an y-slot old copy is considered obsolete.

Then, the discounted Whittle index for a single-copy refresh model is

$$\varphi^{\mathrm{W}}(x) = \begin{cases} \dfrac{(1 - \beta)x - \beta(1 - \beta^x)}{(1 - \beta)^2}r - c, & 1 \leqslant x \leqslant y \\[2ex] \dfrac{(1 - \beta)y - \beta(1 - \beta^y)}{(1 - \beta)^2}r - c, & x > y. \end{cases} \tag{26}$$

whereas the long-run Whittle index has the evaluation

$$\bar{\varphi}^{\mathrm{W}}(x) = \begin{cases} \dfrac{x(x + 1)}{2}r - c, & 1 \leqslant x \leqslant y \\[2ex] \dfrac{y(y + 1)}{2}r - c, & x > y. \end{cases} \tag{27}$$

4 Numerical Study

This section reports on numerical experiments (run in MATLAB) for assessing the quality of the Whittle index policy under the long-run average criterion.

For each instance considered, the long-run average performance objective was evaluated both for the Whittle index policy and for the myopic index policy, using a horizon of $T = 20000$ time slots. Since state dynamics are deterministic, a single run suffices. Further, the upper bounds on the optimal performance objective resulting from Whittle's relaxation were computed.

The experiments aim to test the conjecture stated by Whittle in [8, p. 293], whereby the Whittle index policy is asymptotically optimal as m and n grow to infinity in a fixed proportion, provided the "projects" (web pages in the present model) are drawn from a finite number of types with fixed probabilities. The conjecture's validity has been proven under certain conditions in [14]. Here, the ratio has been set to $m/n = 1/10$, i.e., one agent per 10 copies. The cases of 30 and 40 page types have been considered.

Four sets of experiments were conducted. Experiments 1–2 consider the geometrically decreasing utility model in §3.6, whereas experiments 3–4 consider the linearly decreasing utility model in §3.7. In each set, both the cases of costless and of costly refresh are considered.

4.1 Geometrically Decreasing Utility

The utility parameters α and r in §3.6 were generated for each page type using MATLAB's **rand** pseudo-random number generator (Uniform$[0, 1]$).

Experiment # 1: Costless Refresh. For the costless refresh case ($c = 0$), Figures 1–2 show the results with 30 and 40 page types. In each figure, the variable on the horizontal axis is the number of web pages, n, ranging up to 5000. The variable on the vertical axis is the relative optimality gap, in percent, evaluated as the ratio of the difference between the upper bound on the optimal performance objective and the objective value of the corresponding policy, in the numerator, to the said upper bound in the denominator.

The figures show that, in each instance, the Whittle index policy's performance was so close to the upper bound as to render it indistinguishable from that of an optimal policy. In contrast, the performance of the myopic index policy was substantially suboptimal, with optimality gaps of over 20%.

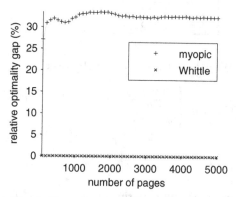

Fig. 1. Experiment # 1 with 30 page types

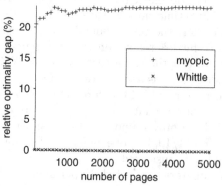

Fig. 2. Experiment # 1 with 40 page types

Experiment # 2: Costly Refresh. The refresh costs are now taken to be 20% of the reward parameters r in §3.6. Figures 3–4 show corresponding results with 30 and 40 page types. The figures show again that, in each instance, the Whittle index policy's performance is optimal or has an exceedingly small optimality gap. As for the performance of the myopic index policy, it is even poorer than in the costless refresh case, showing substantial optimality gaps.

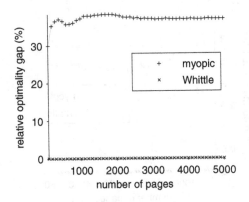

Fig. 3. Experiment # 2 with 30 page types

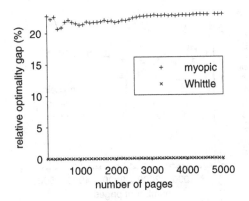

Fig. 4. Experiment # 2 with 40 page types

4.2 Linearly Decreasing Utility

The utility function parameters y and r in §3.7 were generated for each page type using MATLAB's `randi` (Uniform over the integers $1, \ldots, 10$) and `rand` (Uniform$[0, 1]$) pseudo-random number generators.

4.3   Experiment # 3: Costless Refresh

For the costless refresh case, Figures 5–6 show the results with 30 and 40 page types. The figures show that, in each instance, the Whittle index policy's performance was near optimal. As for the performance of the myopic index policy, its performance, though worse than Whittle's, was quite good, with optimality gaps no larger than about 3%.

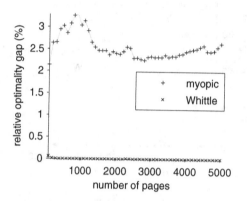

Fig. 5. Experiment # 3 with 30 page types

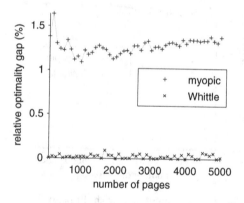

Fig. 6. Experiment # 3 with 40 page types

4.4   Experiment # 4: Costly Refresh

For the costly refresh case, the refresh cost was taken to be $c = ry/4$ for each page type. Figures 7–8 show corresponding results with 30 and 40 page types. In the cases of 30 and 40 page types, the results show a similar pattern to those in the costless refresh case.

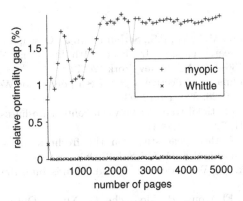

Fig. 7. Experiment # 4 with 30 page types

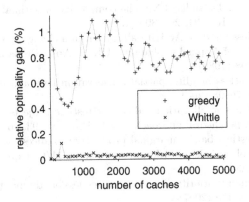

Fig. 8. Experiment # 4 with 40 page types

5   Conclusions

This paper has developed a simple closed-form dynamic policy based on Whittle's priority index for dynamic page-refresh scheduling. Preliminary results of numerical experiments are reported showing that the performance of the proposed policy is consistently near optimal across a range of large-scale randomly generated instances, outperforming the performance of the myopic policy. The approach used to establishing the model's indexability and to evaluate the Whittle index has been based on deploying general sufficient conditions for indexability. The results presented demonstrate the usefulness of the Whittle index, the associated performance bound, and the method of analysis employed to obtain a simple tractable policy with an excellent performance. For future work, it would be interesting to analyze theoretically the performance of such a policy.

Acknowledgment. This work was partially supported by the Spanish Ministry of Science and Innovation project MTM2010-20808.

References

1. Wolf, J.L., Squillante, M.S., Yu, P.S., Sethuraman, J., Ozsen, L.: Optimal crawling strategies for web search engines. In: Proc. 11th Int. Conf. World Wide Web, WWW 2002, pp. 136–147. ACM, New York (2002)
2. Cho, J., García-Molina, H.: Effective page refresh policies for Web crawlers. ACM Trans. Database Syst. 28, 390–426 (2003)
3. Ling, Y., Mi, J.: An optimal trade-off between content freshness and refresh cost. J. Appl. Probab. 41, 721–734 (2004)
4. Lewandowski, D.: A three-year study on the freshness of web search engine databases. J. Information Sci. 34, 817–831 (2008)
5. Olston, C., Najork, M.: Web crawling. Found. Trends Info. Retrieval 4, 175–246 (2010)
6. Raiss-El-Fenni, M., El-Azouzi, R., Menasché, D., Xu, Y.: Optimal sensing policies for smartphones in hybrid networks: A POMDP approach. In: Proc. 6th Int. Conf. Performance Eval. Method. Tools (VALUETOOLS 2012), pp. 89–98. ICST (2012)
7. Papadimitriou, C.H., Tsitsiklis, J.N.: The complexity of optimal queuing network control. Math. Oper. Res. 24, 293–305 (1999)
8. Whittle, P.: Restless bandits: Activity allocation in a changing world. In: Gani, J. (ed.) A Celebration of Applied Probability, UK. J. Appl. Probab. Trust, Sheffield, vol. 25, pp. 287–298 (1988)
9. Niño-Mora, J.: Restless bandits, partial conservation laws and indexability. Adv. Appl. Probab. 33, 76–98 (2001)
10. Niño-Mora, J.: Dynamic allocation indices for restless projects and queueing admission control: A polyhedral approach. Math. Program. 93, 361–413 (2002)
11. Niño-Mora, J.: Restless bandit marginal productivity indices, diminishing returns and optimal control of make-to-order/make-to-stock M/G/1 queues. Math. Oper. Res. 31, 50–84 (2006)
12. Niño-Mora, J.: Dynamic priority allocation via restless bandit marginal productivity indices. Top 15, 161–198 (2007)
13. Bertsekas, D.P.: Nonlinear Programming, 2nd edn. Athena Scientific, Nashua (1999)
14. Weber, R.R., Weiss, G.: On an index policy for restless bandits. J. Appl. Probab. 27, 637–648 (1990)

Dynamic Transmission Scheduling and Link Selection in Mobile Cloud Computing

Huaming Wu and Katinka Wolter

Institut für Informatik,
Freie Universität Berlin,
Takustr. 9, 14195 Berlin, Germany
{huaming.wu,katinka.wolter}@fu-berlin.de

Abstract. Recently, mobile devices have multiple wireless interfaces to use, but how to choose an appropriate network interface? Energy-efficient data transmission is a key issue in mobile cloud computing due to energy-poverty of the mobile devices. In this paper, we study an energy-delay tradeoff and address the issue of energy-efficient offloading that migrates data-intensive but delay-tolerant applications from the mobile devices to a remote cloud. Through dynamic scheduling and link selection based on Lyapunov optimization for data transmission between the mobile devices and the cloud, we are able to reduce battery consumption of the mobile devices for transferring large volumes of data. We derive a control algorithm which determines when and on which network to transmit data so that energy-cost is minimized by leveraging delay tolerance. Further, we propose and compare three kinds of transmission schedulers with energy-efficient link selection policies under heterogeneous wireless network interfaces (e.g., 3G and WiFi), where the average energy consumption is optimized.

Keywords: energy-efficient; transmission scheduling; link selection; optimization; delay-tolerant; mobile cloud computing.

1 Introduction

Mobile cloud computing [1] is emerging as a new computing paradigm that aims to augment resource-poor mobile devices, taking advantage of the abundant resources hosted by clouds. Offloading programs from mobile devices to a remote cloud is becoming an increasingly attractive way to reduce execution time and extend battery life time[2]. It makes running computing/data-intensive applications feasible on resource-constrained mobile devices. Apple's Siri and iCloud [3] are two examples. However, cloud offloading critically depends on a reliable end-to-end communication and on the availability of the cloud. Access to the cloud is usually influenced by uncontrollable factors, such as the instability and intermittency of wireless networks.

Mobile devices often have multiple wireless interfaces, such as 3G/EDGE, 4G LTE and WiFi for data transfer. While in most situations 4G LTE uses most energy and WiFi the least, normally WiFi has the highest bandwidth, 3G/EDGE

B. Sericola, M. Telek, and G. Horváth (Eds.): ASMTA 2014, LNCS 8499, pp. 61–79, 2014.

the lowest. However, the bandwidth, the energy-efficiency and even the availability of these networks can vary significantly, such that the stated ordering does not always hold true. Not only the availability and quality of access points (APs) may vary from place to place, but also the uplink and downlink bandwidths fluctuate frequently due to multiple factors such as weather, mobility, building shield and so on [4]. If we can adaptively select one of the available links in every slot, energy consumption may be reduced.

Energy consumption in mobile devices has become an important issue for network selection. Gribaudo et. al. [5] developed a framework based on the Markovian agent formalism, which could model the dynamics of user traffic and the allocation of the network radio resources. Rahmati et. al. [6] suggested on-the-spot network selection by examining tradeoff between energy consumption for WiFi search and transmission efficiency when the WiFi network was intermittently available. In [7], a power control scheme suitable for a multi-tier wireless network was presented. It maximizes the energy-efficiency of a mobile device transmitting on several communication channels while at the same time ensuring the required minimum quality of service. More recently, "delayed" offloading has been proposed: if there is no WiFi available, traffic can be delayed up to some chosen deadline [8]. Some studies like [9] and [4], suggested energy-efficient delayed network selection by exploring the tradeoff between transmit power of heterogeneous network interfaces (e.g., 3G, WiFi) and transmission delay.

Many mobile applications are dealing with video, audio, sensor data, or are accessing large databases on the Internet. Delay-tolerant applications are less sensitive to network delays. Participatory sensing applications are a good example of data-intensive but delay-tolerant applications. Participatory sensing is the collective sampling of sensor data by a number of sensor nodes. This creates a body of knowledge on parameters such as personal resource consumption, dietary habits and urban documentation [9]. Data is uploaded from a smartphone to a back-end cloud server either through the cellular network or any available WiFi network. Some of the sensor information is not time-critical and its submission to the server may be delayed until the device enters an energy-efficient network. Users can browse or search the obtained archives through a website at the server side.

In this paper, we address the operation of a mobile user terminal equipped with multiple radio access technologies. We focus on energy-efficient offloading of delay-tolerant data to a remote cloud. To this end, we propose a framework based on Lyapunov optimization and contribute the following: (i) minimization of the average energy consumption for the link selection and transmission scheduling problem, and (ii) formulating a number of transmission schedulers when using 3G and WiFi interfaces to transmit data.

The remainder of this paper is organized as follows. Section 2 briefly introduces the link selection problem in mobile cloud computating systems. Section 3 analyzes the energy-delay tradeoff by using Lyapunov optimization. Three kinds of transmission schedulers are proposed and investigated in Section 4. Section 5 gives some simulation results. Finally, the paper is concluded in Section 6.

2   Problem Formulation

We provide a brief introduction of the studied adaptive link selection problem and consider a Markovian queueing model for dynamic transmission scheduling and link selection in mobile cloud computing systems.

2.1   Multiple Wireless Interfaces

Mobile devices usually have multiple wireless interfaces that can be used for data transfer, such as EDGE (Enhanced GPRS), 3G, WiFi and so on. The time intervals of cellular connectivity (EDGE or 3G) are usually much longer than for WiFi. Especially, EDGE has very high coverage. In addition, the data rates differ significantly (from hundreds of Kbps for EDGE, to a few Mbps for 3G, to ten or more Mbps for WiFi). The achievable data rate for different radio transmission depends on the environment and can vary widely. It is sometimes far below the nominal value. Also energy-efficiency of the different technologies is different. The energy usage for transmitting a fixed amount of data can differ by an order of magnitude or more [9]. In general, the WiFi interface is more energy-efficient than the cellular interface, and data transmission using a good connection requires much less energy than under bad conditions [4].

Thus, offloading large data items from a mobile device to the cloud using WiFi can be more energy-efficient than using cellular radio, but WiFi connections are not always available. Therefore it must be decided when to transmit data and across which network interface. However, this decision is not easy to take since we know neither the future availability of APs nor their transmission quality.

2.2   Adaptive Link Selection

The problem when to transmit data and which mobile interface to use can be formulated as an adaptive link selection problem as depicted in Fig. 1. Given a set of available links with energy information, AP availability information as obtained from traces and data system queues, determine whether to use any of the available links (the appropriate network interface) to transfer data, while keeping the transmission delay bounded [9]. In Fig. 1, the 3G interface is chosen for data transfer.

The mobile device selects the link with the best connection quality by running a series of probe-based tests to the cloud. Even after a particular link is selected, the connectivity can still be unstable as it is affected by user mobility, limited coverage of the WiFi APs and other factors. Because it sacrifices delay for energy, the problem of link selection and transmission scheduling for delay-tolerant applications can be naturally formulated using an optimization framework.

Suppose there are M channels available, let $B_j(t)$ denote the bandwidth between the mobile device and the cloud in time slot t when using channel j, where $j \in \{1, \cdots, M\}$. Let $b_j(t)$ or $\hat{b}_j(\alpha(t))$ denote the amount of data transmitted over channel j between the mobile device and the cloud in slot t. It is determined by a

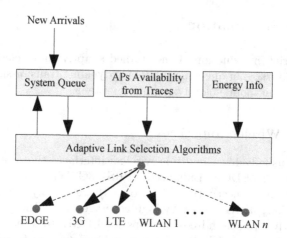

Fig. 1. A mathematical model of adaptive link selection

transmission decision $\alpha(t)$, which is the choice made in slot t, either to transmit data over channel j or not to transfer, and can be expressed as:

$$b_j(t) = \hat{b}_j(\alpha(t)) = \begin{cases} B_j(t) \cdot \tau, & \text{if } \alpha(t) = \text{"Transmit over channel } j\text{"}, \\ 0, & \text{if } \alpha(t) = \text{"Idle"}, \end{cases} \quad (1)$$

where $\alpha(t) = $ "Idle" means that no transmission takes place in slot t and τ is the time duration that the interface is on. For convenience, τ is assumed to be a constant, which is based on the bandwidth estimation and should neither too large to too small [4].

We denote the energy consumption caused by data transmission on the mobile device in time slot t as $E(t) = \hat{E}(\alpha(t))$, which depends on the current link bandwidth and the transmission decision $\alpha(t)$. Over a long time period T, the total amount of transmitted data is $\sum_{t=0}^{T-1} \sum_{j=1}^{M} b_j(t)$, correspondingly, the total energy consumption of the mobile device for transmitting such an amount of data can be denoted as $\sum_{t=0}^{T-1} E(t)$.

Suppose there are N queues of data to be sent from the the mobile device to the cloud, and we define the vector of current queue backlogs by:

$$\boldsymbol{Q}(t) = (Q_1(t), Q_2(t), \cdots, Q_N(t)), \quad \forall t \in \{0, 1, \cdots, T-1\}, \quad (2)$$

where the queues are maintained in the mobile device's memory and for each queue i, $Q_i(t)$ represents its queue backlog of data to be transmitted from the mobile device to the cloud at the beginning of time slot t.

Further, let $A_i(t)$ denote the amount of newly arriving data added to each queue i in time slot t. We assume that each random variable $A_i(t)$ is i.i.d. over time slots with expectation $\mathbb{E}\{A_i(t)\} = \lambda_i$. We call λ_i the arrival rate to queue i.

Therefore, the queue length of queue i in time interval $t + 1$, i.e., $Q_i(t + 1)$ has the following dynamics:

$$Q_i(t+1) = \max\left[Q_i(t) - b_i(t), 0\right] + A_i(t), \forall i \in \{1, 2, \cdots, N\}, \forall t \in \{0, 1, \cdots, T-1\}. \quad (3)$$

Given this notation, we can formally state the queueing constraint that is imposed on our adaptive link selection algorithm. We require all the queues to be stable in the time average sense, i.e.,

$$\bar{Q} \triangleq \limsup_{T \to \infty} \frac{1}{T} \sum_{t=0}^{T-1} \sum_{i=1}^{N} \mathbb{E}\{Q_i(t)\} < \infty, \tag{4}$$

the stability constraint ensures that the average queue length is finite and we should not always defer the transmission.

While maintaining a stable queue we seek to design an adaptive link selection algorithm and dynamic transmission scheduling such that the time-averaged expected transmission energy is minimised [9]:

$$\min \left[\bar{E} \triangleq \limsup_{T \to \infty} \frac{1}{T} \sum_{t=0}^{T-1} \mathbb{E}\{E(t)\} \right], \tag{5}$$

where the required transmission energy $E(t)$ depends on the selected link for transmission during slot t.

3   Energy-Delay Tradeoff

In this section, an optimization model is formulated, with the objective of minimizing the average energy consumption subject to a stability constraint on the queue of data to be transmitted.

3.1   Problem Analysis Using Lyapunov Optimization

To solve the adaptive link selection problem we employ a Lyapunov optimization framework, which enables us to derive a control algorithm that determines when and on which network to transmit our data such that the total energy-cost is minimized. This optimization is not strict with respect to transmission delay.

For each slot t, we define a Lyapunov function [10] as:

$$L(\boldsymbol{Q}(t)) = \frac{1}{2} \sum_{i=1}^{N} Q_i^2(t), \tag{6}$$

which represents a scalar measure of queue length in the network.

We then define the Lyapunov drift as the change in the Lyapunov function from one time slot to the next:

$$
\begin{aligned}
L(\boldsymbol{Q}(t+1)) - L(\boldsymbol{Q}(t)) &= \frac{1}{2} \sum_{i=1}^{N} [Q_i^2(t+1) - Q_i^2(t)] \\
&= \frac{1}{2} \sum_{i=1}^{N} \left[\left(\max[Q_i(t) - b_i(t), 0] + A_i(t) \right)^2 - Q_i^2(t) \right] \\
&\leq \sum_{i=1}^{N} \frac{A_i^2(t) + b_i^2(t)}{2} + \sum_{i=1}^{N} Q_i(t)[A_i(t) - b_i(t)].
\end{aligned} \tag{7}
$$

The conditional Lyapunov drift $\Delta(\boldsymbol{Q}(t))$ is the expected change in the Lyapunov function over one time slot, given that the current state in time slot t is $\boldsymbol{Q}(t)$. That is:

$$\Delta(\boldsymbol{Q}(t)) = \mathbb{E}\{L(\boldsymbol{Q}(t+1)) - L(\boldsymbol{Q}(t))|\boldsymbol{Q}(t)\}. \tag{8}$$

From (7), we have that for a general control policy $\Delta(\boldsymbol{Q}(t))$ satisfies:

$$\Delta(\boldsymbol{Q}(t)) \le \mathbb{E}\left\{ \sum_{i=1}^{N} \frac{A_i^2(t) + b_i^2(t)}{2}|\boldsymbol{Q}(t) \right\} + \sum_{i=1}^{N} Q_i(t)\lambda_i - \mathbb{E}\left\{ \sum_{i=1}^{N} Q_i(t)b_i(t)|\boldsymbol{Q}(t) \right\} \tag{9}$$

where we have used the assumption that arrivals are i.i.d. over slots and hence independent of current queue backlogs, so that $\mathbb{E}\{A_i(t)|\boldsymbol{Q}(t)\} = \mathbb{E}\{A_i(t)\} = \lambda_i$.

Let C be a finite constant that bounds the first term on the right-hand-side of (9), so that for all t, all possible $\boldsymbol{Q}(t)$ and all possible transmission decisions we have:

$$\mathbb{E}\left\{ \sum_{i=1}^{N} \frac{A_i^2(t) + b_i^2(t)}{2}|\boldsymbol{Q}(t) \right\} = \frac{1}{2}\mathbb{E}\left\{ \sum_{i=1}^{N} A_i^2(t) \right\} + \frac{1}{2}\mathbb{E}\left\{ \sum_{i=1}^{N} b_i^2(t)|\boldsymbol{Q}(t) \right\} \le C. \tag{10}$$

There exist constants $A_{\max}^2$ and $b_{\max}^2$ that satisfy the conditions: $\mathbb{E}\{ \sum_{i=1}^{N} A_i^2(t) \}$ $\le A_{\max}^2$ and $\mathbb{E}\{ \sum_{i=1}^{N} b_i^2(t)|\boldsymbol{Q}(t) \} \le b_{\max}^2$, where $A_{\max} \ge A_i(t)$ represents the maximum amount of data that can arrive per time slot, and $b_{\max} \ge b_i(t)$ denotes the maximum amount of data that can be transmitted via the wireless network in a time slot. Hence, we have $C = \frac{1}{2}(A_{\max}^2 + b_{\max}^2)$.

To stabilize the data queue by making sure that there is a balance of arriving data and transmitted data, while minimizing the time-averaged energy $E(t)$, we incorporate the expected energy consumption over one slot t. It can be designed to make transmission decisions that greedily minimize a bound on the following drift-plus-penalty term in each slot t [10]:

$$\Delta(\boldsymbol{Q}(t)) + V\mathbb{E}\{E(t)|\boldsymbol{Q}(t)\}, \tag{11}$$

where $V \ge 0$ is a control parameter that represents an "importance weight" in deciding relative importance among queue backlog, estimated rate, and energy cost. In other words, V can be thought of as a threshold on the queue backlog beyond which the control algorithm decides to transmit, so V controls the energy-delay tradeoff [9].

From (9) and (10) we have:

$$\Delta(\boldsymbol{Q}(t)) + V\mathbb{E}\{E(t)|\boldsymbol{Q}(t)\} \le C + \sum_{i=1}^{N} Q_i(t)\lambda_i + V\mathbb{E}\{E(t)|\boldsymbol{Q}(t)\} - \mathbb{E}\left\{ \sum_{i=1}^{N} Q_i(t)\hat{b}_i(\alpha(t))|\boldsymbol{Q}(t) \right\}$$

$$= C + \sum_{i=1}^{N} Q_i(t)\lambda_i + \mathbb{E}\left\{ \left[VE(t) - \sum_{i=1}^{N} Q_i(t)\hat{b}_i(\alpha(t)) \right] | \boldsymbol{Q}(t) \right\}. \tag{12}$$

Using the concept of opportunistically minimizing an expectation, the optimization of the right-hand-side of (12) is accomplished by greedily minimizing the following term:

$$\arg \min_{\alpha(t)} \left[VE(t) - \sum_{i=1}^{N} Q_i(t)\hat{b}_i(\alpha(t)) \right], \tag{13}$$

where we choose the transmission decision $\alpha(t)$ that will minimize (13).

We denote a decision function as $d(t) = VE(t) - \sum_{i=1}^{N} Q_i(t)\hat{b}_i(\alpha(t))$, which is the decision results that depends on the current link bandwidth and the transmission decision $\alpha(t)$. In order to understand the intuition behind this decision, we would like to see when $d(t)$ can have a low value.

1. **Link with a Good Quality:** $d(t)$ can be small when the link has a high estimated rate. It makes sense that we would like to use any high-quality link to transfer data over a low-quality link.
2. **Queue Backlog is High:** $d(t)$ can achieve a low-value if the queue backlog $Q(t)$ is high. This is also intuitive: when data has been in the queue for long, there should be a higher incentive to transmit.
3. **Link Energy Cost is Low:** $d(t)$ is small when the energy cost $E(t)$ of a link is low (e.g., a WiFi link). Such a link should be preferred over a high-energy cellular link [9].

In other words, the link selection model based on Lyapunov optimization defers transmission until good-quality and low-energy links become available, unless the queue backlog is too high.

Further, considering the decision $\alpha(t)$, the decision function $d(t)$ can be denoted as:

$$d(t) = \begin{cases} VE_i(t) - Q_i(t)b_i(t), & \text{if } \alpha(t)=\text{``Transmit over channel } i\text{''}, \\ 0, & \text{if } \alpha(t)=\text{``Idle''}. \end{cases} \tag{14}$$

3.2   Performance Bounds

For any control parameter $V > 0$, we assume that the data arrival rate λ_i is strictly within the network capacity region, which is defined as the region that can be achieved by the mobile device in communication networks [9]. We can achieve a time-averaged energy consumption and queue backlog satisfying the following constraints [11]:

$$\bar{E} = \limsup_{T \to \infty} \frac{1}{T} \sum_{t=0}^{T-1} \mathbb{E}\{E(t)\} \leq E^* + \frac{C}{V}, \tag{15}$$

$$\bar{Q} = \limsup_{T \to \infty} \frac{1}{T} \sum_{t=0}^{T-1} \sum_{i=1}^{N} \mathbb{E}\{Q_i(t)\} \leq \frac{C + V(E^* - \bar{E})}{\varepsilon}, \tag{16}$$

where $\varepsilon > 0$ is a constant denoting the distance between arrival pattern and the capacity region boundary [9], E^* is a theoretical lower bound on the time-averaged energy consumption using any control policy that achieves queue stability.

Proof. Because the transmission decision $\alpha(t)$ minimizes the right-hand-side of the drift-plus-penalty in inequality (12), in every slot t (given the observed $\boldsymbol{Q}(t)$), we have:

$$\Delta(\boldsymbol{Q}(t)) + V\mathbb{E}\{E(t)|\boldsymbol{Q}(t)\} \leq C + V\mathbb{E}\{\hat{E}(\alpha^*(t))|\boldsymbol{Q}(t)\} + \sum_{i=1}^{N} Q_i(t)\lambda_i$$
$$- \mathbb{E}\left\{\sum_{i=1}^{N} Q_i(t)\hat{b}_i(\alpha^*(t))|\boldsymbol{Q}(t)\right\}, \qquad (17)$$

where $\alpha^*(t)$ is any other (possibly randomized) transmission decision that can be made in slot t. Fixing any value $\varepsilon > 0$ in the capacity region boundary further yields:

$$\Delta(\boldsymbol{Q}(t)) + V\mathbb{E}\{E(t)|\boldsymbol{Q}(t)\} \leq C + V\mathbb{E}\{\hat{E}(\alpha^*(t))|\boldsymbol{Q}(t)\} + \sum_{i=1}^{N} Q_i(t)\lambda_i - \sum_{i=1}^{N} Q_i(t)(\lambda_i + \varepsilon)$$
$$= C + V\mathbb{E}\{\hat{E}(\alpha^*(t))|\boldsymbol{Q}(t)\} - \varepsilon\sum_{i=1}^{N} Q_i(t). \qquad (18)$$

Taking expectations for (18) with respect to $\boldsymbol{Q}(t)$ and using the law of iterated expectations, yields:

$$\mathbb{E}\{L(\boldsymbol{Q}(t+1))\} - \mathbb{E}\{L(\boldsymbol{Q}(t))\} + V\mathbb{E}\{E(t)\} \leq C + VE^* - \varepsilon\sum_{i=1}^{N} \mathbb{E}\{Q_i(t)\}, \qquad (19)$$

where $E^* \triangleq \mathbb{E}\{\hat{E}(\alpha^*(t))\}$.

Summing the above inequality over $t \in \{0, 1, \cdots, T-1\}$ for some positive integer T, yields:

$$\mathbb{E}\{L(\boldsymbol{Q}(T))\} - \mathbb{E}\{L(\boldsymbol{Q}(0))\} + V\sum_{t=0}^{T-1} \mathbb{E}\{E(t)\} \leq CT + VTE^* - \varepsilon\sum_{t=0}^{T-1}\sum_{i=1}^{N} \mathbb{E}\{Q_i(t)\}. \quad (20)$$

Then, dividing (20) by VT and after a simple manipulation we obtain:

$$\frac{1}{T}\sum_{t=0}^{T-1} \mathbb{E}\{E(t)\} \leq \frac{C}{V} + E^* - \frac{\varepsilon\sum_{t=0}^{T-1}\sum_{i=1}^{N} \mathbb{E}\{Q_i(t)\}}{VT} - \frac{\mathbb{E}\{L(\boldsymbol{Q}(T))\}}{VT} + \frac{\mathbb{E}\{L(\boldsymbol{Q}(0))\}}{VT}.$$
$$(21)$$

Since the Lyapunov function is non-negative by definition and so is E^*, neglecting that we subtract non-negative quantities in (21) yields:

$$\frac{1}{T}\sum_{t=0}^{T-1} \mathbb{E}\{E(t)\} \leq P^* + \frac{C}{V} + \frac{\mathbb{E}\{L(\boldsymbol{Q}(0))\}}{VT}. \qquad (22)$$

Similarly, dividing (20) by εT, and after rearranging terms we obtain:

$$\frac{1}{T} \sum_{t=0}^{T-1} \sum_{i=1}^{N} \mathbb{E}\{Q_i(t)\} \leq \frac{C + V(E^* - \frac{1}{T} \sum_{t=0}^{T-1} \mathbb{E}\{E(t)\})}{\varepsilon} + \frac{\mathbb{E}\{L(\boldsymbol{Q}(0))\}}{\varepsilon T}. \quad (23)$$

Finally, taking a lim sup as $T \to \infty$ in inequalities (22) and (23), we can derive (15) and (16), respectively. □

It can be seen that (15) and (16) demonstrate an $[O(1/V), O(V)]$ tradeoff between energy consumption and delay. We can achieve an average energy consumption $\bar{E}$ arbitrarily close to E^* while maintaining queue stability. However, this is achieved at the expense of a larger delay because the average queue backlog $\bar{Q}$ increases linearly with V. Choosing a large value of V can thus push the average energy arbitrarily close to its optimal value. However, this comes by sacrificing average queue backlog or average delay that is $O(V)$ [10]. A good V value is one that achieves a good energy and delay tradeoff, where a unit increase in V yields a very small reduction in $\bar{E}$ with consistently growing delays [9]. In mathematical terms we can choose a $k < 0$ that satisfies:

$$\frac{d(E^* + C/V)}{dV} \geq k \Longrightarrow V \geq \sqrt{\frac{C}{-k}}, \quad (24)$$

where k is the slope of $\bar{E}$ curve.

4 Performance Analysis Models

To understand this link selection algorithm, we consider the two most prominent networks: WiFi and 3G. Typically, the WiFi interface is much more energy-efficient, but its availability is limited while the 3G network is available almost everywhere. Besides, channel quality can be affected by environmental factors and interference. The channel bandwidth can be reduced due to competing users in the same cells. Therefore, for data-intensive but delay-tolerant applications, we can save energy by delaying transmissions until a good-quality or a low-energy interface such as WiFi becomes available, unless the queue backlog is too high.

4.1 Bandwidth Estimation and Energy Models

Since our transmission scheduling model uses the knowledge of current states (i.e, the current network bandwidth is supposed to be known), it closely depends on the bandwidth estimation. We use a predictor proposed in [12], which considers the classical bandwidth predictors (such as Last value, Mean filter, Network weather service forecaster, etc.) synthetically. The framework unifies such decision models by formulating the problem as a statistical decision problem that can either be treated "classically" or using a Bayesian approach. The experimental result shows that the Bayes strategy performs significantly better than the traditional predictors. Thus, this prediction model is more general and

Table 1. Energy model for 3G and WiFi networks

Items	3G	WiFi
Ramp and Transfer Energy $R(x)$	$0.025x + 3.5$	$0.007x + 5.9$
Tail power P	0.62J/s	NA
Tail time T	12.5s	NA

could be used by our offloading system. Further, we assume that the network bandwidth is constant in one time slot.

Table 1 lists the measured energy consumption models according to [13]. The energy needed to transmit x bytes of data over the cellular network can be split into three components: ramp energy, transmission energy and tail energy. $R(x)$ denotes the sum of the ramp and the transfer energy to send x KB, P denotes the tail power and T denotes the tail time.

Obviously, the energy consumption depends on the type of interface that is selected. For the 3G interface, the sum of the ramp and the transfer energy is $R(x) = 0.025x + 3.5$. After transmitting a packet, instead of transitioning from high to low power state, the 3G interface spends substantial time in the high state, which incurs considerable energy, referred to as the tail energy. For the WiFi interface, the sum of the ramp and the transfer energy is $R(x) = 0.007x + 5.9$, and the tail energy is zero. Using WiFi, the data transfer itself is significantly more efficient than using the 3G connection for all transfer sizes. In addition to the transfer cost, the total energy to transmit a packet also depends on the time that the interface is on.

Therefore, the energy consumption for the 3G and WiFi interfaces in time slot t can be expressed as follows:

$$E_{3G}(t) = 0.025 \cdot b_{3G}(t) + 3.5 + 0.62 \cdot 12.5, \tag{25}$$

$$E_{WiFi}(t) = 0.007 \cdot b_{WiFi}(t) + 5.9. \tag{26}$$

4.2   Transmission Scheduler I ($N \neq M$)

The model of the transmission scheduler I for only one queue of arriving jobs is depicted in Fig. 2. The arrival vector $A(t)$ is assumed to be i.i.d over the time slot and $\mathbb{E}\{A(t)\} = \lambda$.

We take decisions of transmission scheduling according to the estimate of the current network bandwidth. In Fig. 2, "$B_{3G}(t)$" represents the estimated 3G bandwidth in slot t, "$B_{WiFi}(t)$" represents the estimated WiFi bandwidth and "Idle" denotes that no transmission takes place in time slot t. If $B_{3G}(t)$ is larger than $B_{WiFi}(t)$, the mobile device will be linked to the 3G interface in time slot t to transmit data, otherwise it will be linked to the WiFi interface. The decision criterion can be denoted as $\max\{B_{3G}(t), B_{WiFi}(t)\}$. Therefore, the bandwidth of the selected interface is as follows:

$$B(t) = \begin{cases} B_{3G}(t), & \text{if } B_{3G}(t) > B_{WiFi}(t), \\ B_{WiFi}(t), & \text{otherwise.} \end{cases} \tag{27}$$

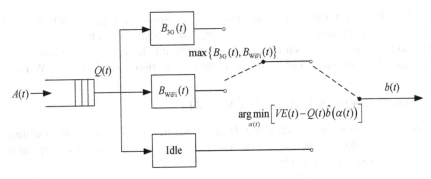

Fig. 2. Model of transmission scheduler I

According to the Lyapunov optimization, the minimization of the average energy consumption is accomplished by greedily minimizing the following criterion:

$$\arg\min_{\alpha(t)}\left[VE(t) - Q(t)\hat{b}(\alpha(t))\right]. \tag{28}$$

Denoting the decision function as $d(t) = VE(t) - Q(t)\hat{b}(\alpha(t))$, when considering the transmission decision $\alpha(t)$ we have:

$$d(t) = \begin{cases} VE(t) - Q(t)B(t)\cdot\tau, & \text{if } \alpha(t) = \text{``transmit''}, \\ 0, & \text{if } \alpha(t) = \text{``idle''}, \end{cases} \tag{29}$$

where $\alpha(t) \in \{\text{``transmit'' and ``idle''}\}$, taking on two possible values and

$$E(t) = \begin{cases} E_{3G}(t), & \text{if } \alpha(t) = \text{``transmit'' and } B_{3G}(t) > B_{WiFi}(t), \\ E_{WiFi}(t), & \text{if } \alpha(t) = \text{``transmit'' and } B_{3G}(t) \le B_{WiFi}(t), \\ 0, & \text{if } \alpha(t) = \text{``idle''}. \end{cases}$$

If the transmission decision is $\alpha(t) = \text{``transmit''}$, we choose to transfer data according to the current channel bandwidth. If $\alpha(t) = \text{``idle''}$, no data is transmitted in slot t, so $E(t) = 0$ and $b(t) = 0$, and then we have $d(t) = 0$. Therefore, transmission takes place only if V satisfies: $VE(t) - Q(t)\hat{b}(\alpha(t)) < 0$. This happens when the bandwidth is high, making a large $\hat{b}(\alpha(t))$, or the queue $Q(t)$ is already congested in time slot t.

Over time, the queuing dynamic is given by:

$$Q(t+1) = \max[Q(t) - b(t), 0] + A(t), \quad \forall t \in \{0, 1, \cdots, T-1\}. \tag{30}$$

By Little's Theorem [14], the average delay can be calculated as:

$$\bar{D} = \frac{\bar{Q}}{\lambda}. \tag{31}$$

The disadvantage of transmission scheduler I is that only the estimated bandwidth of 3G and WiFi in time slot t is considered and the energy usage of

3G and WiFi is not taken into account. For example, if $B_{3G}(t)$=50Kbps and $B_{WiFi}(t)$=49.99Kbps, since $B_{3G}(t)$ is larger than $B_{WiFi}(t)$ we choose the 3G interface to transmit data, even though it consumes much more energy than WiFi. In this situation we should also consider the energy demand of 3G and WiFi.

4.3 Transmission Scheduler II ($N \neq M$)

The model of transmission scheduler II is shown in Fig. 3. There are two links ($M = 2$) available for selection. We also use one queue ($N = 1$) to represent data transmission during each slot.

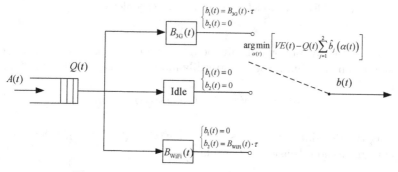

Fig. 3. Model of optimal transmission scheduler II

Using the concept of opportunistically minimizing the expectation, the minimization of average energy consumption is accomplished by greedily minimizing:

$$\arg \min_{\alpha(t)} \left[VE(t) - Q(t) \sum_{j=1}^{M} \hat{b}_j(\alpha(t)) \right]. \tag{32}$$

Similarly, let $d(t) = VE(t) - Q(t) \sum_{j=1}^{M} \hat{b}_j(\alpha(t))$. Since $M = 2$, there are three possible results according to the transmission decision of $\alpha(t)$:

$$d(t) = \begin{cases} VE_{3G}(t) - Q(t)B_{3G}(t) \cdot \tau, & \text{if } \alpha(t)=\text{``transmit via 3G''}, \\ VE_{WiFi}(t) - Q(t)B_{WiFi}(t) \cdot \tau, & \text{if } \alpha(t)=\text{``transmit via WiFi''}, \\ 0, & \text{if } \alpha(t)=\text{``idle''}, \end{cases} \tag{33}$$

where $\alpha(t) \in \{$ "transmit via 3G", "transmit via WiFi" and "idle"$\}$ is the transmission decision in slot t, taking on three possible values.

According to (33), we not only consider the estimated bandwidth but also take into account the energy usage of 3G and WiFi in time slot t. We thus compare the above values and choose the transmission decision corresponding to the smallest outcome. The queuing dynamics and the average delay are given by (30) and (31), respectively.

If the 3G and WiFi interfaces can be used simultaneously, the model of transmission scheduler II in Fig. 3 can be further extended as in Fig. 4.

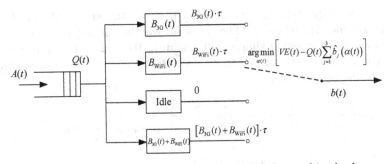

Fig. 4. Model of transmission scheduler II for the combined scheme

Since the combined transmission works just like an extra channel, we have $M = 3$. Thus, there are four possible results in (32) according to the transmission decision of $\alpha(t)$:

$$d(t) = \begin{cases} V E_{3G}(t) - Q(t)B_{3G}(t) \cdot \tau, & \text{if } \alpha(t)=\text{"transmit via 3G"}, \\ V E_{WiFi}(t) - Q(t)B_{WiFi}(t) \cdot \tau, & \text{if } \alpha(t)=\text{"transmit via WiFi"}, \\ V \cdot [E_{3G}(t) + E_{WiFi}(t)] - Q(t) \cdot [B_{3G}(t) + B_{WiFi}(t)] \cdot \tau, & \text{if } \alpha(t)=\text{"transmit via 3G and WiFi"}, \\ 0, & \text{if } \alpha(t)=\text{"idle"}, \end{cases}$$

where $\alpha(t) \in \{$ "transmit via 3G", "transmit via WiFi", "transmit via 3G and WiFi", and "idle"$\}$ is the transmission decision in slot t, taking on four possible values.

4.4 Transmission Scheduler III ($N = M$)

The model of transmission scheduler III is depicted in Fig. 5. To overcome the problem pointed out above and to take more accurate decisions, we divide the data into two queues. The number of channels is equal to the number of queues, that is $N = M = 2$.

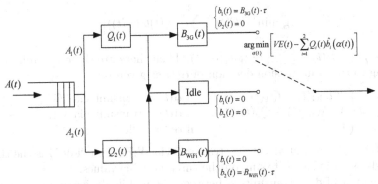

Fig. 5. Model of transmission scheduler III

It can be seen from Fig. 5 that $A_1(t)$ is only transmitted through the 3G interface while $A_2(t)$ is only transmitted through the WiFi interface. We assume that $A_1(t)$ and $A_2(t)$ take integer units of packets, the arrival vector $A(t)$ is i.i.d over slot and $\mathbb{E}\{A(t)\} = \lambda$. The question whether or not to allocate $A(t)$ to $A_1(t)$ and $A_2(t)$ in equal shares still remains. To analyze this problem, we simplify the model as shown in Fig. 6, such that it involves routing decisions besides scheduling decisions.

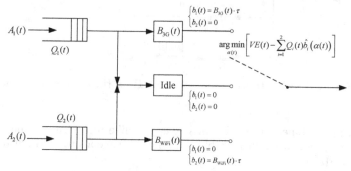

Fig. 6. Equivalent model of transmission scheduler III

There are two separate queues depicted in Fig. 6, the arrival vectors $A_1(t)$ and $A_2(t)$ are i.i.d over slot, $\mathbb{E}\{A_1(t)\} = \lambda_1$ and $\mathbb{E}\{A_2(t)\} = \lambda_2$. Since $A_1(t)+A_2(t) = A(t)$, according to the property of the Poisson distribution we have:

$$\lambda_1 + \lambda_2 = \lambda, \tag{34}$$

where $\lambda_1 = \rho\lambda$, $\lambda_2 = (1-\rho)\lambda$, and $0 \leq \rho \leq 1$ is the ratio of arrival rate to queue 1. There are two extreme cases: when $\rho = 0$, the mobile device only uses the WiFi interface to transmit data and when $\rho = 1$, the mobile device only uses the 3G interface.

Similarly, using the concept of opportunistic minimization of the expectation, the minimization of the average energy consumption is accomplished by greedily minimizing:

$$\arg \min_{\alpha(t)} \left[VE(t) - \sum_{i=1}^{2} Q_i(t)\hat{b}_i(\alpha(t)) \right]. \tag{35}$$

Let $d(t) = VE(t) - \sum_{i=1}^{2} Q_i(t)\hat{b}_i(\alpha(t))$. Then there are three possible results according to the transmission decision of $\alpha(t)$ as given by:

$$d(t) = \begin{cases} VE_{3G}(t) - Q_1(t)b_1(t), & \text{if } \alpha(t)=\text{``transmit via 3G''}, \\ VE_{WiFi}(t) - Q_2(t)b_2(t), & \text{if } \alpha(t)=\text{``transmit via WiFi''}, \\ 0, & \text{if } \alpha(t)=\text{``idle''}, \end{cases} \tag{36}$$

where $\alpha(t) \in \{$ "transmit via 3G", "transmit via WiFi" and "idle"$\}$ is the transmission decision in slot t, taking on the three possible values.

The amount of data transmitted between the mobile device and the cloud in slot t is as follows:

$$\{b_1(t), b_2(t)\} = \begin{cases} \{B_{3G}(t) \cdot \tau, 0\}, & \text{if } \alpha(t)=\text{``transmit via 3G''}, \\ \{0, B_{WiFi}(t) \cdot \tau\}, & \text{if } \alpha(t)=\text{``transmit via WiFi''}, \\ \{0, 0\}, & \text{if } \alpha(t)=\text{``idle''}, \end{cases}$$

and the queuing dynamics are given by:

$$Q_i(t+1) = \max[Q_i(t) - b_i(t), 0] + A_i(t), \ \forall i \in \{1, 2\}, \ \forall t \in \{0, 1, \cdots, T-1\}. \quad (37)$$

Similarly, the average delay for this system is:

$$\bar{D} = \frac{\overline{Q_1 + Q_2}}{\lambda_1 + \lambda_2}. \quad (38)$$

Furthermore, the transmission scheduler III can be extended in the same way to more general scenarios as depicted in Fig. 1, where several traffic queues can be concurrently distributed over several communication channels.

5 Simulation Results

As for parameter setting, we assume that data arrivals follow Possion Process with $\lambda = 4$ packets/minute and the size of each packet is 100 KB. Suppose that the network bandwidths stay the same during each time slot. Our algorithms are simulated in 1,000 time slots for each of the V value ranging from 1 to 300. We study the impact of parameter V on time-averaged energy consumption, queue backlog, delay and transmit data. The energy consumption models are according to (25) and (26) for the 3G and WiFi interfaces, respectively.

We first estimate the achievable network bandwidth $B(t)$ at the beginning of every time slot t. Since data communication time between the mobile device and the cloud depends on the network bandwidth and the bandwidth of wireless LAN is remarkably higher than the bandwidth provided by radio access on a mobile device, we suppose that the bandwidth for the 3G interface follows a uniform distribution on $[1, 100]$ KB/s and the bandwidth for the WiFi interface follows a uniform distribution on the interval $[1, 300]$ KB/s. We set the length of each slot $\tau = 60$, and the bandwidth in the corresponding time slot t is used for every 60 seconds.

It can be seen from Fig. 7 (transmission scheduler I, refer to Fig. 2) that the time-average energy consumption and transmit data fall quickly at the beginning and then tend to descend slowly while the time-average queue backlog grows linearly with V. This finding confirms the $[O(1/V), O(V)]$ tradeoff as captured in (15) and (16). According to different delay-tolerant and data-intensive applications, we can adjust the value of V to control the energy-delay tradeoff. Especially, there exists a sweet spot of V, and at this point, the marginal energy conservation is not worth the consistently growing delay with increasing of V. For example, when V increases from 100 to 200, it shows a negligible decrease of the average energy consumption while the average delay increases significantly, thus we should not trade energy with delay. Further, according to (24), the slope of the curve is $k \approx 0$ at this point.

The numerical results of using transmission scheduler II are depicted in Fig. 8. We compare the scheme that combines 3G and WiFi (refer to Fig. 4) with the one that transmit separately (refer to Fig. 3). It can be seen that the average

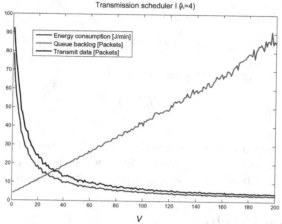

Fig. 7. The impact of V on time-averaged energy consumption, queue backlog and transmit data for transmission scheduler I

number of transmitted packets and average delay in both schemes almost coincide with each other while the combined scheme achieves a lower average energy consumption than both individual schemes when the control factor V is small (e.g., $V \leq 50$).

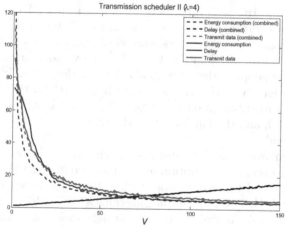

Fig. 8. Comparison of different schemes for transmission scheduler II

The numerical results of using transmission scheduler III for the scenario (refer to Fig. 6) are depicted in Figs. 9-11.

It is known that when $\rho = 0$ (ρ is defined before as the dispatching ratio of arrival rate to queue 1), the mobile device only uses the WiFi interface to transmit data and when $\rho = 1$, it only uses the 3G interface. As shown in Fig. 9, when V is small, it has the minimum energy consumption when only using 3G for data transfer, while it has the maximum energy consumption when only

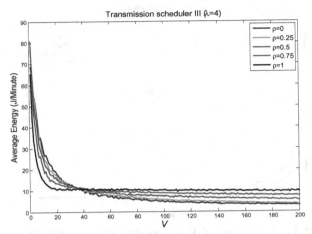

Fig. 9. The impact of V on time-averaged energy consumption

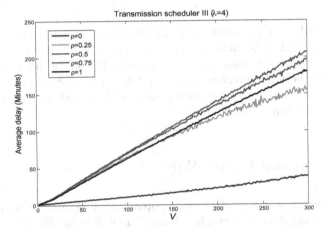

Fig. 10. The impact of V on time-averaged delay

using WiFi. The average energy increases with the increase of ρ when $V \leq 37$. However, when V arrives to a certain value ($V \approx 37$), the scheme that only uses WiFi for data transfer has the minimum energy consumption while the one that only uses 3G has the maximum energy consumption. The time-averaged energy consumption increases with the increment of ρ when $V > 37$. Therefore, the energy consumption for such a transmission scheduler closely depends on the value of ρ.

The impact of V on the time-averaged delay is shown in Fig. 10. It is found that the average delay is minimal when only using WiFi to transmit data. With the increase of ρ, the average delay at first increases, but it then decreases after ρ arrives at some value, for example, the average delay is smaller for $\rho = 1$ than for $\rho = 0.75$.

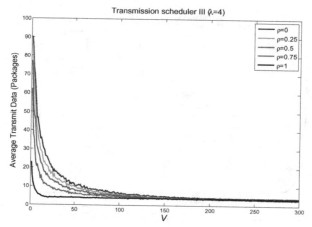

Fig. 11. The impact of V on time-averaged transmit data

The impact of V on time-averaged transmitted data is depicted in Fig. 11. It can be seen that when V is small, the average transmit data decreases with the increment of ρ, thus the mobile device can transfer the largest amount of data when only using the WiFi interface to transmit data due to its high bandwidth. However, when V is large, the average transmitted data is almost the same and does not change with increasing of V.

6   Conclusion and Future Work

In this paper, we present a fundamental approach for designing an online algorithm for the energy-delay tradeoff in "delayed" mobile data offloading through the Lyapunov optimization framework. Considering the changing landscape of network connectivity, the problem of link selection and data transmission scheduling can be formulated as an optimization problem, in which a significant amount of energy can be saved without sacrificing on the transmission delay too much. Three types of transmission schedulers are proposed and compared based on simulation results. These energy-efficient transmission schedulers consider several factors: data backlog, channel quality and energy consumption of the wireless interface, when making transmission decisions. They will choose to transmit data when the connectivity is good enough or when the queues in the mobile device are congested.

So far the validation of the approach is based on simulation under simplifying assumptions. For future work, validation based on real workloads and more realistic application examples will be provided to gain insights about efficiency of the proposed algorithm in practice. Besides, a mobile-cloud offloading middleware will be developed to apply those schedulers to reduce energy consumption for delay-tolerant applications on mobile devices.

References

1. Armbrust, M., Fox, A., Griffith, R., Joseph, A., Katz, R., Konwinski, A., Lee, G., Patterson, D., Rabkin, A., Stoica, I., Zaharia, M.: Above the Clouds: A Berkeley View of Cloud Computing. Technical Report No. UCB/EECS-2009-28.University of California at Berkley, USA (2009)
2. Niu, D.: Gearing Resource-Poor Mobile Devices with Powerful Clouds: Architectures, Challenges, and Applications. IEEE Wireless Communications 20(3), 14–22 (2013)
3. Apple Company, http://www.apple.com/icloud/
4. Shu, P., Liu, F., Jin, H., Chen, M., Wen, F., Qu, Y., Li, B.: eTime: Energy-Efficient Transmission between Cloud and Mobile Devices. In: Proc. of IEEE Infocom, pp. 195–199. IEEE Press, New York (2013)
5. Gribaudo, M., Manini, D., Chiasserini, C.: Studying Mobile Internet Technologies with Agent based Mean-Field Models. In: Dudin, A., De Turck, K. (eds.) ASMTA 2013. LNCS, vol. 7984, pp. 112–126. Springer, Heidelberg (2013)
6. Rahmati, A., Zhong, L.: Context-for-Wireless: Context-Sensitive Energy-Efficient Wireless Data Transfer. In: Proc. of the 5th International Conference on Mobile Systems, Applications and Services, pp. 165–178. ACM, New York (2007)
7. Galinina, O., Trushanin, A., Shumilov, V., Maslennikov, R., Saffer, Z., Andreev, S., Koucheryavy, Y.: Energy-Efficient Operation of a Mobile User in a Multi-tier Cellular Network. In: Dudin, A., De Turck, K. (eds.) ASMTA 2013. LNCS, vol. 7984, pp. 198–213. Springer, Heidelberg (2013)
8. Mehmeti, F., Spyropoulos, T.: Performance Analysis of On-the-Spot Mobile Data Offloading. In: Proc. of IEEE Globecom. IEEE Press, New York (2013)
9. Ra, M., Paek, J., Sharma, A., Govindan, R., Krieger, M., Neely, M.: Energy-Delay Tradeoffs in Smartphone Applications. In: 8th International Conference on Mobile Systems, Applications, and Services, pp. 255–270. ACM, New York (2010)
10. Neely, M.J.: Stochastic Network Optimization with Application to Communication and Queueing Systems. In: Synthesis Lectures on Communication Networks, vol. 3(1), pp. 1–211. Morgan & Claypool Publishers (2010)
11. Georgiadis, L., Neely, M.J., Tassiulas, L.: Resource Allocation and Cross-Layer Control in Wireless Networks. Now Publishers Inc. (2006)
12. Wolski, R., Gurun, S., Krintz, C., Nurmi, D.: Using Bandwidth Data to Make Computation Offloading Decisions. In: IEEE International Symposium on Parallel and Distributed Processing, IPDPS 2008, pp. 1–8. IEEE Press, New York (2008)
13. Balasubramanian, N., Balasubramanian, A., Venkataramani, A.: Energy Consumption in Mobile Phones: A Measurement Study and Implications for Network Applications. In: 9th ACM SIGCOMM Conference on Internet Measurement Conference, pp. 280–293. ACM, New York (2009)
14. Bertsekas, D.P., Gallager, R.G., Humblet, P.: Data Networks. Prentice-Hall International, New Jersey (1992)

Non-linear Programming Method for Buffer Allocation in Unreliable Production Lines

Yassine Ouazene, Alice Yalaoui, Farouk Yalaoui, and Hicham Chehade

Institut Charles Delaunay
Laboratoire d'Optimisation des Systèmes Industriels (UMR-CNRS 6281)
Université de Technologie de Troyes, 12 rue Marie Curie, CS 42060
10004 TROYES, France
yassine.ouazene@utt.fr

Abstract. This paper proposes a new algorithm based on a non-linear programming approach to deal with the buffer allocation problem in the case of unreliable production lines. Processing, failure and repair times are assumed to be random variables exponentially distributed. The proposed approach can be used to solve the different versions of the buffer allocation problem: primal, dual and generalized.

This method is based on the modeling and the analysis of the serial production line using an equivalent machines method. The idea is to model the different possible states of each buffer using dedicated birth-death Markov processes to calculate the blockage and starvation probabilities of each machine. Then, each original machine is replaced by an equivalent one taking into account these probabilities.

A comparative study based on different test instances issued from the literature is presented and discussed. The obtained results show the effectiveness and the accuracy of the proposed approach.

Keywords: Buffer allocation, Unreliable production lines, Equivalent Machine Method, Non-Linear Programming, Birth-death processes.

1 Introduction

The buffer allocation problem is one of the major issues in the production systems design because of the great effect that buffers can do on improving the performance and the efficiency of a production system. However, the use of these intermediate stocks generates investment costs in terms of space or other physical means. They also increase work-in-process inventories through the production line. That is why, the buffer allocation is a major optimization problem faced by manufacturing systems designers as well as by researchers. The problem is complicated and critical since it introduces computational complexity and involves trade-offs between the constraints and the objectives posed by the problem itself [18].

Due to its complexity and importance, the buffer allocation problem has been widely studied in the literature. One of the earliest works dealing with this

B. Sericola, M. Telek, and G. Horváth (Eds.): ASMTA 2014, LNCS 8499, pp. 80–94, 2014.

problem was presented by Koenigsberg [5] more than fifty years ago. The author proposed an analysis of the fundamental problems combining the allocation of storage areas and the efficiency of production systems.

Vergara and Kim [17] stated that the buffer allocation research literature can, in general, be separated into two main categories. The first category focuses on general design rules that have been developed after extensive computational experimentations. For example, Conway et al. [1] did extensive simulation studies of open production lines and identified general buffer allocation principles and design rules for balanced unbalanced and unbalanced production lines. A similar study was presented by Matta [6].The author proposes explicit mathematical programming representations for jointly simulating and optimizing discrete event systems for solving the buffer allocation problem in flow lines with finite buffer capacities.

The second category of buffer allocation literature focuses on algorithms for buffer allocation optimization in serial production lines and is directly related to the topic of this paper. In general, these algorithms possess some method for evaluating the performance of each candidate solution, and include some search procedure to select candidate solutions. For example, Nahas et al. [7] described a new local search approach for solving the buffer allocation problem in unreliable production lines. The authors used a variant of a local search metaheuristic, called the degraded ceiling approach to select the t candidate solutions. Then an analytical decomposition-type approximation is used to estimate the production line throughput.

Other studies, such as those of Dallery and Gershwin [2] and Papadopoulos and Heavey [11], were interested in the analysis of various analytical methods and mathematical models describing the effects of intermediate buffers on different kinds of production systems. Recently, Papadopoulos et al. [10] proposed a classification of the different optimization methods taking into account the length (size) of the production lines.

More details about the buffer allocation problem literature can be found on the comprehensive literature survey proposed by Demir et al. [3]. To provide a systematic review of current relevant research, first studies are grouped in two categories: reliable production lines and unreliable production lines. Next, the studies in each group are reviewed based on topology of the production line, the solution methodologies suggested and the objective function employed.

In this paper we address the problem of buffer allocation in serial production lines. Here the focus is on assigning a fixed number of buffer spaces to maximize system throughput. The proposed approach can then be used to find the minimum of buffer sizes required to meet or exceed a target average throughput, sometimes referred to as a dual buffer placement optimization problem. The main contribution presented in this paper is the development of a new non-linear method that can rapidly identify a near-optimal buffer allocation solution. This optimization method is an adaptation of the equivalent machine method proposed by Ouazene et al. [8] to evaluate the throughput of serial production lines.

The originality is how to transform a performance evaluation method into an optimization algorithm.

The remainder of this paper is organized as follows: Section 2 introduces the problem studied with the different assumptions and notations. In Section 3 and Section 4, the analytical approach and the non-linear programming model proposed in this paper are developed. An experimental study is presented and discussed in Section 5. Finally, a conclusion section summarizes the guidelines of our contribution.

2 Problem Description and Assumptions

The production line considered in this paper consists of K unreliable machines or stations separated by $(K-1)$ intermediate buffers (see Figure 1). The products flow continuously in a fixed sequence from the first machine to the last one. It is assumed that the first machine is never starved and the last one is never blocked. It follows that there is an infinite number of products at the input of the system and unlimited storage capacity at the output of the last machine.

The failures are operation-dependent, a machine cannot fail while it is starved, blocked or idle. Times to failure and to repair are independent and exponentially distributed. The processing times of each machine are assumed to be independent random variables following the exponential distribution. The mean values of these variables are not necessarily identical.

If no products are available in the upstream buffer B_{j-1} the machine M_j will be starved and when the downstream buffer B_j is full, it will be blocked.

The capacities of the different intermediate buffers are finite but unknown. In fact, the aim of this paper is to propose an efficient method to determine these capacities in order to maximize the throughput of the production line.

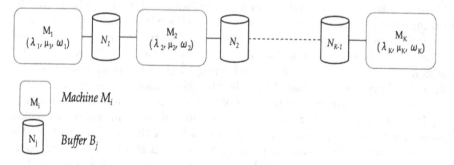

Machine M_i

Buffer B_j

Fig. 1. A K-machine $(K-1)$-buffer serial production line

The notations used in this paper are defined below.

λ_i Equivalent failure rate of the machine M_i
μ_i Equivalent repair rate of the machine M_i
ω_i Processing rate of the machine M_i
$N_j, j \in \{1...K - 1\}$ Capacity of the intermediate buffer B_j
$\overline{N_j}, j \in \{1...K - 1\}$ average inventory level in buffer j
$\xi_i, i \in \{1...K\}$ Probability of both blockage and starvation of machine M_i
$\alpha_j, j \in \{1...K - 1\}$ Processing rates fraction related to the buffer B_j
$\rho_i, i \in \{1...K\}$ Equivalent production rate of the machine M_i
$P_j^s, s \in \{0...N_j\}$ Steady state probability to have s products in the buffer B_j
ψ Production line throughput

In mathematical terms, the buffer allocation problem can be stated as a non-linear programming problem. Gershwin and Schor [4] presented two versions of this problem.

Primal Problem: the objective of the primal formulation is to minimize the total buffer space required for the line to meet or exceed a given average production rate. This formulation is appropriate if either storage space or the buffering mechanism is expensive, if work-in-process inventory is inexpensive, and if an average production rate is mandated.

$$
\begin{cases}
Min & \sum_{j=1}^{K-1} N_j \\
Constraints: \\
& \psi \geq \psi_{required} \\
& \sum_{j=1}^{K-1} N_j \leq N_{available} \\
& N_j \in \mathbb{N}, \forall j = 1...K - 1
\end{cases}
\tag{1}
$$

Dual Problem: the objective of the dual formulation is to maximize the production rate achievable with a given total buffer space. This is appropriate in cases where the total storage space is limited, where the number of buffer locations is fixed.

$$
\begin{cases}
Max & \psi \\
Constraints: \\
& \sum_{j=1}^{K-1} N_j \leq N_{available} \\
& N_j \in \mathbb{N}, \forall j = 1...K - 1
\end{cases}
\tag{2}
$$

Seong et al. [16] introduced the **generalized version** of the buffer allocation problem. In this case, the objective is to maximize the difference between the profit obtained from the throughput and the holding cost incurred by work-in-process inventory.

$$\begin{cases} Max & f_1 \times \psi - f_2 \times \sum_{j=1}^{K-1} \overline{N_j} \\ \\ \text{Constraints:} \\ & \sum_{j=1}^{K-1} N_j \leq N_{available} \\ & N_j \in \mathbb{N}, \forall j = 1...K-1 \\ \overline{N_j} & \text{average inventory level in buffer } B_j \\ f_1 & \text{profit per unit} \\ f_2 & \text{holding cost per unit} \end{cases} \quad (3)$$

The mathematical programming approach, proposed in this paper, can be used to solve the buffer allocation problem in the different cases cited above.

This approach is inspired by the analytical method proposed by Ouazene et al. [8] to evaluate the throughput of a serial production line while the capacities of the different buffers are assumed to be known. This method is based on the analysis of the different buffers states using dedicated birth-death Markov processes. The idea of this paper is to transform this analytical method into an optimization method to deal with the buffer allocation problem where the capacities of the different buffers are variables to be determined.

3   Analytical Method Description

The main idea of the proposed approach is to replace each machine by an equivalent one that has only up and down states. The blockage and starvation are integrated in the up state of the machine. This formulation is a generalization of two-machine-one-buffer production line proposed by Ouazene et al. [9].

Based on the analysis of the buffers steady states using birth-death Markov processes, the probabilities of starvation and blockage of each buffer are determined. Then, these probabilities are used to model and analyze the system behavior in its steady state.

3.1   Two-Machine-One-Buffer Model

Before detailing the general formulation, we introduce the simple system which consists of two machines separated by one buffer. This model is used as a building block to construct the general model. To analyze the steady states of the buffer, we consider a birth-death Markov process with $(N+1)$ states $\{0, 1...N\}$ such as N is the capacity of the intermediate buffer and ω_1 and ω_2 are respectively the birth and death transition rates.

The differential equations for the probability that the system is in state j at time t are:

$$\begin{cases} \frac{\partial P_0(t)}{\partial t} = -\omega_1 \times P_0(t) + \omega_2 \times P_1(t) \\ \\ \frac{\partial P_j(t)}{\partial t} = \omega_1 \times P_{j-1}(t) - (\omega_1 + \omega_2) \times P_j(t) + \omega_2 \times P_{j+1}(t) \\ \\ \frac{\partial P_N(t)}{\partial t} = \omega_1 \times P_{N-1}(t) + \omega_2 \times P_N(t) \end{cases} \quad (4)$$

In the steady state of the system, all the differential terms are equal to zero (see Equation 5).

$$\begin{cases} 0 = -\omega_1 \times P_0 + \omega_2 \times P_1 \\ 0 = \omega_1 \times P_{j-1} - (\omega_1 + \omega_2) \times P_j + \omega_2 \times P_{j+1} \\ 0 = \omega_1 \times P_{N-1} + \omega_2 \times P_N \end{cases} \tag{5}$$

So, by simplifying the system above and considering the normalization equation: $\sum_{j=0}^{N} P_j = 1$ we obtain the different steady state probabilities. We are especially interested in the starvation and blockage probabilities respectively represented by empty and full buffer states given by the following equations:

$$P_0 = \begin{cases} \frac{1-\alpha}{1-\alpha^{N+1}} & \text{if } \alpha \neq 1 \\ \frac{1}{N+1} & \text{if } \alpha = 1 \end{cases} \tag{6}$$

$$P_N = \begin{cases} \frac{\alpha^N \times (1-\alpha)}{1-\alpha^{N+1}} & \text{if } \alpha \neq 1 \\ \frac{1}{N+1} & \text{if } \alpha = 1 \end{cases} \tag{7}$$

Based on these two probabilities, the effective production rate of each work station is defined as function of machine processing rate, machine and the buffer availabilities (8).

$$\rho_i = \omega_i \times \frac{\mu_i \times \xi_i}{\mu_i + \xi_i \times \lambda_i} \tag{8}$$

Such as: $\xi_1 = 1 - P_N$ and $\xi_2 = 1 - P_0$.
The system throughput ψ is defined as the bottleneck between the two effective production rates ρ_1 and ρ_2.

$$\psi = \begin{cases} \omega_1 \times \min\{\frac{\mu_1 \times (1-\alpha^N)}{\mu_1 \times (1-\alpha^{N+1}) + \lambda_1 \times (1-\alpha^N)}; \\ \frac{\mu_2 \times (1-\alpha^N)}{\mu_2 \times (1-\alpha^{N+1}) + \lambda_2 \times \alpha \times (1-\alpha^N)}\} \\ \text{if } \alpha \neq 1 \\ \\ \omega \times \min\{\frac{N \times \mu_1}{N \times \lambda_1 + (N+1) \times \mu_1}, \frac{N \times \mu_2}{N \times \lambda_2 + (N+1) \times \mu_2}\} \\ \text{if } \alpha = 1 \end{cases} \tag{9}$$

More implicitly:

$$\psi = \begin{cases} \omega_1 \times \min\{\frac{1}{\frac{1-\alpha^{N+1}}{1-\alpha^N} + \frac{\lambda_1}{\mu_1}}; \frac{1}{\frac{1-\alpha^{N+1}}{1-\alpha^N} + \alpha \times \frac{\lambda_2}{\mu_2}}\} \\ \text{if } \alpha \neq 1 \\ \\ \omega \times \min\{\frac{1}{\frac{N}{N+1} + \frac{\lambda_1}{\mu_1}}; \frac{1}{\frac{N}{N+1} + \frac{\lambda_2}{\mu_2}}\} \\ \text{if } \alpha = 1 \end{cases} \tag{10}$$

3.2   General Model with K Machines and $(K-1)$ Buffers

The two-machine-one-buffer presented above is used as a building block to analysis larger production lines. Therefore, the states of each intermediate buffer B_j are analyzed using a dedicated birth-death Markov process. These Markov processes differ in terms of number of states because the buffers are not identical. They also differ in terms of birth and death transition rates because each buffer B_j is differently influenced by the machines and the other buffers (see Figure 2).

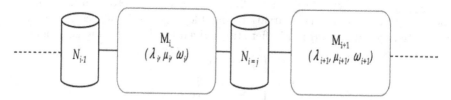

Fig. 2. Sub-system of the original production line

In the simple case of two machines and one buffer, each available machine processes ω_i products per time unit. For this reason, ω_1 and ω_2 are respectively the birth and death transition rates. But in the general case, the machines M_i and M_{i+1} related to the buffer $B_{j=i}$ are subject to starvation and blockage. So their effective processing rates are affected by the availabilities of the buffers B_{i-1} and B_{i+1}. The upstream machine M_i can process products if the buffer B_{i-1} is not empty and the downstream machine M_{i+1} can process products when the buffer B_{i+1} is not full.

The first buffer and the last one should be considered as particular cases because the first machine cannot be starved and the last machine cannot be blocked. The birth-death Markov process, related to the buffer B_i, is represented in Figure 3.

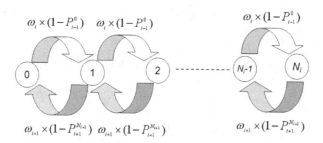

Fig. 3.   Birth-death Markov process related to the buffer B_i

The different states of the $(K-1)$ buffers are modeled by $(K-1)$ related but different birth-death Markov processes. Each stochastic process is defined by its processing rates ratio α_j. The different ratios α_j are defined by the following equations:

$$\alpha_1 = \frac{\omega_1 \times (1 - P_{N_2}^2)}{\omega_2}. \tag{11}$$

$$\alpha_{K-1} = \frac{\omega_{K-1} \times (1 - P_0^{K-1})}{\omega_K}. \tag{12}$$

$$\forall j = 2...K - 2, \alpha_j = \frac{\omega_i \times (1 - P_0^{j-1})}{\omega_{i+1} \times (1 - P_{N_{j+1}}^{j+1})}. \tag{13}$$

Equations (11) and (12) consider respectively the particular case of the first and the last buffer because the first machine cannot be starved and the last one cannot be blocked.

Based on the analysis presented in Section 2, the probabilities of empty and full states of each buffer are calculated using Equations (10) and (11).

$$P_0^j = \begin{cases} \frac{1-\alpha_j}{1-\alpha_j^{N_{j+1}}} & \text{if } \alpha_j \neq 1 \\ \\ \frac{1}{N_j+1} & \text{if } \alpha_j = 1 \end{cases} \tag{14}$$

$$P_{N_i}^j = \begin{cases} \frac{\alpha_j^{N_j} \times (1-\alpha_j)}{1-\alpha_j^{N_{j+1}}} & \text{if } \alpha_j \neq 1 \\ \\ \frac{1}{N_j+1} & \text{if } \alpha_j = 1 \end{cases} \tag{15}$$

The resolution of Equations (11) to (15) allows the determination of the processing rates ratio α_j and empty and full states probabilities $(P_0^j, P_{N_i}^j)$ of each buffer. So, based on these information and the two-machine-one-buffer model presented above, we can calculate the effective production rate of each machine considering the influence of the buffers availabilities using Equation (16).

$$\forall i = 1...K, \rho_i = \omega_i \times \frac{\mu_i \times \xi_j}{\mu_i + \xi_j \times \lambda_i}. \tag{16}$$

Such as:

$$\begin{cases} \xi_1 = 1 - P_{N_1}^1 \\ \\ \xi_K = 1 - P_0^{K-1} \\ \\ \forall j = 2...K - 1, \xi_j = (1 - P_0^{j-1}) \times (1 - P_{N_{j+1}}^{j+1}) \end{cases} \tag{17}$$

Similarly to the two-machine-one-buffer model, the throughput of the production line ψ is defined as the bottleneck between the effective production rates of all machines:

$$\psi = \min\{\omega_i \times \frac{\mu_i \times \xi_i}{\mu_i + \xi_i \times \lambda_i}\}, i = 1...K .$$ (18)

4   Non-linear Programming Model

The transformation of the analytical approach for performance evaluation proposed above into an optimization approach is obtained by considering that the capacities of the different buffers $(N_1, N_2,..., N_{K-1})$ are finite but variables. So the different characteristics (probabilities of the blockage, starvation, birth and death ratios and equivalent machines ratios) are expressed as function dependent on the buffer capacities.

The obtained non-linear programming model is solved using LINGO software as illustrated by algorithm 1. Without loss of generality, algorithm 1 presents the dual version of the buffer allocation problem. It can be easily adapted to the generalized case by adding the formula for calculating the average storage level of the production line.

The average inventory level of the system can be defined as the sum of the average inventory levels of the different buffers.

$$\overline{N} = \sum_{j=1}^{K-1} \overline{N_j}$$ (19)

Or,

$$\overline{N_j} = E[P_{B_j}]$$

$$= \sum_{s=1}^{N_j} s \times P_j^s$$

$$= \begin{cases} \sum_{s=1}^{N_j} s \times \frac{\alpha_j^s \times (1-\alpha_j)}{1-\alpha_j^{N_j+1}} & \text{si } \alpha_j \neq 1 \\ \\ \sum_{s=1}^{N_j} s \times \frac{1}{1+N_j} & \text{si } \alpha_j = 1 \end{cases}$$ (20)

$$\overline{N_j} = \begin{cases} \frac{N_j \times \alpha_j^{N_j+2} - (N_j+1) \times \alpha_j^{N_j+1} + \alpha_j}{\alpha_j^{N_j+2} - \alpha_j^{N_j+1} - \alpha_j + 1} & \text{si } \alpha_j \neq 1 \\ \\ \frac{N_j}{2} & \text{si } \alpha_j = 1 \end{cases}$$ (21)

Remark:

The formulation above is based on the following finite series sum:

$$\sum_{s=1}^{N_j} s \times \frac{\alpha_j^s(1-\alpha_j)}{1-\alpha_j^{N_j+1}} = \frac{1-\alpha_j}{1-\alpha_j^{N_j+1}} \times \sum_{s=1}^{N_j} s \times \alpha_j^s \tag{22}$$

Such as:

$$\sum_{s=1}^{N_j} s \times \alpha_j^s = \alpha_j \times \frac{\partial \sum_{s=1}^{N_j} \alpha_j^s}{\partial \alpha_j}$$

$$= \alpha_j \times \frac{\partial \frac{\alpha_j \times (1-\alpha_j^{N_j})}{1-\alpha_j}}{\partial \alpha_j} \tag{23}$$

$$= \frac{\alpha_j - (N_j+1) \times \alpha_j^{N_j+1} + N_j \times \alpha_j^{N_j+2}}{(1-\alpha_j)^2}$$

So,

$$\overline{N_j} = \frac{1-\alpha_j}{1-\alpha_j^{N_j+1}} \times \frac{\alpha_j - (N_j+1) \times \alpha_j^{N_j+1} + N_j \times \alpha_j^{N_j+2}}{(1-\alpha_j)^2}$$

$$= \frac{N_j \times \alpha_j^{N_j+2} - (N_j+1) \times \alpha_j^{N_j+1} + \alpha_j}{\alpha_j^{N_j+2} - \alpha_j^{N_j+1} - \alpha_j + 1} \tag{24}$$

5   Numerical Experiments

In this section, the proposed non-linear programming model is solved using LINGO software (version 12.0). The optimization algorithm was implemented on a laptop with a Core Duo processor running at 2.00 GHz. Some numerical results are given concerned with the buffer allocation problem in unbalanced production lines based on different instances proposed in the literature.

This benchmark has been widely reported by various authors, such as Seong et al. [15], Gershwin and Schor [4], Papadopoulos and Vidalis [12] and Sabuncuoglu et al. [13]. It is consists of ten different instances representing production lines with 4, 5, 6, 8, 9 and 10 machines and total storage capacity varying from 10 to 315 units. Each of the configurations considers the general case of asynchronous non-homogeneous line with machines having different reliability parameters and different processing times. Table 2 summarizes the design of these instances.

The proposed non-linear programming method is compared with six methods issued from the literature.

SEVA, Non-SEVA1 and **Non-SEVA2** proposed by Seong et al. [15]. The authors formulated the buffer allocation as a nonlinear multidimensional search problem. They developed two algorithms called SEVA (standard exchange vector algorithm) and Non-SEVA (non-standard exchange vector algorithm).

Algorithm 1. Non-linear programming model coded using LINGO Software

---

	K	Number of machines
	$K-1$	Number of buffers
	λ_i	Failure rate of machine M_i
Require:	μ_i	Repair rate of machine M_i
	ω_i	Processing rate of machine M_i
	$P_0^0 = 0$	first machine is never starved
	$P_K^{N_K} = 0$	Last machine never blocked

for all each buffer B_j do
  for all each machine M_i do
    Max $\psi = \min_{i=1\ldots K}\{\rho_i\}$
    $N_j \in \mathbb{N}$ and $N_j \neq 0$
    $\alpha_j = \dfrac{\min_{i=1\ldots j} \rho_i}{\min_{i=j+1\ldots K} \rho_i}$
    $\rho_i = \omega_i \times \dfrac{\mu_i \times (1-P_{j-1}^0) \times (1-P_j^{N_j})}{\mu_i + (1-P_{j-1}^0) \times (1-P_j^{N_j}) \times \lambda_i}$
    if $\alpha_j = 1$ then
      $P_j^{N_j} = P_j^0 = \dfrac{1}{N_j+1}$
    else
      $P_j^{N_j} = \dfrac{\alpha_i^{N_j} \times (1-\alpha_j)}{1-\alpha_i^{N_j+1}}$
      $P_j^0 = \dfrac{1-\alpha_j}{1-\alpha_j^{N_j+1}}$
    end if
  end for
end for
return ψ and $(N_1, N_2, ..., N_{K-1})$

Both methods are based on the principle of local search by exploring a specific neighborhood. In the first algorithm, the neighborhood search is identified according to the concept of standard exchange vector whereas the in the second algorithm, the authors introduced the concepts of pseudo-gradient and gradient projection to accelerate exploration the search space. The authors considered two different versions of this algorithm: Non- SEVA1 and Non- SEVA2 which differ in terms of heuristic used for the approximation of the gradient of the objective function.

LIBA is a simulation-based heuristic procedure proposed by Selvi [14] and reproduced by Sabuncuoglu et al. [13]. The main feature of the procedure is to minimize the difference between the throughput values of two sub-lines created by dividing the line around each buffer and to transfer the buffers from the faster sub-line to the slower one.

H-1 and **H-2** are two heuristic approaches proposed by Sabuncuoglu et al. [13]. The first heuristic starts with a uniform allocation as initial solution, whereas the second heuristic uses the solution obtained by the method **LIBA**.

Table 2. Parameters of production lines reported in Seong et al. [15] and Sabuncuoglu et al [13]

Instance	K	$\sum_{j=1}^{K-1} N_j$	$(\lambda_i, \mu_i, \omega_i)$
1	4	10	(0.07,0.17,3.7); (0.11,0.37,1.5); (0.49,0.78,1.1); (0.19,0.50,3.0)
2	4	30	(0.38,0.45,3.0); (0.30,0.55,1.0); (0.35,0.50,2.0); (0.45,0.40,3.6)
3	5	10	(0.10,0.30,1.2); (0.30,0.50,1.0); (0.50,0.20,3.0); (0.40,0.30,2.0); (0.20,0.10,1.8)
4	5	15	(0.30,0.64,2.8); (0.40,0.83,1.7); (0.45,0.75,2.5); (0.35,0.85,3.4); (0.10,0.74,1.9)
5	6	115	(0.08,0.40,2.6); (0.24,0.40,3.0); (0.20,0.60,3.4); (0.17,0.50,4.7); (0.10,0.30,1.5)
6	6	130	(0.30,0.20,3.0); (0.50,0.50,1.0); (0.10,0.30,1.2); (0.20,0.10,1.8); (0.30,0.20,1.5); (0.40,0.30,2.0)
7	8	125	(0.25,0.52,1.0); (0.18,0.48,3.6); (0.23,0.58,1.7); (0.32,0.50,1.4); (0.19,0.47,2.8); (0.35,0.46,2.7); (0.26,0.66,1.6); (0.20,0.41,1.2)
8	9	200	(0.20,0.70,2.5); (0.10,0.60,1.5); (0.30,0.80,2.8); (0.20,0.80,3.6); (0.10,0.70,2.1); (0.10,0.60,1.9); (0.30,0.80,2.7); (0.20,0.50,3.0); (0.30,0.60,2.0)
9	10	310	(0.20,0.70,2.5); (0.10,0.60,1.5); (0.30,0.80,2.8); (0.20,0.80,3.6); (0.10,0.70,2.1); (0.10,0.60,1.9); (0.30,0.80,2.7); (0.20,0.50,3.0); (0.30,0.60,2.0);(0.10,0.70,2.1)
10	10	315	(0.365,0.465,2.4); (0.215,0.565,1.7); (0.305,0.485,2.8); (0.375,0.455,2.2);(0.340,0.455,2.1); (0.390,0.390,2.5); (0.265,0.500,1.1); (0.285,0.490,1.3);(0.255,0.495,1.6); (0.240,0.505,0.8)

Tables 3 and 4 summarize the performance of the different approaches investigated in this comparative study. For each instance, the buffer allocation solution, the corespondent throughput and the computational times are given.

The results of the comparative study show the efficiency of the proposed method to solve the buffer allocation problem. In fact, we note that for the small instances (1 to 4), the proposed approach and the method **SEVA** obtain the optimal solutions except for the first instance. This instance is particular case because the optimal solution contain a zero-buffer solution (the first buffer does not exist). The proposed method does not consider the case of zero-buffer capacity since it is based on the analysis of the different buffers states.

For some large instances, the results of the method **SEVA** are not reported because it failed to obtain solutions because of the computational complexity.

We note also that for the rest of the instance the different methods obtain different solutions (buffer allocations) but the values of the objective function are very close. In fact, in the buffer allocation problem the optimal solution is not unique especially when the total capacity to be allowed is hight.

As conclusion, the proposed non-linear programming approach is computationally faster than all other methods and obtain high-quality solutions.

Table 3. Results of the comparative study (1)

Instance	Method	buffers allocation	ψ	min:sec
1	**SEVA** (Seong et al. [15])	(0,7,3)	0.6363	00:06
	Non-SEVA1 (Seong et al. [15])	(0,6,4)	0.6363	00:03
	Non-SEVA2 (Seong et al. [15])	(0,7,3)	0.6363	00:04
	LIBA (Sabuncuoglu et al. [13])	(0,7,3)	0.6258	-
	H-1 (Sabuncuoglu et al. [13])	(0,7,3)	0.6258	-
	H-2 (Sabuncuoglu et al. [13])	(0,7,3)	0.6258	-
	Proposed method	(2,5,3)	0.6327	00:00
2	**SEVA** (Seong et al. [15])	(11,16,3)	0.6394	02:53
	Non-SEVA1 (Seong et al. [15])	(11,16,3)	0.6394	01:06
	Non-SEVA2 (Seong et al. [15])	(11,16,3)	0.6394	00:38
	LIBA (Sabuncuoglu et al. [13])	(11,16,3)	0.6384	-
	H-1 (Sabuncuoglu et al. [13])	(11,15,4)	0.6384	-
	H-2 (Sabuncuoglu et al. [13])	(11,15,4)	0.6384	-
	Proposed method	(11,16,3)	0.6460	00:00
3	**SEVA** (Seong et al. [15])	(1,3,4,2)	0.3484	00:13
	Non-SEVA1 (Seong et al. [15])	(1,3,4,2)	0.3484	00:13
	Non-SEVA2 (Seong et al. [15])	(1,3,4,2)	0.3484	00:13
	LIBA (Sabuncuoglu et al. [13])	(1,3,4,2)	0.3468	-
	H-1 (Sabuncuoglu et al. [13])	(1,3,3,3)	0.3475	-
	H-2 (Sabuncuoglu et al. [13])	(1,3,3,3)	0.3475	-
	Proposed method	(1,3,4,2)	0.3662	00:01
4	**SEVA** (Seong et al. [15])	(4,6,3,2)	0.9460	00:14
	Non-SEVA1 (Seong et al. [15])	(4,6,4,1)	0.9439	00:15
	Non-SEVA2 (Seong et al. [15])	(4,6,4,1)	0.9439	00:15
	LIBA (Sabuncuoglu et al. [13])	(4,7,3,1)	0.9714	-
	H-1 (Sabuncuoglu et al. [13])	(4,7,3,1)	0.9714	-
	H-2 (Sabuncuoglu et al. [13])	(4,7,3,1)	0.9714	-
	Proposed method	(4,6,3,2)	0.9940	00:04
5	**Non-SEVA1** (Seong et al. [15])	(29,15,29,42)	1.1250	00:43
	Non-SEVA2 (Seong et al. [15])	(22,11,22,60)	1.1250	00:35
	LIBA (Sabuncuoglu et al. [13])	(17,17,22,59)	1.1248	-
	H-1 (Sabuncuoglu et al. [13])	(21,27,16,51)	1.1248	-
	H-2 (Sabuncuoglu et al. [13])	(20,31,11,53)	1.1248	-
	Proposed method	(17,22,4,56)	1.1250	00:01
6	**Non-SEVA1** (Seong et al. [15])	(26,45,26,16,17)	0.4995	07:37
	Non-SEVA2 (Seong et al. [15])	(23,38,47,14,8)	0.4989	09:19
	LIBA (Sabuncuoglu et al. [13])	(23,38,47,14,8)	0.4797	-
	H-1 (Sabuncuoglu et al. [13])	(23,38,47,14,8)	0.4799	-
	H-2 (Sabuncuoglu et al. [13])	(23,38,47,14,8)	0.4801	-
	Proposed method	(21,78,9,15,7)	0.4999	0:01
7	**Non-SEVA1** (Seong et al. [15])	(33,2,18,0,21,19,32)	0.6756	83:49
	Non-SEVA2 (Seong et al. [15])	(33,2,18,0,21,19,32)	0.6756	92:22
	LIBA (Sabuncuoglu et al. [13])	(57,8,17,17,5,8,13)	0.6755	-
	H-1 (Sabuncuoglu et al. [13])	(48,2,18,0,21,19,32)	0.6755	-
	H-2 (Sabuncuoglu et al. [13])	(54,13,18,15,4,10,17)	0.6755	-
	Proposed method	(87,6,18,17,5,8,17)	0.6753	00:05

Table 3. *(Continued)*

	Non-SEVA1 (Seong et al. [15])	(42,42,8,25,25,25,8,25)	1.2857	61:26
8	Non-SEVA2 (Seong et al. [15])	(70,44,15,17,16,16,6,16)	1.2857	01:36
	LIBA (Sabuncuoglu et al. [13])	(28,47,20,15,21,20,22,27)	1.2780	-
	H-1 (Sabuncuoglu et al. [13])	(25,44,4,18,18,24,18,49)	1.2783	-
	H-2 (Sabuncuoglu et al. [13])	(28,41,12,10,20,25,18,46)	1.2784	-
	Proposed method	(29,108,2,4,7,8,2,40)	1.2818	00:05

Table 4. Results of the comparative study (2)

Instance	Method	buffers allocation	ψ	min:sec
9	Non-SEVA1 (Seong et al. [15])	(34,34,34,2,34,34,34,34,70)	0.9864	24:12
	Non-SEVA2 (Seong et al. [15])	(20,15,16,18,25,15,20,112,69)	0.9863	89:20
	LIBA (Sabuncuoglu et al. [13])	(6,22,15,15,21,20,41,112,58)	0.9860	-
	H-1 (Sabuncuoglu et al. [13])	(6,25,14,22,22,20,39,110,52)	0.9860	-
	H-2 (Sabuncuoglu et al. [13])	(6,25,14,20,25,23,39,106,52)	0.9860	-
	Proposed method	(5,10,4,23,8,6,25,174,55)	0.9863	00:07
10	Non-SEVA1 (Seong et al. [15])	(35,35,35,35,18,52,118,35,52)	0.5423	05:28
	Non-SEVA2 (Seong et al. [15])	(65,20,5,5,5,65,5,5,140)	0.5423	03:18
	LIBA (Sabuncuoglu et al. [13])	(29,18,26,31,30,38,49,40,54)	0.5417	-
	H-1 (Sabuncuoglu et al. [13])	(35,22,35,35,35,35,35,35,48)	0.5417	-
	H-2 (Sabuncuoglu et al. [13])	(29,24,26,31,30,38,49,40,48)	0.5417	-
	Proposed method	(4,11,7,13,21,42,110,23,94)	0.5423	00:08

6 Conclusion

This paper proposes an efficient non-linear programming method for the buffer allocation problem in serial unreliable production lines. This approach can be used to deal with the three versions of this problem: primal, dual and generalized. Numerical analysis shows that the computational time required to solve the buffer allocation problem can be significantly reduced by using the proposed formulation. It shows also the efficiency of the proposed approach on solving both small-size and big-size problems.

The originality of this method is its reduced number of variables used to model the behavior of the system because it considers only the empty and full states of each buffer. Therefore, to solve the buffer allocation problem in the case K-machine $(K-1)$-buffer production line, we have to solve a non-linear programming model with $(5 \times K - 3)$ variables.

Future extension of this work may be the adaptation of this approach to deal with other production systems types such as series-parallel structures or assembly/disassembly systems. The main challenges in such extensions are more in the modeling and the analysis than the optimization algorithm.

Acknowledgments. The authors would like to thank the anonymous reviewers for their valuable remarks, comments and suggestions that helped to improve this paper. They would like to thank Andras Horvath for his assistance on the revision of this paper.

References

1. Conway, R., Maxwell, W., McClain, J.O., Thomas, L.J.: The role of work-in-process inventory in serial production lines. Operations Research 36(2), 229–241 (1988)
2. Dallery, Y., Gershwin, S.B.: Manufacturing flow line systems: A review of models and analytical results. Queueing Systems 12(1-2), 3–94 (1992)
3. Demir, L., Tunali, S., Eliiyi, D.T.: The state of the art on buffer allocation problem: A comprehensive survey. Journal of Intelligent Manufacturing, 1–22 (2012)
4. Gershwin, S.B., Schor, J.E.: Efficient algorithms for buffer space allocation. Annals of Operations Research 93, 117–144 (2000)
5. Koenigsberg, E.: Production lines and internal storage-a review. Management Science 5(4), 410–433 (1959)
6. Matta, A.: Simulation optimization with mathematical programming representation of discrete event systems. In: Simulation Conference, WSC 2008. Winter, pp. 1393–1400 (December 008)
7. Nahas, N., Ait-Kadi, D., Nourelfath, M.: A new approach for buffer allocation in unreliable production lines. International Journal of Production Economics 103(2), 873–881 (2006)
8. Ouazene, Y., Chehade, H., Yalaoui, A., Yalaoui, F.: Equivalent machine method for approximate evaluation of buffered unreliable production lines. In: 2013 IEEE Workshop on Computational Intelligence in Production and Logistics Systems (CIPLS), 2013, pp. 33–39 (April 2013)
9. Ouazene, Y., Yalaoui, A., Chehade, H.: Analysis of a buffered two-machine production line with unreliable machines. Journal of Multiple-Valued Logic and Soft Computing, (accepted: July 25, 2013)
10. Papadopoulos, C.T., Vidalis, M.J., OKelly, M.E.J., Spinellis, D.: The buffer allocation problem. In: Analysis and Design of Discrete Part Production Lines. Springer Optimization and Its Applications, vol. 31, pp. 131–159. Springer, New York (2009)
11. Papadopoulos, H.T., Heavey, C.: Queueing theory in manufacturing systems analysis and design: A classification of models for production and transfer lines. European Journal of Operational Research 92(1), 1–27 (1996)
12. Papadopoulos, H.T., Vidalis, M.I.: A heuristic algorithm for the buffer allocation in unreliable unbalanced production lines. Computers & Industrial Engineering 41(3), 261–277 (2001)
13. Sabuncuoglu, I., Erel, E., Gocgun, Y.: Analysis of serial production lines: characterisation study and a new heuristic procedure for optimal buffer allocation. International Journal of Production Research 44(13), 2499–2523 (2006)
14. Selvi, O.: The line balancing algorithm for optimal buffer allocation in production lines. MS thesis. Bilkent University, Turkey (2002)
15. Seong, D., Chang, S.Y., Hong, Y.: Heuristic algorithms for buffer allocation in a production line with unreliable machines. International Journal of Production Research 33(7), 1989–2005 (1995)
16. Seong, D., Chang, S.Y., Hong, Y.: An algorithm for buffer allocation with linear resource constraints in a continuous-flow unreliable production line. Asia-Pacific Journal of Operational Research 17, 169–180 (2000)
17. Vergaraand, H.A., Kim, D.S.: A new method for the placement of buffers in serial production lines. International Journal of Production Research 47(16), 4437–4456 (2009)
18. Vouros, G.A., Papadopoulos, H.T.: Buffer allocation in unreliable production lines using a knowledge based system. Computers & Operations Research 25(12), 1055–1067 (1998)

The $PH/PH/1$ Multi-threshold Queue

Niek Baer, Richard J. Boucherie, and Jan-Kees van Ommeren

Stochastic Operations Research, Department of Applied Mathematics, University of
Twente, Drienerlolaan 5, 7500 AE Enschede, The Netherlands
{n.baer,r.j.boucherie,j.c.w.vanommeren}@utwente.nl

Abstract. We consider a $PH/PH/1$ queue in which a threshold policy
determines the stage of the system. The arrival and service processes
follow a Phase-Type (PH) distribution depending on the stage of the
system. Each stage has both a lower and an upper threshold at which
the stage of the system changes, and a new stage is chosen according to a
prescribed distribution. The $PH/PH/1$ multi-threshold queue is a Quasi-
Birth-and-Death process with a tri-diagonal block structured boundary
state which we model as a Level Dependent Quasi-Birth-and-Death process. An efficient algorithm is presented to obtain the stationary queue
length vectors using Matrix Analytic methods.

Keywords: $PH/PH/1$ queue, multiple thresholds, Matrix Analytic
methods, Quasi-Birth-and-Death process, tri-diagonal block structured
boundary state.

1 Introduction

We consider a $PH/PH/1$ queue in which a threshold policy determines the
stage of the system. The arrival and service processes follow a Phase-Type (PH)
distribution depending on the stage of the system. Each stage has both a lower
and an upper threshold at which the stage of the system changes. At these
thresholds a new stage is chosen according to a prescribed distribution.

In literature, threshold policies are often used to activate or deactivate servers
when the queue length reaches certain thresholds. The $M/M/2$ queue in which
the second server is activated when the queue length reaches an upper threshold and deactivated when it reaches a lower threshold is studied in [11], where
a closed form expression is obtained for the steady-state probabilities. In [13],
see also Section 4.2, closed form expressions are obtained for the steady-state
distributions for the $M/M/c$ with c heterogeneous servers. Using Green's function, Ibe and Keilson [9] studied the $M/M/c$ queue with homogeneous servers
and the $M/M/2$ queue with heterogeneous servers. The $M/M/c$ with heterogeneous servers is also studied in [14] where the steady-state probabilities are
obtained using a stochastic complement analysis for uncoupling Markov Chains.
A $MAP/M/c$ with homogeneous servers is analysed in [4] and the $PH/M/2$
queue with heterogeneous servers is studied by Neuts [16]. In [5], see also Section 4.3, a very general setting is studied in which the generator of the queueing

B. Sericola, M. Telek, and G. Horváth (Eds.): ASMTA 2014, LNCS 8499, pp. 95–109, 2014.
© Springer International Publishing Switzerland 2014

system forms a nested Quasi-Birth-and-Death process. In this model a threshold policy controls the stage of the system which, in turn, determines the arrival process and the service process. An upper threshold increases the stage by one whereas the the lower threshold decreases the stage by one, creating a *staircase* threshold policy. In [12] an $M/M/2$ queue is studied with two heterogeneous servers in which the second server is exponentially delayed before activation.

Threshold policies are also used to send servers to a certain queue, as is shown in [7]. In this paper, a system is studied containing two queues and two servers where both interarrival times and service times are exponentially distributed. After each service completion, the server chooses a queue to serve according to a threshold policy. A generalisation of this model is analysed in [6] where customers from multiple classes arrive according to a Poisson process and require an exponential amount of service. The queueing system contains a fixed number of servers which are allocated to a customer class according to a threshold policy. Each server experiences an exponential delay once it is assigned to a different customer class. In [17], the joint queue length distribution is obtained for an $M/G/1$ queue with multiple customer classes in which customers from higher class are blocked when thresholds are reached.

Motivating Example. The queueing system in this paper is motivated by the hysteretic relation between density and speed of traffic flows observed on a highway, see Helbing [8]. In [8] it is stated that this hysteretic behaviour is controlled by two critical densities, denoted by ρ_1 and ρ_2. When the density of cars on the highway increases vehicles are more and more affected by each other and the driving speeds decrease. Once the density reaches ρ_2 the highway becomes congested and driving speeds decrease drastically. The density must reduce to ρ_1 for the highway to become non-congested. In Baer, Boucherie and van Ommeren [2], an $M/M/1$ threshold queue was used to model a particular highway section. In [2], the arrival rates were kept constant, whereas the service rates where altered according to a 2-stage threshold policy. When the queue length surpasses an upper threshold the service rates decreased. The service rates were increased again when the queue length dropped below a lower threshold. In [2], the mean sojourn time is determined. Since a single queue represents a highway section, this directly gives the average time to cross the highway section and the mean speed of a vehicle. The motivating example in Figure 1 is an extension to the model in [2], where not only the service rates are controlled by a threshold policy, but also the arrival rates. This models the hysteretic relation within a highway section, but also between two consecutive highway sections. We will, get back to this example in Section 4.1.

Contribution. This paper generalises the model of [5] to an arbitrary threshold policy and introduces a novel dedicated solution method based on the Level Dependent Quasi-Birth-and-Death process of [3]. In particular, a class of $PH/PH/1$ multi-threshold queueing systems is described for which the solution method in [3] can be decomposed to find the stationary queue length vector for each stage separately. The stationary distribution of the $PH/PH/1$ multi-threshold queue

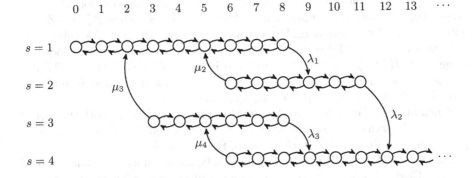

Fig. 1. State Diagram

can be obtained using the results in [3] but for a large number of stages, this may result in computational demanding calculations. In this paper we use the structure of the $PH/PH/1$ multi-threshold queue to form, based on the results in [3], smaller and easier equations to obtain the stationary distribution.

Overview. Section 2 introduces the $PH/PH/1$ multi-threshold queue and presents the queueing system as a Level Dependent Quasi-Birth-and-Death process. In Section 3 we analyse the multi-threshold queue using Matrix Analytic methods and obtain the stationary queue length probabilities. Furthermore, we present a decomposition theorem for a class of multi-threshold queues providing an explicit description of the stationary queue length probability vectors. In Section 4 we illustrate our results via three multi-threshold queues obtained from literature. Section 5 gives concluding remarks.

2   Model Description

Consider a $PH/PH/1$ queue, controlled by a threshold policy. The system can be in different stages $s = 1 \ldots, S$, where every stage s is associated with a set of feasible queue lengths $\{L_s, \ldots, U_s\}$. The quantities L_s and U_s are the lower, respectively upper thresholds for stage s. In case $U_s = \infty$, we say that stage s has no upper threshold. For each queue length $n = 0, 1, \ldots$, a stage s is a *potential* stage when $L_s \le i \le U_s$. If the system is in stage s and a departure or arrival causes the queue length to drop below L_s or to exceed U_s, the stage of the system changes (the threshold policy). If the queue length increases to $U_s + 1$ the stage changes from s to t with probability $p_{s,t}$. Note that $p_{s,t} > 0$ implies that t is a potential stage for queue length $U_s + 1$. If the queue length decreases to $L_s - 1$ the stage changes from s to t with probability $q_{s,t}$. See Figure 1 for an illustration with exponential service times and Poisson arrivals.

The arrival process in stage s follows a $PH(\Lambda_s, \lambda_s)$ distribution of $v_s + 1$ phases (v_s transient phases and 1 absorbing phase). We define $\Lambda_s^0 = -\Lambda_s e_{v_s}$, with e_{v_s} a $v_s \times 1$ vector of ones. Furthermore we assume that the absorbing state is never chosen as initial state, i.e. $\lambda_s e_{v_s} = 1$. Similarly, the service process in stage s is $PH(M_s, \mu_s)$ distributed with $w_s + 1$ phases. We define $M_s^0 = -M_s e_{w_s}$ and assume $\mu_s e_{w_s} = 1$. The mean interarrival times and mean service time if given by $-\lambda_j \Lambda_j^{-1} e_{v_j}$ and $-\mu_j M_j^{-1} e_{w_j}$, see Neuts [15].

When an arrival or departure changes the stage of the system both the arrival process and service process are reset by choosing a new initial phase for both processes according to the distributions of the new stage.

This $PH/PH/1$ multi-threshold queue can be modelled as a four-dimensional Markov Chain (i, s, x, y) where i and s represent the queue length and stage of the system, $x = 1, \ldots, v_s$ the phase of the arrival process and $y = 1, \ldots, w_s$ the phase of the service process. This queueing system is a Quasi-Birth-and-Death process (QBD) [10] in which the levels are represented by the queue length i, with $i > \max_s\{U_s\}$. Modelling the system as a QBD-process results in a boundary level (level 0) containing the entire threshold policy. By ordering the states lexicographically a tri-diagonal block structure emerges in the boundary level. This structure is utilised by modelling the queueing system as a Level Dependent Quasi-Birth-and-Death process (LDQBD) [3] in which the levels of the LDQBD are the queue length i. We stress that, from here on, we refer to the queue lenght as the level of the LDQBD. The other three variables represent the phase within a level. The states are ordered lexicographically in (i, s, x, y).

The generator Q for this LDQBD is:

$$
Q = \begin{bmatrix}
L^{(0)} & F^{(0)} & 0 & \cdots & & \\
B^{(1)} & L^{(1)} & F^{(1)} & \ddots & & \\
0 & B^{(2)} & L^{(2)} & \ddots & & \\
\vdots & \ddots & \ddots & \ddots & F^{(i-1)} & \\
& & & B^{(i)} & L^{(i)} & \ddots \\
& & & & \ddots & \ddots
\end{bmatrix}, \tag{1}
$$

where $B^{(i)}$ denotes the *backward* transitions (departures) from level i to level $i - 1$, $L^{(i)}$ the *local* transitions within level i and $F^{(i)}$ the *forward* transitions (arrivals) from level i to level $i + 1$.

If the number of potential stages for level $i - 1$, i and $i + 1$, are ℓ, m and n respectively, $B^{(i)}$ is a $m \times \ell$ matrix of submatrices $B_{(j,k)}^{(i)}$, $L^{(i)}$ is a $m \times m$ matrix of submatrices $L_{(j,k)}^{(i)}$ and $F^{(i)}$ is a $m \times n$ matrix of submatrices $F_{(j,k)}^{(i)}$, describing the backward, local and forward transition rates from stage j to stage k. Let I_t denote the $t \times t$ identity matrix and let $\otimes$ denote the Kronecker product. For $s = 1, \ldots, S$, the forward, local and backward submatrices are given by:

$$F_{(s,j)}^{(i)} = \begin{cases} \Lambda_s^0 \otimes \lambda_s \otimes I_{w_s}, & \text{if } j = s \text{ and } L_s \leq i < U_s, \\ p_{s,j} \cdot \Lambda_s^0 \otimes e_{w_s} \otimes \lambda_j \otimes \mu_j, & \text{if } i = U_s, \\ 0, & \text{otherwise.} \end{cases} \tag{2}$$

$$L_{(s,j)}^{(i)} = \begin{cases} \Lambda_s \otimes I_{w_s} + I_{v_s} \otimes M_s, & \text{if } j = s, i > 0 \text{ and } L_s \leq i \leq U_s, \\ \Lambda_s \otimes I_{w_s}, & \text{if } j = s, i = 0 \text{ and } L_s = 0, \\ 0, & \text{otherwise.} \end{cases} \tag{3}$$

$$B_{(s,j)}^{(i)} = \begin{cases} I_{v_s} \otimes M_s^0 \otimes \mu_s, & \text{if } j = s \text{ and } L_s < i \leq U_s, \\ q_{s,j} \cdot e_{v_s} \otimes M_s^0 \otimes \lambda_j \otimes \mu_j, & \text{if } i = L_s, \\ 0, & \text{otherwise.} \end{cases} \tag{4}$$

These formulas can be obtained by closely observing the queueing system. Consider, for instance, the forward transition matrices $F_{(s,j)}^{(i)}$. When $L_s \leq i < U_s$ the stage cannot change upon an arrival, so $j = s$. Now, with rate Λ_s^0 an arrival occurs at which an initial state is chosen with probability λ_s, independent of the phase of the service process. The stage will change when an arrival occurs when $i = U_s$. Now, with rate Λ_s^0, independent of the phase of the service process, an arrival occurs and the stage changes from s to j with probability $p_{s,j}$. During this event an initial phase is chosen for both the arrival process *and* the service process respectively probability λ_j and μ_j. Similar reasoning gives the relations for $L_{(s,j)}^{(i)}$ and $B_{(s,j)}^{(i)}$.

Remark 1. Note that modelling the queueing system as a QBD-process results in the following generator

$$\tilde{Q} = \begin{bmatrix} \tilde{Q}_{00} & \tilde{Q}_{01} & 0 & \cdots & \cdots \\ \tilde{Q}_{10} & L & F & \ddots & \\ 0 & B & L & F & \ddots \\ \vdots & \ddots & B & L & \ddots \\ \vdots & & & \ddots & \ddots & \ddots \end{bmatrix}. \tag{5}$$

The threshold policy, in the LDQBD-process described by the levels $0, \ldots, U_{max}$, is now described in the submatrix $\tilde{Q}_{00}$ with

$$U_{max} = 1 + \max\{U_s \ : \ s = 1, \ldots, S, \ U_s < \infty\}.$$

Finding the stationary distribution for the QBD-process, i.e. solving $\pi \tilde{Q} = 0$, would also include solving

$$\pi_0 \tilde{Q}_{00} + \pi_1 \tilde{Q}_{10} = 0,$$

with π_0 and π_1 denoting the stationary distribution of the entire threshold policy and of the first level in the QBD-process respectively. By modelling the queueing system as the LDQBD-process in (1) we split up level 0 in the QBD-process (5) into smaller blocks such that the stationary distribution π is easier obtained.

3   Steady-State Analysis

In the previous section we modelled the $PH/PH/1$ multi-threshold queue as a LDQBD. In this section, following the analysis in [3] we obtain the steady-state probabilities of the Markov Chain using Matrix Analytic methods. The special structure of our generator allows us to obtain an efficient algorithm for the R-matrices.

We assume the queueing system is stable, i.e., the mean service time is less than the mean interarrival time, see [15], in stages without upper threshold:

$$-\mu_j M_j^{-1} e_{w_j} < -\lambda_j \Lambda_j^{-1} e_{v_j}, \qquad \text{for } j \text{ such that } U_j = \infty.$$

The equilibrium distribution $\pi = [\pi_0, \pi_1, \pi_2, \dots]$ is then given, see Bright and Taylor [3], by

$$\pi_n = \pi_0 \prod_{i=0}^{n-1} R^{(i)},$$

where $R^{(i)}$ is the minimal non-negative solution to

$$F^{(i)} + R^{(i)} L^{(i+1)} + R^{(i)} R^{(i+1)} B^{(i+2)} = 0, \tag{6}$$

with 0 the zero matrix, see [3]. The element $[R^{(i)}]_{(r,t)}$ describes the mean sojourn time in state $(i+1,t)$ per unit sojourn time in the state (i,r) before returning to level i, given that the process started in state (i,r) see p. 499 in [3]. The $R^{(i)}$-matrices can be obtained using the algorithm for LDQBD's by Bright and Taylor [3]. For later convenience, by analogy of $F^{(i)}_{(j,k)}$, $L^{(i)}_{(j,k)}$ and $B^{(i)}_{(j,k)}$, we define the submatrix $R^{(i)}_{(j,k)}$ of $R^{(i)}$ in which the element $[R^{(i)}_{(j,k)}]_{(r,t)}$ describes the mean sojourn time in state $(i+1,t)$ and stage k per unit sojourn time in state (i,r) and stage j before returning returning to level i, given that the process started in state (i,r) and stage j.

We obtain π_0 by solving the boundary condition:

$$\pi_0 L^{(0)} + \pi_1 B^{(1)} = \pi_0 \left(L^{(0)} + R^{(0)} B^{(1)} \right) = 0,$$

and the normalising equation:

$$1 = \sum_{n=0}^{\infty} \pi_n e = \pi_0 \left(I + \sum_{n=1}^{\infty} \prod_{i=0}^{n-1} R^{(i)} \right) e.$$

Above level $U_{\max}$ only stages without upper threshold are active and we may define $F = F^{(i)}$, $L = L^{(i)}$ and $B = B^{(i)}$, $i \geq U_{max}$, i.e., the LDQBD is level independent from level U_{max} upwards. We have $R^{(i)} = R$, $i \geq U_{max}$, where R is the minimal nonnegative solution of

$$F + RL + R^2 B = 0. \tag{7}$$

The LDQBD is level independent from level U_{max}. Therefore, the matrices F, L, B and R are diagonal block matrices. As a consequence, (7) reduces to the matrix equation for the submatrices $R_{(s,s)}$ of R

$$F_{(s,s)} + R_{(s,s)} L_{(s,s)} + R^2_{(s,s)} B_{(s,s)} = 0, \qquad \text{for } s \text{ such that } U_s = \infty. \tag{8}$$

For $i < U_{max}$, the matrices $\boldsymbol{R}^{(i)}$ are obtained from (6) by iteration

$$\boldsymbol{R}^{(i)} = -\boldsymbol{F}^{(i)} \left[\boldsymbol{L}^{(i+1)} + \boldsymbol{R}^{(i+1)} \boldsymbol{B}^{(i+2)} \right]^{-1}, \qquad i = 0, 1, \ldots, U_{max} - 1. \qquad (9)$$

Following the appendix in [3] the inverse exists and has only non-positive elements so that $\boldsymbol{R}^{(i)}$, given by (9), is the unique non-negative solution to (6).

Notice that, unlike [3], we do not need to truncate the iteration for large i, as the structure of our multi-threshold queue guarantees the existence of $U_{max} < \infty$, or for $U_{max} = \infty$ reduces to a single stage.

For a special class of multi-threshold queue the submatrices $\boldsymbol{R}^{(i)}_{(j,k)}$ of $\boldsymbol{R}^{(i)}$ can be obtained efficiently by considering the block elements of the l.h.s. of (6). This result is presented in Theorem 1.

Theorem 1. *For a multi-threshold queue consisting of S stages such that*

(i) $\boldsymbol{F}^{(i)}_{(j,k)} = \boldsymbol{0}$, *for $k < j$ and $i = 0, 1, \ldots$, and*
(ii) *if $\boldsymbol{B}^{(i)}_{(j,k)} \neq \boldsymbol{0}$, for $k < j$, then $\boldsymbol{L}^{(i-1)}_{(x,x)} = \boldsymbol{0}$, for $k < x \leq j$,*

the submatrices $\boldsymbol{R}^{(i)}_{(j,k)}$ of $\boldsymbol{R}^{(i)}$ are given by

$$\boldsymbol{R}^{(i)}_{(j,j)} = -\boldsymbol{F}^{(i)}_{(j,j)} \left[\boldsymbol{L}^{(i+1)}_{(j,j)} + \sum_{b=j}^{S} \boldsymbol{R}^{(i+1)}_{(j,b)} \boldsymbol{B}^{(i+2)}_{(b,j)} \right]^{-1}, \qquad (10)$$

$$\boldsymbol{R}^{(i)}_{(j,k)} = \begin{cases} \boldsymbol{0}, & \text{if } k < j, \\[2mm] -\left[\boldsymbol{F}^{(i)}_{(j,k)} + \sum_{a=j}^{k-1} \sum_{b=a}^{S} \boldsymbol{R}^{(i)}_{(j,a)} \boldsymbol{R}^{(i+1)}_{(a,b)} \boldsymbol{B}^{(i+2)}_{(b,k)} \right] \\[2mm] \quad \cdot \left[\boldsymbol{L}^{(i+1)}_{(k,k)} + \sum_{b=k}^{S} \boldsymbol{R}^{(i+1)}_{(k,b)} \boldsymbol{B}^{(i+2)}_{(b,k)} \right]^{-1}, & \text{if } k > j. \end{cases} \qquad (11)$$

and

$$\boldsymbol{R}^{(i)}_{(x,y)} = \boldsymbol{0} \text{ if } \boldsymbol{B}^{(i+1)}_{(j,k)} \neq \boldsymbol{0} \text{ for } k < x \leq y \leq j. \qquad (12)$$

Proof. Assuming $\boldsymbol{R}^{(i+1)}$ is an upper triangular block matrix one can verify that the unique solution to the block elements of the l.h.s. of (6), i.e.

$$\boldsymbol{0} = \boldsymbol{F}^{(i)}_{(j,k)} + \sum_{a=1}^{S} \boldsymbol{R}^{(i)}_{(j,a)} \boldsymbol{L}^{(i+1)}_{(a,k)} + \sum_{a=1}^{S} \sum_{b=1}^{S} \boldsymbol{R}^{(i)}_{(j,a)} \boldsymbol{R}^{(i+1)}_{(a,b)} \boldsymbol{B}^{(i+2)}_{(b,k)}$$

$$= \boldsymbol{F}^{(i)}_{(j,k)} + \boldsymbol{R}^{(i)}_{(j,k)} \boldsymbol{L}^{(i+1)}_{(k,k)} + \sum_{a=1}^{S} \sum_{b=a}^{S} \boldsymbol{R}^{(i)}_{(j,a)} \boldsymbol{R}^{(i+1)}_{(a,b)} \boldsymbol{B}^{(i+2)}_{(b,k)}.$$

is given by (10), (11) and (12). Since $\boldsymbol{R}$ is a diagonal block matrix this proves by induction that $\boldsymbol{R}^{(i)}$, $i = 0, 1, \ldots$, is an upper triangular block matrix and that its submatrices are uniquely determined by (10), (11) and (12).

The conditions of Theorem 1 can be interpreted as (i) at upper thresholds the stage of the system can only change to higher stages, and (ii) at lower thresholds the stage of the system can change to higher stages and to at most one lower stage. If at level i the stage of the system changes from s to t, with $t < s$, then all stages, $r = t + 1, \ldots, s - 1$ must not be potential stage for level $i - 1$.

Remark 2 (Upper triangularity of $\boldsymbol{R}^{(i)}$). Note that under the conditions of Theorem 1, $\boldsymbol{R}^{(i)}$ must be an upper triangular block matrix for all i. This implies that only stage 1 has no lower threshold.

To prove this, we extend the interpretation of $\boldsymbol{R}^{(i)}$ to the product $\boldsymbol{R}^{(i)}\boldsymbol{R}^{(i+1)}$. Observe that the element $\left[\boldsymbol{R}^{(i)}\boldsymbol{R}^{(i+1)}\right]_{(r,t)}$ describes the mean sojourn time in state $(i + 2, t)$ per unit sojourn time in state (i, r) before returning to level i, given that the process started in state (i, r). If the element $\left[\boldsymbol{R}^{(i)}\boldsymbol{R}^{(i+1)}\right]_{(r,t)} = 0$ then state $(i + 2, t)$ cannot be reached from state (i, r) without visiting level i. The same interpretation holds for the submatrices of the product

$$R(n) = \prod_{i=0}^{n-1} \boldsymbol{R}^{(i)}.$$

If the submatrix $\boldsymbol{R}(n)_{(j,k)}$ of $\boldsymbol{R}(n)$ is $\boldsymbol{0}$, then stage k at level n can never be reached from stage j at level 0. Under the conditions of Theorem 1, $\boldsymbol{R}^{(i)}$ is an upper triangular block matrix for $i \geq 0$, therefore, $\boldsymbol{R}(n)$ is also an upper triangular block matrix for $n \geq 0$. Suppose now that stage $j \neq 1$ has no lower threshold, then stages $k < j$ can never be reached from stage j since $\boldsymbol{R}(n)_{(j,k)} = \boldsymbol{0}$ for $k < j$ and $n \geq 0$. This implies that stages $k < j$ can be removed from the threshold policy. Since the Markov Chain is irreducible, $j = 1$.   ∎

In Corollary 1, we provide an efficient algorithm to compute the stationary queue length vectors $\boldsymbol{\pi}_i$, $i = 0, 1, \ldots$, using the submatrices of $\boldsymbol{R}^{(i)}$ defined in Theorem 1 and equation (8).

Corollary 1. *Define the vector* $\boldsymbol{p}_i = \begin{bmatrix} p_i^1 \ p_i^2 \cdots \ p_i^S \end{bmatrix}$ *for* $i = 0, 1, \ldots$ *such that*

$$p_i^j = \begin{cases} \displaystyle\sum_{a=1}^{j} \boldsymbol{p}_{i-1}^a \boldsymbol{R}_{(a,j)}^{(i-1)}, & i = 1, \ldots, U_{max}, \\ \boldsymbol{p}_{U_{max}}^j \left[\boldsymbol{R}_{(j,j)}\right]^{i - U_{max}}, & i = U_{max} + 1, U_{max} + 2, \ldots, \end{cases} \tag{13}$$

with $\boldsymbol{p}_0^1$ *the solution to*

$$\boldsymbol{p}_0^1 \left[\boldsymbol{L}_{(1,1)}^{(0)} + \sum_{a=1}^{S} \boldsymbol{R}_{(i,a)}^{(0)} \boldsymbol{B}_{(a,i)}^{(1)} \right] = \boldsymbol{0}, \tag{14}$$

such that

$$\boldsymbol{p}_0^1 \boldsymbol{e} = 1, \tag{15}$$

and $p_0^j = 0$ for $j = 2, \ldots, S$. Under the conditions of Theoren 1, the stationary probability vector, $\boldsymbol{\pi}_i = [\pi_i^1 \ \pi_i^2 \ \cdots \ \pi_i^S]$, is given by

$$\pi_i^j = \frac{p_i^j}{\sum_{k=1}^{S} \beta_k}, \tag{16}$$

with

$$\beta_k = \begin{cases} \displaystyle\sum_{i=L_k}^{U_k} p_i^k e, & \text{if } U_k < \infty, \\ \displaystyle\sum_{i=L_k}^{U_{max}-1} p_i^k e + p_{U_{max}}^k \left[I - R_{(k,k)}\right]^{-1} e, & \text{if } U_k = \infty, \end{cases}$$

where e is a vector of ones and $\boldsymbol{I}$ the identity matrix of appropriate size.

Proof. From (13) is follows directly that

$$\boldsymbol{p}_i = \boldsymbol{p}_{i-1} \boldsymbol{R}^{(i-1)},$$

and from (16)

$$\boldsymbol{\pi}_i = \boldsymbol{\pi}_{i-1} \boldsymbol{R}^{(i-1)}.$$

At level 0, only stage 1 is active (see Remark 1), it then follows from (14) that

$$\boldsymbol{p}_0 \left[\boldsymbol{L}^{(0)} + \boldsymbol{R}^{(0)} \boldsymbol{B}^{(1)}\right] = \boldsymbol{0},$$

and that

$$\boldsymbol{\pi}_0 \left[\boldsymbol{L}^{(0)} + \boldsymbol{R}^{(0)} \boldsymbol{B}^{(1)}\right] = \boldsymbol{0}.$$

Stability of the multi-threshold queue guarantees that

$$\sum_{j=1}^{S} \sum_{i=0}^{\infty} p_i^j e = \sum_{\{j \,:\, U_j < \infty\}} \sum_{i=L_j}^{U_j} p_i^j e + \sum_{\{j \,:\, U_j = \infty\}} \left\{ \sum_{i=L_j}^{U_{max}-1} p_i^j e + \sum_{i=U_{max}}^{\infty} p_i^j e \right\}$$

$$= \sum_{\{j \,:\, U_j < \infty\}} \beta_j + \sum_{\{j \,:\, U_j = \infty\}} \left\{ \sum_{i=L_j}^{U_{max}-1} p_i^j e + p_{U_{max}}^j \sum_{i=0}^{\infty} [R_{(j,j)}]^i e \right\}$$

$$= \sum_{\{j \,:\, U_j < \infty\}} \beta_j + \sum_{\{j \,:\, U_j = \infty\}} \left\{ \sum_{i=L_j}^{U_{max}-1} p_i^j e + p_{U_{max}}^j [I - R_{(j,j)}]^{-1} e \right\}$$

$$= \sum_{j=1}^{S} \beta_j < \infty,$$

and that $\boldsymbol{\pi}$ is the stationary queue length distribution.

Remark 3 (Permutations of stages). Consider a multi-threshold queue with S stages. If there exists a permutation of the S stages such that the conditions of Theorem 1 hold, its stationary queue length vector can efficiently be obtained using this permutation and the results from Theorem 1 and Corollary 1.

4   Examples

In this section expressions for $\boldsymbol{R}^{(i)}_{(j,k)}$ and the stationary queue length distribution π^j_i are obtained for three multi-threshold queueing systems. These expressions follow using Theorem 1 and Corollary 1 and are obtained by straightforward but tedious derivations. The three multi-threshold queueing systems we will consider are the multi-threshold queue from Figure 1, the staircase multi-threshold with exponential service and arrival rates from [13] and the staircase multi-threshold queue in a general setting from [5].

4.1   Extended Traffic Model

Consider the multi-threshold queue in Figure 1. Observe that the threshold policy in Figure 1 satisfies both conditions of Theorem 1. In this multi-threshold queueing system, inspired by the traffic model in [2], we assume that

$$0 = L_1 < L_3 < L_2 = L_4 < U_1 = U_3 < U_2 < U_4 = \infty$$

and we define $\rho_i = \frac{\lambda_i}{\mu_i}$. Note that by assuming exponential arrival and service rates, each submatrix $\boldsymbol{R}^{(i)}_{(j,k)}$ reduces to a single element. Therefore, the solution to equation (8) is ρ_4 and each submatrix $\boldsymbol{R}^{(i)}_{(j,k)}$ is given by:

$$\boldsymbol{R}^{(i)}_{(1,1)} = \begin{cases} \rho_1, & i = 0, \ldots, L_3 - 2, \\[2mm] \rho_1 \dfrac{\left(1-\rho_1^{U_1-i}\right)\left(\rho_2^{U_2-U_1}-\rho_2^{U_2-L_2+2}\right)+\left(1-\rho_1^{U_1-L_2+2}\right)\left(1-\rho_2^{U_2-U_1}\right)}{\left(1-\rho_1^{U_1+1-i}\right)\left(\rho_2^{U_2-U_1}-\rho_2^{U_2-L_2+2}\right)+\left(1-\rho_1^{U_1-L_2+2}\right)\left(1-\rho_2^{U_2-U_1}\right)}, & \\[2mm] & i = L_3 - 1, \ldots, L_2 - 2, \\[2mm] \dfrac{\rho_1-\rho_1^{U_1+1-i}}{1-\rho_1^{U_1+1-i}}, & i = L_2 - 1, \ldots, U_1 - 1, \end{cases}$$

$$\boldsymbol{R}^{(i)}_{(1,2)} = \frac{\lambda_1}{\mu_2} \frac{\left(\rho_1^{U_1-i}-\rho_1^{U_1-i+1}\right)\left(1-\rho_2^{U_2-U_1}\right)}{\left(1-\rho_1^{U_1+1-i}\right)\left(1-\rho_2^{U_2+1-i}\right)}, \qquad i = L_2 - 1, \ldots, U_1,$$

$$\boldsymbol{R}^{(i)}_{(1,3)} = \frac{\lambda_1}{\mu_3} \frac{\left(\rho_1^{U_1-i}-\rho_1^{U_1+1-i}\right)\left(\rho_2^{U_2-U_1}-\rho_2^{U_2-L_2+2}\right)}{\left(1-\rho_1^{U_1+1-i}\right)\left(\rho_2^{U_2-U_1}-\rho_2^{U_2-L_2+2}\right)+\left(1-\rho_1^{U_1-L_2+2}\right)\left(1-\rho_2^{U_2-U_1}\right)},$$
$$i = L_3 - 1, \ldots, L_4 - 2,$$

$$\boldsymbol{R}^{(i)}_{(1,4)} = \frac{\lambda_1}{\mu_4} \frac{\left(\rho_1^{U_1-i}-\rho_1^{U_1+1-i}\right)\left(\rho_2^{U_2-U_1}-\rho_2^{U_2+1-i}\right)}{\left(1-\rho_1^{U_1+1-i}\right)\left(1-\rho_2^{U_2+1-i}\right)}, \qquad i = L_4 - 1, \ldots, U_1,$$

$$\boldsymbol{R}^{(i)}_{(2,2)} = \frac{\rho_2-\rho_2^{U_2+1-i}}{1-\rho_2^{U_2+1-i}}, \qquad i = L_2, \ldots, U_2 - 1,$$

$$\boldsymbol{R}^{(i)}_{(2,3)} = 0, \qquad \forall i,$$

$$\boldsymbol{R}^{(i)}_{(2,4)} = \frac{\lambda_2}{\mu_4} \frac{\rho_2^{U_2-i}-\rho_2^{U_2+1-i}}{1-\rho_2^{U_2+1-i}}, \qquad i = L_2, \ldots, U_2,$$

$$\boldsymbol{R}^{(i)}_{(3,3)} = \begin{cases} \rho_3, & i = L_3, \ldots, L_4 - 2, \\[2mm] \dfrac{\rho_3-\rho_3^{U_3+1-i}}{1-\rho_3^{U_3+1-i}}, & i = L_4 - 1, \ldots, U_3 - 1, \end{cases}$$

$$R_{(3,4)}^{(i)} = \frac{\lambda_3}{\mu_4} \frac{\rho_3^{U_3-i} - \rho_3^{U_3+1-i}}{1 - \rho_3^{U_3+1-i}}, \qquad i = L_4 - 1, \ldots, U_3,$$

$$R_{(4,4)}^{(i)} = \rho_4, \qquad i = L_4, L_4 + 1, \ldots.$$

The stationary queue length probability of i customers in stage j, π_i^j, follows from Corollary 1 by normalising p_i^j. For $i = 0$:

$$p_0^j = \begin{cases} 1, & j = 1, \\ 0, & j \neq 1, \end{cases}$$

and for $i > 0$:

$$p_i^1 = p_{i-1}^1 R_{(1,1)}^{(i-1)}, \qquad 0 < i \leq U_1,$$

$$p_i^2 = \begin{cases} p_{i-1}^1 R_{(1,2)}^{(i-1)}, & i = L_2, \\ p_{i-1}^1 R_{(1,2)}^{(i-1)} + p_{i-1}^2 R_{(2,2)}^{(i-1)}, & L_2 < i \leq U_1 + 1, \\ p_{i-1}^2 R_{(2,2)}^{(i-1)}, & U_1 + 1 < i \leq U_2, \end{cases}$$

$$p_i^3 = \begin{cases} p_{i-1}^1 R_{(1,3)}^{(i-1)}, & i = L_3, \\ p_{i-1}^1 R_{(1,3)}^{(i-1)} + p_{i-1}^3 R_{(3,3)}^{(i-1)}, & L_3 < i \leq L_4 - 1, \\ p_{i-1}^3 R_{(3,3)}^{(i-1)}, & L_4 - 1 < i \leq U_3, \end{cases}$$

$$p_i^4 = \begin{cases} p_{i-1}^1 R_{(1,4)}^{(i-1)} + p_{i-1}^3 R_{(3,4)}^{(i-1)}, & i = L_4, \\ p_{i-1}^1 R_{(1,4)}^{(i-1)} + p_{i-1}^2 R_{(2,4)}^{(i-1)} + p_{i-1}^3 R_{(3,4)}^{(i-1)} + p_{i-1}^4 R_{(4,4)}^{(i-1)}, & \\ & L_4 < i \leq U_1 + 1, \\ p_{i-1}^2 R_{(2,4)}^{(i-1)} + p_{i-1}^4 R_{(4,4)}^{(i-1)}, & U_1 + 1 < i \leq U_2 + 1, \\ p_{U_2+1}^4 \left[R_{(4,4)}^{(U_2+1)} \right]^{i-U_2-1}, & U_2 + 1 < i. \end{cases}$$

4.2   Le Ny and Tuffin [13]

Consider a multi-threshold queue of S stages as analysed by Le Ny and Tuffin in [13]. In each stage i arrivals are Poisson distributed with rate λ_i, service times are exponentially distributed with rate μ_i and we define $\rho_i = \frac{\lambda_i}{\mu_i}$. An arrival changes the stage from j to $j+1$ at U_j and a departure changes the stage from j to $j-1$ at L_j. We assume

$$0 = L_1 < L_2 < \cdots < L_S \leq U_1 < \cdots < U_{S-1} < U_S = \infty.$$

The state diagram created by this threshold policy forms a *staircase* as schematically shown in Figure 2.

As in Section 4.1 each submatrix $R_{(j,k)}^{(i)}$ consists of a single element and equation (7), and in particular (8), gives

$$R_{(S,S)}^{(U_{max})} = \rho_S.$$

Both conditions of Theorem 1 are satisfied by the threshold policy and $R_{(j,k)}^{(i)}$ is given by:

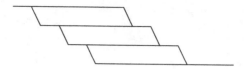

Fig. 2. Schematic representation of the state diagram of a staircase threshold policy with 4 stages.

$$
\boldsymbol{R}^{(i)}_{(j,j)} =
\begin{cases}
\rho_j, & L_j \le i \le L_{j+1} - 2, \\[2mm]
\dfrac{\rho_j - \rho_j^{U_j+1-i}}{1 - \rho_j^{U_j+1-i}}, & L_{j+1} - 1 \le i \le U_j,
\end{cases}
$$

$$
\boldsymbol{R}^{(i)}_{(S,S)} = \rho_S, \qquad L_S \le i,
$$

$$
\boldsymbol{R}^{(i)}_{(j,k)} =
\begin{cases}
\dfrac{\lambda_j}{\mu_k} \dfrac{\rho_j^{U_j-i} - \rho_j^{U_j+1-i}}{1 - \rho_j^{U_j+1-i}} \\[2mm]
\quad \cdot \prod_{a=j+1}^{k-1} \dfrac{\rho_a^{U_a-U_{a-1}} - \rho_a^{U_a+1-i}}{1 - \rho_a^{U_a+1-i}}, & L_k - 1 \le i \le L_{k+1} - 2, \\[4mm]
\dfrac{\lambda_j}{\mu_k} \dfrac{\left(\rho_j^{U_j-i} - \rho_j^{U_j+1-i}\right)\left(1 - \rho_k^{U_k-U_{k-1}}\right)}{\left(1-\rho_j^{U_j+1-i}\right)\left(1-\rho_k^{U_k+1-i}\right)} \\[2mm]
\quad \cdot \prod_{a=j+1}^{k-1} \dfrac{\rho_a^{U_a-U_{a-1}} - \rho_a^{U_a+1-i}}{1 - \rho_a^{U_a+1-i}}, & L_{k+1} - 1 \le i \le U_j,
\end{cases}
$$

$$
\boldsymbol{R}^{(i)}_{(j,S)} = \dfrac{\lambda_j}{\mu_S} \dfrac{\rho_j^{U_j-i} - \rho_j^{U_j+1-i}}{1 - \rho_j^{U_j+1-i}} \prod_{a=j+1}^{S-1} \dfrac{\rho_a^{U_a-U_{a-1}} - \rho_a^{U_a+1-i}}{1 - \rho_a^{U_a+1-i}}, \quad L_S - 1 \le i.
$$

The stationary queue length distribution π_i^j follows from Corollary 1 by normalising p_i^j. For $i = 0$:

$$
p_0^j =
\begin{cases}
1, & j = 1, \\
0, & j \ne 1,
\end{cases}
$$

for $i > 0$ and $j = 1$ or $j = 2$:

$$
p_i^1 = p_{i-1}^1 \boldsymbol{R}^{(i-1)}_{(1,1)}, \qquad 0 < i \le U_1, \tag{17}
$$

$$
p_i^2 =
\begin{cases}
p_{i-1}^1 \boldsymbol{R}^{(i-1)}_{(1,2)}, & i = L_2, \\
p_{i-1}^1 \boldsymbol{R}^{(i-1)}_{(1,2)} + p_{i-1}^2 \boldsymbol{R}^{(i-1)}_{(2,2)}, & L_2 < i \le U + 1, \\
p_{i-1}^2 \boldsymbol{R}^{(i-1)}_{(2,2)}, & U + 1 < i \le U_2,
\end{cases} \tag{18}
$$

for $i > 0$ and $j = 3, \ldots, S - 1$:

$$
p_i^j =
\begin{cases}
\sum_{a=1}^{j-1} p_{i-1}^a \boldsymbol{R}^{(i-1)}_{(a,j)}, & i = L_j, \\
\sum_{a=1}^{j} p_{i-1}^a \boldsymbol{R}^{(i-1)}_{(a,j)}, & L_j < i \le U_1 + 1, \\
\sum_{a=k}^{j} p_{i-1}^a \boldsymbol{R}^{(i-1)}_{(a,j)}, & U_{k-1} + 1 < i \le U_k + 1, \ k = 2, \ldots, j-1, \\
p_{i-1}^j \boldsymbol{R}^{(i-1)}_{(j,j)}, & U_{j-1} + 1 < i \le U_j.
\end{cases} \tag{19}
$$

and for $i > 0$ and $j = S$

$$p_i^S = \begin{cases} \sum_{a=1}^{S-1} p_{i-1}^a R_{(a,S)}^{(i-1)}, & i = L_S, \\ \sum_{a=1}^{S} p_{i-1}^a R_{(a,S)}^{(i-1)}, & L_S < i \le U_1 + 1, \\ \sum_{a=k}^{S} p_{i-1}^a R_{(a,S)}^{(i-1)}, & U_{k-1} + 1 < i \le U_k + 1, \ k = 2, \dots, S-1, \\ p_{i-1}^S \left[R_{(S,S)}^{(i-1)} \right]^{i - U_{max}}, & U_{max} < i. \end{cases}$$

$$(20)$$

4.3 Choi et al [5]

Consider the multi-threshold queue of S stages as analysed by Choi et al [5]. This model generalises the staircase model of [13] to $PH(\Lambda_s, \lambda_s)$ arrivals and $PH(M_s, \mu_s)$ services in stage s. The forward, local and backward transition matrices are given by (2), (3) and (4) respectively. In this case, the submatrices $R_{(j,k)}^{(i)}$ are not single elements and the matrix equation (8) must be solved numerically. The submatrices $R_{(j,k)}^{(i)}$, $i = 0, \dots, U_{max} - 1$, are iteratively given, following Theorem 1, by

$$R_{(j,j)}^{(i)} = \begin{cases} - F_{(j,j)}^{(i)} \left[L_{(j,j)}^{(i+1)} + R_{(j,j)}^{(i+1)} B_{(j,j)}^{(i+2)} \right]^{-1}, & L_j \le i < U_j - 1, \ i \ne L_{j+1} - 2, \\ - F_{(j,j)}^{(i)} \left[L_{(j,j)}^{(i+1)} + \sum_{b=j}^{j+1} R_{(j,b)}^{(i+1)} B_{(b,j)}^{(i+2)} \right]^{-1}, & \\ & i = L_{j+1} - 2, \\ - F_{(j,j)}^{(i)} \left[L_{(j,j)}^{(i+1)} \right]^{-1}, & i = U_j - 1, \\ 0, & \text{otherwise}, \end{cases}$$

$$R_{(j,k)}^{(i)} = \begin{cases} - \left[\sum_{a=j}^{k-1} R_{(j,a)}^{(i)} R_{(a,k)}^{(i+1)} B_{(k,k)}^{(i+2)} \right] \\ \quad \cdot \left[L_{(k,k)}^{(i+1)} + R_{(k,k)}^{(i+1)} B_{(k,k)}^{(i+2)} \right]^{-1}, & L_k - 1 \le i < U_j, \ i \ne L_{k+1} - 2, \\ - \left[\sum_{b=k}^{k+1} \sum_{a=j}^{k-1} R_{(j,a)}^{(i)} R_{(a,b)}^{(i+1)} B_{(b,k)}^{(i+2)} \right] \\ \quad \cdot \left[L_{(k,k)}^{(i+1)} + \sum_{b=k}^{k+1} R_{(k,b)}^{(i+1)} B_{(b,k)}^{(i+2)} \right]^{-1}, & i = L_{k+1} - 2, \\ - \left[F_{(j,k)}^{(i)} \mathbb{1}_{\{k = j+1\}} + \sum_{a=j+1}^{k-1} R_{(j,a)}^{(i)} R_{(a,k)}^{(i+1)} B_{(k,k)}^{(i+2)} \right] \\ \quad \cdot \left[L_{(k,k)}^{(i+1)} + R_{(k,k)}^{(i+1)} B_{(k,k)}^{(i+2)} \right]^{-1}, & i = U_j, \\ 0, & \text{otherwise}, \end{cases}$$

for $j = 1, \dots, S - 1$, and

$$R_{(S,S)}^{(i)} = \begin{cases} R_{(S,S)}, & L_S \le i, \\ 0, & \text{otherwise}. \end{cases}$$

The stationary queue length distribution π_i^j follows from Corollary 1 by normalising p_i^j. The vectors p_i^j, $i > 0$, are given by equations (17), (18), (19) and (20). Finally, p_0^1 is obtained from (14) and (15) and $p_0^j = 0$, $j > 1$.

5   Summary and Conclusion

We introduced the $PH/PH/1$ multi-threshold queue where the arrival process and service process are controlled by a threshold policy. The threshold policy determines, based on the queue length, the stage of system, and the stage determines the arrival and service processes. We modelled this queue as a Level Dependent Quasi-Birth-and-Death process and obtained the stationary queue length probabilities using Matrix Analytic methods.

A special class of multi-threshold queues is presented and explicit description of the R-matrices has been obtained in terms of its submatrices. This decomposition theorem allows an efficient computation of each R-submatrix as well as the stationary queue length probability vectors.

Future work consists of a network of PH/PH/1 threshold queues in which the threshold policy can control the service rates of previous queue, see Baer, Al Hanbali, Boucherie and van Ommeren [1].

Acknowledgement. This research is supported by the Centre for Telematics and Information Technology (CTIT) of the University of Twente. The authors would like to thank the anonymous reviewers for their helpful suggestions.

References

1. Baer, N., Al Hanbali, A., Boucherie, R.J., van Ommeren, J.C.W.: A successive censoring algorithm for a system of connected qbd-processes. Memorandum 2030. Department of Applied Mathematics, University of Twente, Enschede, The Netherlands (2013)
2. Baer, N., Boucherie, R.J., van Ommeren, J.C.W.: Threshold queueing describes the fundamental diagram of uninterrupted traffic. Memorandum 2000. Department of Applied Mathematics, University of Twente, Enschede, The Netherlands (2012)
3. Bright, L., Taylor, P.G.: Calculating the equilibrium distribution in level dependent quasi-birth-and-death processes. Communications in Statistics - Stochastic Models 11(3), 497–525 (1995)
4. Chakravarthy, S.R.: A multi-server queueing model with markovian arrivals and multiple thresholds. Asia-Pacific Journal of Operational Research 24(2), 223–243 (2007)
5. Choi, S.H., Kim, B., Sohraby, K., Choi, B.D.: On matrix-geometric solution of nested QBD chains. Queueing Systems 43, 5–28 (2003)
6. Chou, C.F., Golubchik, L., Lui, J.C.S.: Multiclass multiserver threshold based systems: A study of noninstantaneous server activation. IEEE Transactions on Parallel and Distributed Systems 18(1), 96–110 (2007)
7. Feng, W., Adachi, K., Kowada, M.: A two-queue and two-server model with a threshold-based control service policy. European Journal of Operational Research 137, 593–611 (2002)
8. Helbing, D.: Traffic and related self-driven many-particle systems. Reviews of Modern Physics 73 (2001)
9. Ibe, O.C., Keilson, J.: Multi-server threshold queues with hysteresis. Performance Evaluation 21, 185–213 (1995)

10. Latouche, G., Ramaswami, V.: A logarithmic reduction algorithm for quasi-birth-death processes. Journal of Applied Probability 30, 650–674 (1993)
11. Le Ny, L.M.: Probabilité stationnaire d'une file d'attente á 2 seuils. Technical report. IRISA (1987)
12. Le Ny, L.M.: Exact analysis of a threshold-based queue with hysteresis and a delayed additional server. Technical report, IRISA (2000)
13. Le Ny, L.M., Tuffin, B.: A simple analysis of heterogeneous multi-server threshold queues with hysteresis. In: Proceedings of the Applied Telecommunication Symposium (2002)
14. Lui, J.C.S., Golubchik, L.: Stochastic complement analysis of multi-server threshold queues with hysteresis. Performance Evaluation 35, 185–213 (1999)
15. Neuts, M.F.: Matrix-Geometric Solutions in Stochastic Models - An Algorithmic Approach. Dover Publications, Inc., New York (1981)
16. Neuts, M.F., Rao, B.M.: On the design of a finite-capacity queue with phase-type service times and hysteretic control. European Journal of Operational Research 62, 221–240 (1992)
17. Zhu, Y., Yang, P., Basu, K.: Overload control in finite buffer message system with general distributed service time. In: Proceedings Internation Conference on Communication Technology, vol. 1, pp. 408–411 (1996)

Time-Dependent Behavior of Queueing Delay in $GI/M/1/K$-type Model with N-policy: Analytical Study with Applications to WSNs

Wojciech M. Kempa

Silesian University of Technology, Institute of Mathematics,
ul. Kaszubska 23, 44-100 Gliwice, Poland
wojciech.kempa@polsl.pl

Abstract. The transient queueing delay distribution in the $GI/M/1/K$-type model with the N-policy is investigated. After finishing each busy period the service is being initialized simultaneously with the Nth packet arrival occurrence. Using the approach based on the idea of embedded Markov chain, the formula of total probability, renewal theory and linear algebra, the compact-form representation for the Laplace transform of the tail of delay distribution is obtained. The results can be useful in modeling the operation of wireless sensor networks (WSNs) with energy saving mechanism based on "queued" waking up of nodes. A network-motivated numerical example is attached.

Keywords: Finite-buffer queue, N-policy, Queueing delay, Transient state, Wireless sensor network (WSN).

1 Introduction

Wireless sensor networks (WSNs) are commonly used nowadays in controlling of different real-life phenomena and in the risk alerting. The WSN technology is apllied in observation of air and water pollution, monitoring of patients' health condition in hospitals, road traffic analysis, fire prevention systems, military operations and many others. In typical WSN nodes (sensors) are equipped with a non-rechargable battery so the implementation of efficient energy saving mechanism is one of the most essential challenges. Since the information is transmitted, according to a proper routing algorithm, from node to node to the gateway sensor node which is connected with the Internet, then the problem of power saving is extremely important for sensors which are situated closer to the gateway, and for which the probability of their using in the processing is greater than for nodes situated more distant from the gateway. Moreover, in WSN nodes are often located in hardly accessible places and hence the frequent replacement of "discharged" sensors with new ones can be difficult. In relation to the above-mentioned applications of wireless sensor networks, one of the most important stochastic characteristics helpful in performance evaluation of each WSN is a

B. Sericola, M. Telek, and G. Horváth (Eds.): ASMTA 2014, LNCS 8499, pp. 110–124, 2014.

node queueing delay $v(t)$, that at fixed epoch t gives the time needed for initialization of the processing of a packet joining the queue exactly at time t. Evidently, the total queueing delay of the packet depends on the number of nodes located on the route between the source of the signal and the gateway sensor node, and on the power saving mechanism applied in successive nodes.

In the paper we study the transient behavior of queueing delay distribution in the $GI/M/1/K$-type model with power saving mechanism based on the N-policy. After the busy period of the node, during which the information is being processed continuously, the radio transmitter/receiver is being turned off and becomes active simultaneously with the Nth arrival occurrence only. So, the nodes are "queued" waked up after each busy period. This model of server operation allows to achieve optimal energy management level in conjunction with adequate quality of the network. In the case of data with high validity (e.g. monitoring the fire risk), you can choose a lower threshold value, while for the less important data (e.g. monitoring the road traffic or parking places), the threshold can be set higher. As one can note, most results obtained for stochastic characteristics of different-type finite-buffer queues relate to the stationary state of the system. However, time-dependent analysis is often desired or even necessary, e.g. during the observation of the system just after its opening after a failure, or together with application of the new control mechanism. Moreover, in practice, the stochastic behavior of the system may be destabilized by the phenomena like fade-out or interference which occur in wireless communication. Using the approach based on the idea of embedded Markov chain, the continuous version of the formula of total probability, renewal theory and linear algebra, we find the closed-form representation for the LT (=Laplace transform) of the tail of d.f. (=distribution function) of queueing delay in the considered system. The formula can be efficiently numerically treated and we attach a network-motivated numerical example.

An infinite-buffer $M/G/1$-type queueing system with the N-policy is considered in [5] as a model of WSN's node with battery saving. Some similar power saving models based on different-type server vacations can be found e.g. in [3], [4] and [14], however the obtained analytical results relate only to performance measures of the stationary state of the system, e.g. mean queueing delay or mean queue size. In [7] and [9] time-dependent analysis of the queue-size distribution in the $M^X/G/1$-type system with batch arrivals, infinite buffer and the N-policy can be found. Some new results for transient queueing delay distribution are obtained e.g. in [6] and [10]. In particular, in [10] (see also [8]) the representation for the double LT of the d.f. of time-dependent queueing delay in the $M/G/1/K$-type model with single server vacation is derived.

The remaining part of the paper is organized as follows. In the next Section 2 we give the precise mathematical description of the considered queueing model. In Section 3 we find the formulae for LTs of d.fs of queueing delay in the buffer loading period and the period duration. Section 4 is devoted to the analysis of queueing delay distribution in the busy cycle. The main result, utilizing the representations obtained in Sections 3 and 4, is presented in Section 5. In Section 6 one can find a numerical example and the last Section 7 contains short conclusion.

2   Mathematical Description of the Model

Let us consider the finite-buffer $GI/M/1/K$ queue in which interarrival times are independent and generally distributed random variables with a d.f. $F(\cdot)$, and the incoming packets are being served individually with an exponential service time with mean μ^{-1}, according to the FIFO service discipline. The total number of packets present in the system is bounded by K i.e. there is a finite buffer with capacity K and one place for service. It is assumed that the system starts the operation together with the first packet arrival and the server initializes the processing simultaneously with the arrival epoch of the Nth packet (the N-policy), where $1 \leq N \leq K$. When the server becomes idle it is being turned off and begins the service again if it finds N packets accumulated in the buffer, and so on. In consequence, the evolution of the system can be observed on successive buffer loading periods $L_1, L_2, \ldots$ followed by untypically defined busy cycles $B_1, B_2, \ldots$, consisting of a busy period during which the queue empties (beginning with the initial level N) and the idle time which ends at the arrival epoch of the first packet after the busy period. Since successive arrival epochs ar Markov moments in the $GI/M/1/K$-type queue (see e.g. [2]), then (L_k) and (B_k), $k = 1, , 2, \ldots$, are sequences of totally independent random variables having the same d.fs "inside" each sequence separately. In the article buffer loading periods and busy cycles will often be identified with their durations.

　　In the next sections we obtain the representations for the LTs of queueing delay distributions during a single buffer loading period and a busy cycle. Besides, we find the formulae for LTs of busy cycle and buffer loading time durations. Next, by using the renewal-theory approach, basing on the fact that successive L_k's and B_k's are independent and identically distributed separately, we obtain the formula in the general case.

3   Analysis of a Buffer Loading Period

Let us analyze, firstly, the queueing delay distribution at arbitrary time epoch t during the first buffer loading period L_1, which begins at $t = 0$ together with the first arrival occurrence. Let $F^{j*}(\cdot)$ denote the j-fold convolution of the d.f. $F(\cdot)$ with itself and let $\mathbb{E}_{k,\mu}(\cdot)$ be the k-Erlang with parameter μ d.f. i.e.

$$\mathbb{E}_{k,\mu}(x) = 1 - e^{-\mu x} \sum_{i=0}^{k-1} \frac{(\mu x)^i}{i!}. \tag{1}$$

Moreover, let $\overline{F}(\cdot)$ and $\overline{\mathbb{E}}_{k,\mu}(\cdot)$ be tails of d.fs $F(\cdot)$ and $\mathbb{E}_{k,\mu}(\cdot)$, respectively.

Let us note that the following representation is true:

$$\mathbf{P}\{(v(t) > x) \cap (t \in L_1)\} = \sum_{i=0}^{N-2} \int_{y=0}^{t} dF^{i*}(y) \int_{u=t-y}^{\infty} dF(u) \times$$

$$\times \int_{v=x-y-u+t}^{\infty} \overline{\mathbb{E}}_{i+1,\mu}(x - y - u - v + t) dF^{(N-i-1)*}(v). \qquad (2)$$

On the right side of the formula above i indicates the number of packets entering before t (excluding the packet entering at $t = 0$) and v denotes the last inter-arrival time during the buffer loading period (the interarrival time between the $(N-1)$th and the Nth packet). The "virtual" packet joining the system exactly at time t will wait more than x, if and only if the accumulated service time of N packets plus the time distant from t to the completion epoch of the buffer loading period will be greater than x.

Introducing now the following notation:

$$\tilde{v}^L(s, x) = \int_0^{\infty} e^{-st} \mathbf{P}\{(v(t) > x) \cap (t \in L_1)\} dt, \quad \mathrm{Re}(s) > 0, \qquad (3)$$

we obtain from (2)

$$\tilde{v}^L(s, x) = \sum_{i=0}^{N-2} \int_{t=0}^{\infty} e^{-st} dt \int_{y=0}^{t} dF^{i*}(y) \int_{u=t-y}^{\infty} dF(u) \qquad (4)$$

$$\times \int_{v=x-y-u+t}^{\infty} \overline{\mathbb{E}}_{i+1,\mu}(x - y - u - v + t) dF^{(N-i-1)*}(v)$$

$$= \sum_{i=0}^{N-2} f^{i+1}(s) \int_{z=0}^{\infty} e^{-sz} dz \int_{v=0}^{x+z} \overline{\mathbb{E}}_{i+1,\mu}(x - v + z) dF^{(N-i-1)*}(v), \qquad (5)$$

where $f(s) = \int_0^{\infty} e^{-st} dF(t)$, $\mathrm{Re}(s) > 0$.

Let us note that, since each buffer loading period L_k, $k \geq 1$, begins with one packet present, then its duration equals the waiting time for the $(N-1)$th arrival. Thus, we have

$$\tilde{d}^L(s) = \tilde{d}_k^L(s) = \int_0^{\infty} e^{-st} d\mathbf{P}\{L_k < t\} dt$$

$$= \int_0^{\infty} e^{-st} dF^{(N-1)*}(t) = f^{N-1}(s). \qquad (6)$$

4   Queueing Delay in a Busy Cycle

Suppose temporarily, for the needs of this section, that the system may start a busy cycle not necessarily with N packets accumulated in the buffer queue but with arbitrary number of n packets, where $1 \leq n \leq K$. Besides, let us identify the start epoch of the first busy cycle B_1 with $t = 0$, and introduce the following notation:

$$V_n^B(t, x) = \mathbf{P}\{(v(t) > x) \cap (t \in B_1) \,|\, X(0) = n\}, \; 1 \le n \le K, \, x > 0, \, t > 0. \quad (7)$$

So, $V_n^B(t, x)$ is the tail distribution function of the queueing delay on the first busy cycle B_1, conditioned by the number of packets present in the buffer at the initial moment of B_1.

Since successive arrival epochs are Markov moments then, applying the formula of total probability with respect to the first arrival instant after $t = 0$, we get the following system of integral equations:

$$V_n^B(t, x) = \sum_{k=0}^{n-1} \int_0^t \frac{(\mu y)^k}{k!} e^{-\mu y} V_{n-k+1}^B(t - y, x) dF(y)$$

$$+ \sum_{k=n}^{\infty} \int_0^{\infty} \frac{(\mu t)^k}{k!} e^{-\mu y} V_1^B(t - y, x) dF(y) + \overline{F}(t) e^{-\mu(t+x)} \frac{[\mu(t + x)]^n}{n!}, \quad (8)$$

where $1 \le n \le K - 1$, and

$$V_K^B(t, x) = \int_0^t e^{-\mu y} V_K^B(t - y, x) dF(y)$$

$$+ \sum_{k=1}^{K-1} \int_0^t \frac{(\mu y)^k}{k!} e^{-\mu y} V_{K-k+1}^B(t - y, x) dF(y)$$

$$+ \sum_{k=K}^{\infty} \int_0^{\infty} \frac{(\mu t)^k}{k!} e^{-\mu y} V_1^B(t - y, x) dF(y) + \overline{F}(t) e^{-\mu(t+x)} \frac{[\mu(t + x)]^K}{K!}. \quad (9)$$

Indeed, te first summand on the right side of (8) relates to the situation in which the first arrival (after $t = 0$) occurs at time $y < t$. If the buffer does not empty before this moment (there are $0 \le k \le n - 1$ departures before y), then, including the first arrival, the system contains exactly $n - k + 1$ packets at the renewal (Markov) moment y. In the second summand the buffer empties before the first arrival and the last summand describes the situation in which there are no arrivals before t. On the right side of (9) the first sum is taken from $k = 1$ since if before the first arrival occurring at time $y < t$ no services are finished, the packet entering at time y is lost (compare the first summand on the right side of (9)).

Denote

$$\widetilde{v}_n^B(s, x) = \int_0^{\infty} e^{-st} V_n^B(t, x) dt, \; n \ge 1, \quad (10)$$

$$a_n(s) = \int_0^{\infty} e^{-(s+\mu)t} \frac{(\mu t)^n}{n!} dF(t), \; n \ge 0, \quad (11)$$

$$b_{n,x}(s) = \int_0^{\infty} e^{-(s+\mu)t - \mu x} \overline{F}(t) \frac{[\mu(t + x)]^n}{n!} dt, \; n \ge 1, \quad (12)$$

where $\mathrm{Re}(s) > 0$. Introducing (10)–(12) into the system (8)–(9), we get

$$\widetilde{v}_n^B(s,x) = \sum_{k=-1}^{n-2} a_{k+1}(s)\widetilde{v}_{n-k}^B(s,x) - \widetilde{v}_n^B(s,x) = \phi_n(s,x), \quad 1 \le n \le K-1, \quad (13)$$

and

$$[1 - f(s+\mu)]\widetilde{v}_K^B(s,x) = \sum_{k=1}^{K-1} a_k(s)\widetilde{v}_{K-k+1}^B(s,x) + \widetilde{v}_1^B(s,x)\sum_{k=K}^{\infty} a_k(s), \quad (14)$$

where

$$\phi_n(s,x) = -\widetilde{v}_1^B(s,x)\sum_{k=n}^{\infty} a_k(s) - b_n(s,x), \quad n \ge 1. \quad (15)$$

In [11] (see also [12]) the infinite-sized system of type (13) is considered, where $n \ge 2$ and is not bounded. It is proved there that each solution of (13) can be written as

$$\widetilde{v}_n^B(s,x) = C_1(s,x)R_{n-1}(s) + \sum_{k=2}^{n} R_{n-k}(s)\phi_k(s,x), \quad n \ge 2, \quad (16)$$

where $C_1(s,x)$ does not depend on n, and successive terms of the sequence $(R_n(s))$ can be found recursively in the following way:

$$R_0(s) = 0, \quad R_1(s) = a_0^{-1}(s),$$

$$R_{n+1}(s) = R_1(s)\big(R_n(s) - \sum_{i=0}^{n} a_{i+1}(s)R_{n-i}(s)\big), \quad (17)$$

where $n \ge 1$.

Let us note that the representation for $\widetilde{v}_1^B(s,x)$ must be found in another way, since the formula (16) is valid only for $n \ge 2$. Since in the system (13) the number of equations is finite, we can efficiently use the identity (14) as a boundary condition and find $C_1(s,x)$ (and next $\widetilde{v}_1^B(s,x)$) explicitly. Indeed, substituting $n = 2$ into (16), we get

$$\widetilde{v}_2^B(s,x) = C_1(s,x)a_0^{-1}(s). \quad (18)$$

Next, taking $n = 2$ in (13) and referring to (15), we obtain

$$a_0(s)\widetilde{v}_2^B(s) - \widetilde{v}_1^B(s,x) = -\widetilde{v}_1^B(s,x)\sum_{k=1}^{\infty} a_k(s) - b_1(s,x), \quad (19)$$

and hence, since $\sum_{k=0}^{\infty} a_k(s) = f(s)$ and $a_0(s) = f(s+\mu)$, substituting (18) into (19), we get

$$\widetilde{v}_1^B(s,x) = \big(1 + f(s+\mu) - f(s)\big)^{-1}\big(C_1(s,x) + b_1(s,x)\big). \quad (20)$$

At this stage we have all tools necessary for finding $C_1(s,x)$ explicitly. Indeed, substituting (16) and (20) into (14) we obtain

$$[1 - f(s+\mu)]\left\{C_1(s,x)R_{K-1}(s) + \sum_{k=2}^{K} R_{K-k}(s)\left[-(1 + f(s+\mu) - f(s))^{-1}\right.\right.$$

$$\times (C_1(s,x) + b_1(s,x))\sum_{i=k}^{\infty} a_i(s) - b_k(s,x)\Big]\Big\}$$

$$= \sum_{k=1}^{K-1} a_k(s)\left\{C_1(s,x)R_{K-k}(s) + \sum_{i=2}^{K-k+1} R_{K-k+1-i}(s)\right.$$

$$\times \left[-(1 + f(s+\mu) - f(s))^{-1}(C_1(s,x) + b_1(s,x))\sum_{j=i}^{\infty} a_j(s) - b_i(s,x)\right]\Big\}$$

$$+ (1 + f(s+\mu) - f(s))^{-1}(C_1(s,x) + b_1(s,x))\sum_{k=K}^{\infty} a_k(s). \qquad (21)$$

From (21) we eliminate $C_1(s,x)$ as follows:

$$C_1(s,x) = \frac{\Delta_1(s,x) - \Delta_2(s,x)}{\Pi_1(s,x) - \Pi_2(s,x)}, \qquad (22)$$

where

$$\Delta_1(s,x) = (1 - f(s+\mu))$$

$$\times \sum_{k=2}^{K} R_{K-k}(s)\left[(1 + f(s+\mu) - f(s))^{-1}b_1(s,x)\sum_{i=k}^{\infty} a_i(s) + b_k(s,x)\right]$$

$$+ (1 + f(s+\mu) - f(s))^{-1}b_1(s,x)\sum_{k=K}^{\infty} a_k(s), \qquad (23)$$

$$\Delta_2(s,x) = \sum_{k=1}^{K-1} a_k(s)\sum_{i=2}^{K-k+1} R_{K-k+1-i}(s)\left[(1 + f(s+\mu) - f(s))^{-1}b_1(s,x)\right.$$

$$\times \sum_{j=i}^{\infty} a_j(s) + b_i(s,x)\Big], \qquad (24)$$

$$\Pi_1(s,x) = (1 - f(s+\mu))\left[R_{K-1}(s) - (1 + f(s+\mu) - f(s))^{-1}\right.$$

$$\times \sum_{k=2}^{K} R_{K-k}(s)\sum_{i=k}^{\infty} a_i(s)\Big] - (1 + f(s+\mu) - f(s))^{-1}\sum_{k=K}^{\infty} a_k(s) \qquad (25)$$

and

$$\Pi_2(s, x) = \sum_{k=1}^{K-1} a_k(s) \Big[R_{K-k}(s) - \big(1 + f(s + \mu) - f(s)\big)^{-1}$$

$$\times \sum_{i=2}^{K-k+1} R_{K-k+1-i}(s) \sum_{j=i}^{\infty} a_j(s) \Big]. \tag{26}$$

Thus, the following theorem can be written:

Theorem 1. *The representations for the LT $\widetilde{v}_n^B(s, x)$ of the conditional queueing delay tail distribution on the busy cycle are following:*

$$\widetilde{v}_1^B(s, x) = \big(1 + f(s + \mu) - f(s)\big)^{-1} \Big(\frac{\Delta_1(s, x) - \Delta_2(s, x)}{\Pi_1(s, x) - \Pi_2(s, x)} + b_1(s, x) \Big) \tag{27}$$

and

$$\widetilde{v}_n^B(s, x) = \frac{\Delta_1(s, x) - \Delta_2(s, x)}{\Pi_1(s, x) - \Pi_2(s, x)} R_{n-1}(s) + \sum_{k=2}^{n} R_{n-k}(s) \phi_k(s, x) \tag{28}$$

where $\mathrm{Re}(s) > 0$, $2 \leq n \leq K$, *and the formulae for* $a_k(s)$, $b_k(s)$, $\phi_k(s, x)$, $R_k(s)$, $\Delta_1(s, x)$, $\Delta_2(s, x)$, $\Pi_1(s, x)$ *and* $\Pi_2(s, x)$ *can be found in (11), (12), (15), (17), (23), (24), (25) and (26), respectively.*

Before we will prove the main theorem we need yet a representation for the LT of the busy cycle B_k, $k \geq 1$, duration, conditioned by the initial buffer state. Introduce the following notation:

$$\widetilde{d}_n^B(s) = \widetilde{d}_n^{B_k}(s) = \int_0^{\infty} e^{-st} d\mathbf{P}\{B_k < t \mid X(0) = n\} dt, \tag{29}$$

where $\mathrm{Re}(s) > 0$ and $1 \leq n \leq K$.

Applying the formula of total probability we obtain now (compare (8)–(9))

$$\widetilde{d}_n^B(s) = \sum_{k=0}^{n-1} \widetilde{d}_{n-k+1}^B(s) \int_0^{\infty} e^{-sx} \frac{(\mu x)^k}{k!} e^{-\mu x} dF(x)$$

$$+ \sum_{k=n}^{\infty} \int_0^{\infty} e^{-sx} \frac{(\mu x)^k}{k!} e^{-\mu x} dF(x), \quad 1 \leq n \leq K - 1 \tag{30}$$

and

$$\widetilde{d}_K^B(s) = \widetilde{d}_K^B(s) \int_0^{\infty} e^{-sx} e^{-\mu x} dF(x)$$

$$+ \sum_{k=1}^{K-1} \widetilde{d}_{K-k+1}^B(s) \int_0^{\infty} e^{-sx} \frac{(\mu x)^k}{k!} e^{-\mu x} dF(x) + \sum_{k=K}^{\infty} \int_0^{\infty} e^{-sx} \frac{(\mu x)^k}{k!} e^{-\mu x} dF(x). \tag{31}$$

Denoting

$$\Psi_n(s) = -\sum_{k=n}^{\infty} a_k(s), \qquad (32)$$

where $a_k(s)$ was defined in (11), we can rewrite (30)–(31) as

$$\sum_{k=-1}^{\infty} a_{k+1}(s)\widetilde{d}^B_{n-k}(s) - \widetilde{d}^B_n(s) = \Psi_n(s), \qquad (33)$$

where $1 \le n \le K - 1$, and

$$\widetilde{d}^B_K(s)\big(1 - f(s+\mu)\big) = \sum_{k=1}^{K-1} a_k(s)\widetilde{d}^B_{K-k+1}(s) + \sum_{k=K}^{\infty} a_k(s). \qquad (34)$$

The representation for the solution of the system (33)–(34) has the following form (see (16)):

$$\widetilde{d}^B_n(s) = C_2(s)R_{n-1}(s) + \sum_{k=2}^{n} R_{n-k}(s)\Psi_k(s), \quad 2 \le n \le K, \qquad (35)$$

where the formula for $R_k(s)$ was given in (17). To find the representations for $C_2(s)$ and $\widetilde{d}^B_1(s)$, substitute $n = 2$ into (33) and (35). We get

$$a_0(s)\widetilde{d}^B_2(s) - \widetilde{d}^B_1(s) = \Psi_1(s) \qquad (36)$$

and

$$\widetilde{d}^B_2(s) = C_2(s)a_0^{-1}(s). \qquad (37)$$

Combining (36) and (37), since $a_0(s) = f(s+\mu)$, gives

$$\widetilde{d}^B_1(s) = C_2(s) + f(s) - f(s+\mu). \qquad (38)$$

Substituting now the representation (35) into (34) we obtain

$$\big(1 - f(s+\mu)\big)\Big[C_2(s)R_{K-1}(s) - \sum_{k=2}^{K} R_{K-k}(s)\sum_{i=k}^{\infty} a_i(s)\Big]$$

$$= \sum_{k=1}^{K-1} a_k(s)\Big[C_2(s)R_{K-k}(s) - \sum_{i=2}^{K-k+1} R_{K-k+1-i}(s)\sum_{j=i}^{\infty} a_j(s)\Big] + \sum_{k=K}^{\infty} a_k(s) \qquad (39)$$

and hence we find the formula of $C_2(s)$ in the form

$$C_2(s) = \big(\Lambda_1(s) - \Lambda_2(s)\big)\big(\Theta(s)\big)^{-1}, \qquad (40)$$

where

$$\Lambda_1(s) = \big(1 - f(s + \mu)\big) \sum_{k=2}^{K} R_{K-k}(s) \sum_{i=k}^{\infty} a_i(s) + \sum_{k=K}^{\infty} a_k(s), \tag{41}$$

$$\Lambda_2(s) = \sum_{k=1}^{K-1} a_k(s) \sum_{i=2}^{K-k+1} R_{K-k+1-i}(s) \sum_{j=i}^{\infty} a_j(s) \tag{42}$$

and

$$\Theta(s) = \big(1 - f(s + \mu)\big) R_{K-1}(s) - \sum_{k=1}^{K-1} a_k(s) R_{K-k}(s). \tag{43}$$

Collecting the formulae (32), (35), (38) and (40)–(43) we obtain the following theorem that gives the compact-form representation for the LT $\widetilde{d}_n^B(s)$ of the d.f. of the conditioned busy cycle duration:

Theorem 2. *For* $\mathrm{Re}(s) > 0$ *the following formulae hold true:*

$$\widetilde{d}_1^B(s) = \Big(\Lambda_1(s) - \Lambda_2(s)\Big)\big(\Theta(s)\big)^{-1} + f(s) - f(s + \mu) \tag{44}$$

and

$$\widetilde{d}_n^B(s) = \Big(\Lambda_1(s) - \Lambda_2(s)\Big)\big(\Theta(s)\big)^{-1} R_{n-1}(s) - \sum_{k=2}^{n} R_{n-k}(s) \sum_{i=k}^{\infty} a_i(s), \tag{45}$$

where $2 \le n \le K$, *and the representations for* $a_k(s)$, $R_k(s)$, $\Lambda_1(s)$, $\Lambda_2(s)$ *and* $\Theta(s)$ *are found in (11), (17), (41), (42) and (43), respectively.*

5   General Result

In this section we will use the renewal-theory approach and analytical results obtained in Sections 3 and 4 to prove the following main theorem:

Theorem 3. *The representation for the LT of the tail of the d.f. of queueing delay* $v(t)$ *in the* $GI/M/1/K$-*type system with* N-*policy ("queued" wake up of the server) is following:*

$$\int_0^{\infty} e^{-st} \mathbf{P}\{v(t) > x\} dt = \frac{\widetilde{v}^L(s,x) + \widetilde{d}^L(s)\widetilde{v}_N^B(s,x)}{1 - \widetilde{d}_N^B(s)\widetilde{d}^L(s)}, \tag{46}$$

where the formulae for $\widetilde{v}^L(s,x)$, $\widetilde{d}^L(s)$, $\widetilde{v}_N^B(s,x)$ *and* $\widetilde{d}_N^B(s)$ *were obtained in (4), (6), (28) and (45), respectively.*

Proof. It is easy to note that

$$\mathbf{P}\{v(t) > x\} = \sum_{k=1}^{\infty} \Big(\mathbf{P}\{(v(t) > x) \cap (t \in L_k)\}$$

$$+ \mathbf{P}\{(v(t) > x) \cap (t \in B_k)\}\Big). \qquad (47)$$

Now, since all terms of sequences (L_k) and (B_k), $k \geq 1$, are independent and identically distributed random variables "inside" each sequence separately, then we obtain

$$\mathbf{P}\{(v(t) > x) \cap (t \in L_k)\}$$

$$= \int_0^t \mathbf{P}\{(v(t-y) > x) \cap (t-y \in L_1)\} d(D^L * D_N^B)^{(k-1)*}(y) \qquad (48)$$

and, similarly,

$$\mathbf{P}\{(v(t) > x) \cap (t \in B_k)\}$$

$$= \int_0^t \mathbf{P}\{(v(t-y) > x) \cap (t-y \in B_1)\} d[(D^L)^{k*} * (D_N^B)^{(k-1)*}](y), \qquad (49)$$

where $D^L(\cdot)$ and $D^B(\cdot)$ denote d.fs of buffer loading period and busy cycle duration, respectively. Introducing Laplace transforms into equations (48)–(49), by virtue of (47), we get the conclusion (46). ∎

6 Numerical Example

In this section we present numerical examples in which the behavior of time-dependent queueing delay distribution is shown for two examples of packet traffic. In numerical analysis we use for the main representation (46) from Theorem 3 the algorithm of numerical Laplace transform inversion described in details in [1]. The algorithm is based on the Bromwich integral which is used for finding the value of the function $q(\cdot)$ at fixed $t > 0$ from its transform $\widehat{q}(\cdot)$, namely

$$q(t) = \frac{1}{2\pi i} \int_{\epsilon-i\infty}^{\epsilon+i\infty} e^{st} \widehat{q}(s) ds, \qquad (50)$$

where $\epsilon \in \mathbf{R}$ is located on the right to all singularities of $\widehat{q}(\cdot)$. From (50) we obtain the following approximation $q_\Delta(t)$ of $q(t)$:

$$q_\Delta(t) = \frac{\Delta e^{\epsilon t}}{2\pi} \widehat{q}(\epsilon) + \frac{\Delta e^{\epsilon t}}{\pi} \sum_{k=1}^{\infty} \mathrm{Re}[e^{ik\Delta t} \widehat{q}(\epsilon + ik\Delta t)]. \qquad (51)$$

Substituting $\Delta = \frac{\pi}{Lt}$ and $\epsilon = \frac{A}{2Lt}$, we get the series representation

$$q_\Delta(t) = q_{A,L}(t) = \sum_{k=0}^{\infty} (-1)^k u_k(t), \qquad (52)$$

where

$$u_k(t) = \frac{e^{A/2L}}{2Lt}\omega_k(t), \quad k \geq 0, \tag{53}$$

$$\omega_0(t) = \widehat{q}\left(\frac{A}{2Lt}\right) + 2\sum_{j=1}^{L} \mathrm{Re}\left[\widehat{q}\left(\frac{A}{2Lt} + \frac{ij\pi}{Lt}\right)e^{ij\pi/L}\right] \tag{54}$$

and

$$\omega_k(t) = 2\sum_{j=1}^{L} \mathrm{Re}\left[\widehat{q}\left(\frac{A}{2Lt} + \frac{ij\pi}{Lt} + \frac{ik\pi}{t}\right)e^{ij\pi/L}\right], \quad k \geq 1. \tag{55}$$

Applying the following Euler formula:

$$\sum_{k=0}^{\infty}(-1)^k c_k \approx \sum_{k=0}^{m}\binom{m}{k}\frac{1}{2^m}\sum_{j=0}^{n+k}(-1)^j c_j, \tag{56}$$

where m is usually of order of dozen and n - of order of several dozen (see [1]), we derive the final formula in the form

$$q(t) \approx \sum_{k=0}^{m}\binom{m}{k}\frac{1}{2^m}\sum_{j=0}^{n+k}(-1)^j u_j(t), \tag{57}$$

where typical values of parameters are following (see [1]): $m = 38, n = 11, A = 19$ and $L = 1$. As it was pointed out in [1], to evaluate the estimation error in (57), it is recommended to execute the calculation twice, changing one of the parameters m, n, A or L by one. The difference between the obtained results, e.g. for $n = 11$ and $n = 12$, gives a good evaluation of the error.

Let us take into consideration the stream of packets of average sizes 100 B arriving at the WSN node with the "queued" wake up energy saving mechanism, and are being transmitted with speed 300 kb/s. Fix $K = 6$ as the buffer size and analyze separately two different arrival rates: 180 kb/s ($\rho = 0.6$) and 300 kb/s ($\rho = 1$). Assume that interarrival times of entering packets have hyperexponential distribution with the following probability density function:

$$dF(t) = \left(pe^{-\lambda_1 t} + (1-p)e^{-\lambda_2 t}\right)dt, \quad t > 0.$$

Suppose that $p = 0.5$ and $\lambda_1 = 2\lambda_2$. Then for the arrival rate 180 kb/s we obtain $\lambda_1 = 336.5$ and $\lambda_2 = 168.75$ packets/sec. Similarly, for the intensity 300 kb/s we get $\lambda_1 = 562.5$ and $\lambda_2 = 281.25$ packets/sec. For the exponentially distributed processing time we obtain $\mu = 375$ packets/sec. In consequence, mean interarrival times are 4.444 ms for $\rho = 0.6$ and 2.667 ms for $\rho = 1$, and mean processing time equals 2.667 ms.

In Figures 1 and 2 probabilities $\mathbf{P}\{v(t) > 0.01\}$ for $K = 2, 3$ and 4 are presented for the cases of $\rho = 0.6$ and $\rho = 1$, respectively.

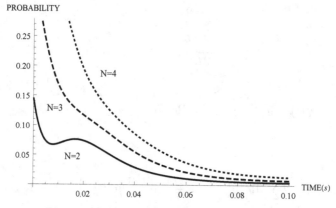

Fig. 1. $\mathbf{P}\{v(t) > 0.01\}$ for $N = 2, 3, 4$ and $\rho = 0.6$

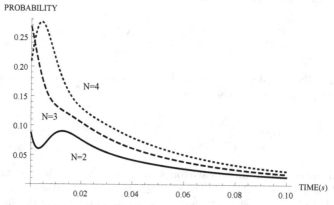

Fig. 2. $\mathbf{P}\{v(t) > 0.01\}$ for $N = 2, 3, 4$ and $\rho = 1$

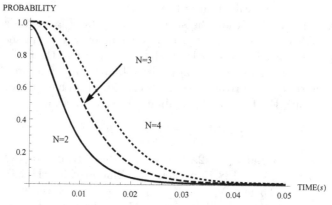

Fig. 3. $\mathbf{P}\{v(t) > 0.0001\}$ for $N = 2, 3, 4$ and $\rho = 1$

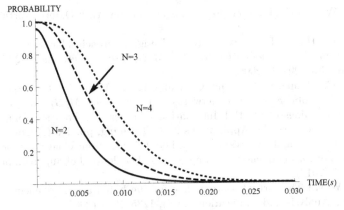

Fig. 4. $\mathbf{P}\{v(t) > 0.0001\}$ for $N = 2, 3, 4$ and $\rho = 1$

Similarly, in Figures 3 and 4 we present the time-dependent behavior of the probability $\mathbf{P}\{v(t) > 0.0001\}$. As one can note, the considered conditional probabilities are the smallest for the threshold $N = 2$ and the greatest for $N = 4$ and with the passage of time, all characteristics are close to each other. Indeed, just after the opening of the system, the probability of long delay will be the lowest for the lowest threshold of accumulation, while in the long-term time horizon the value N of the threshold becomes less and less essential.

7 Conclusion

In the paper the explicit representation for the LT of time-dependent queueing delay d.f. in the finite-buffer $GI/M/1/K$-type queueing model with "queued" wake up of the server (N-policy) was obtained, using the approach based on embedded Markov chain paradigm, renewal theory and linear algebra results. The considered system can be used as a good theoretical model in performance evaluation of WSN with the mechanism of energy saving based on "queued" wake up of sensors. The algebraic form of analytical results allows for their numerical treatment. A network-motivated numerical example was attached.

Acknowledgment. The project was financed with subsidies from the National Science Centre in Poland, granted by virtue of the decision number DEC-2012/07/B/ST6/01201.

References

1. Abate, J., Choudhury, G.L., Whitt, W.: An introduction to numerical transform inversion and its application to probability models. In: Grassmann, W. (ed.) Computational Probability, pp. 257–323. Kluwer, Boston (2000)

2. Cohen, J.W.: The single server queue. Amsterdam-New York-Oxford, North-Holland (1982)
3. Jiang, F.-C., Huang, D.-C., Wang, K.-H.: Design approaches for optimizing power consumption of sensor node with N-policy $M/G/1$. In: Proceedings of QTNA 2009, Singapore, July 29–31 (2009)
4. Jiang, F.-C., Huang, D.-C., Tang, C., Wang, K.-H.: Mitigation techniques for the energy hole problem in sensor networks using N-policy $M/G/1$ queueing models. In: Proceedings of the IET International Conference: Frontier Computinging Theory, Technologies and Applications 2010, Taichung, pp. 4–6 (August 2010)
5. Jiang, F.-C., Huang, D.-C., Tang, C.-T., Leu, F.Y.: Lifetime elongation for wireless sensor network using queue-based approaches. The Journal of Supercomputing 59, 1312–1335 (2012)
6. Kempa, W.M.: The virtual waiting time for the batch arrival queueing systems. Stochastic Analysis and Applications 22(5), 1235–1255 (2004)
7. Kempa, W.M.: The transient analysis of the queue-length distribution in the batch arrival system with N-policy, multiple vacations and setup times. In: AIP Conference Proceedings, vol. 1293, pp. 235–242 (2010)
8. Kempa, W.M.: Departure process in finite-buffer queue with batch arrivals. In: Al-Begain, K., Balsamo, S., Fiems, D., Marin, A. (eds.) ASMTA 2011. LNCS, vol. 6751, pp. 1–13. Springer, Heidelberg (2011)
9. Kempa, W.M.: On transient queue-size distribution in the batch arrival system with the N-policy and setup times. Mathematical Communications 17, 285–302 (2012)
10. Kempa, W.M.: The virtual waiting time in a finite-buffer queue with a single vacation policy. In: Al-Begain, K., Fiems, D., Vincent, J.-M. (eds.) ASMTA 2012. LNCS, vol. 7314, pp. 47–60. Springer, Heidelberg (2012)
11. Korolyuk, V.S.: Boundary-value problems for compound Poisson processes. Naukova Dumka, Kiev (1975) (in Russian)
12. Korolyuk, V.S., Bratiichuk, N.S., Pirdzhanov, B.: Boundary-value problems for random walks. Ylym, Ashkhabad (1987) (in Russian)
13. Lee, H.W., Lee, S.S., Park, J.O., Chae, K.-C.: Analysis of the $M^X/G/1$ queue with N-policy and multiple vacations. Journal of Applied Probability 31(2), 476–496 (1994)
14. Mancuso, V., Alouf, S.: Analysis of power saving with continuous connectivity. Computer Networks 56, 2481–2493 (2012)

The Impact of Class Clustering on a System with a Global FCFS Service Discipline

Willem Mélange, Joris Walraevens, Dieter Claeys,
Bart Steyaert, and Herwig Bruneel

Department of Telecommunications and Information Processing,
Ghent University - UGent
{wmelange,jw,dc,bs,hb}@telin.Ugent.be
http://telin.ugent.be/

Abstract. This paper considers a continuous-time queueing model with two types (classes) of customers each having their own dedicated server with exponential service times. The system adopts a "global FCFS" service discipline, i.e., all arriving customers are accommodated in one single FCFS queue, regardless of their types. "Class clustering", i.e., the fact that customers of any given type may (or may not) have a tendency to "arrive back-to-back", is a concept that we believe is often neglected in literature. As it is clear that customers of different types hinder each other more as they tend to arrive in the system more clustered according to class, the major aim of this paper is to estimate the impact of the degree of class clustering on the system performance. In this paper both classes of customers have an own "cluster parameter". The motivation of our work are systems where this kind of blocking is encountered, such as input-queueing network switches, security checkpoints or road splits.

Keywords: queueing, blocking, global FCFS, Markov, non-workconserving, class clustering.

1 Introduction

In general, queueing phenomena occur when some kind of customers, desiring to receive some kind of service, compete for the use of a service facility (containing one or multiple servers) able to deliver the required service. Most queueing models assume that a service facility delivers exactly one type of service and that all customers requiring this type of service are accommodated in one common queue. If more than one service is needed, multiple different service facilities can be provided, i.e., one service facility for each type of service, and individual separate queues are formed in front of these service facilities. In all such models, customers are only hindered by customers that require exactly the same kind of service, i.e., that compete for the same resources.

In some applications, it may not be physically feasible or desirable to provide separate queues for each type of service that customers may require, and it may be necessary or desirable to accommodate different types of customers

B. Sericola, M. Telek, and G. Horváth (Eds.): ASMTA 2014, LNCS 8499, pp. 125–139, 2014.

(i.e., customers requiring different types of service) in the same queue. In such cases, customers of one type (i.e., requiring a given type of service) may also be hindered by customers of other types. For instance, if a road or a highway is split in two or more subroads leading to different destinations, cars on that road heading for destination A may be hindered or even blocked by cars heading for destination B, even when the subroad leading to destination A is free, simply because cars that go to B are in front of them. In other words, there is a first-come-first-served (FCFS) order on the main road. This blocking also takes place in weaving sections on highways albeit to a lesser extent [1,2]. We refer to [3,4] for a general overview and validation of modelling traffic flows with queueing models. Similarly, in switching nodes of telecommunication networks, information packets with a given destination of node A may have to wait for the transmission of packets destined to node B that arrived earlier, even when the link to node A is free, if the arriving packets are accommodated in so-called input queues according to the source from which they originate (the well-known HOL-blocking effect, see [5,6,7,8,9]). Analogously, at a security checkpoint (e.g., at an international airport or train station) people are usually body-searched by someone of the same gender. As a result, when a group of friends of the same gender arrive, the people of the opposite gender behind them may have to wait until the whole group has been checked, even when the other security person is available, at least when it is not allowed to overtake at the security checkpoint (which is often the case for security reasons). In general, these applications can be modelled by a queueing system with different types of traffic, servers which are dedicated to these different classes, and a FCFS scheduling in the shared queue. Therefore, customers of one type can be blocked by customers of the other type that are waiting in front of them, even when their server is available. We will refer to this scheduling as "global FCFS" in the remainder. A lot of work has already been done on multi-type queues ([10,11,12]). However, the novelty of this paper lies in the global FCFS service discipline (and thus no longer the single server case).

In [13], we already got some insight in the impact of this kind of phenomenon on the performance of the involved systems. In this paper we want to extend our model to introduce class clustering, i.e., the fact that customers of any given type may (or may not) have a tendency to "arrive back-to-back". Class clustering is a concept that is often neglected in literature to keep the model as simple as possible, but in this paper we want to demonstrate that it is not always possible to ignore this concept and this is especially true for our system. It is already intuitively clear that when the customers arrive with alternating types, less blocking will occur than when the types alternate only very rarely.

In [14], we already briefly studied the concept of class clustering with a very simple "cluster model" characterized by one simple "cluster parameter" indicating the probability the next customer in the arrival stream has the same type as the previous customer. Consequently, in [14], we could only study cases where both classes of customers were equiprobable and thus both types of customers accounted for half of the total load of the system. In this paper, we got rid of this restriction as both classes no longer have the same cluster parameter. In this paper, it is also possible that one class has a big tendency to arrive back-to-back, while the other class has the complete opposite tendency.

The structure of the rest of the paper is as follows: in section 2, we start by giving a brief description of the mathematical model. In section 3, we first discuss the stability condition of our system. It is clear that only when the stability condition is met, our analysis is justified. In section 4, the distribution of the system occupancy, i.e., the number of customers in the system, is determined and some related performance measures are calculated. Section 5 is devoted to a discussion of the results derived in previous sections and some numerical examples are provided. Some conclusions and directions for future work are given in section 6.

2   Mathematical Model

We consider a continuous-time queueing model with infinite waiting room. Two servers have exponential service times; server 1 has a service rate of μ_1 and server 2 a service rate of μ_2. The servers are dedicated to a given class of customers. Server 1 only serves customers of one type (say type 1) and server 2 serves customers of the other type (type 2). The customers are served in their order of arrival, regardless of the class they belong to (global FCFS).

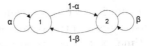

Fig. 1. 2-state Markov chain to determine the type of an arriving customer

The customers enter the system according to a Poisson arrival process with arrival rate λ. The type of the arriving customer is determined by a two-state Markov chain (see Fig. 1). If the previous customer is of type 1, then the customer is of type 1 with probability α and of type 2 with probability $(1 - \alpha)$. If the previous customer is of type 2, then the current customer is of type 1 with probability $(1 - \beta)$ and of type 2 with probability β. Notice here already that we can transform α and β in two other parameters σ and K that have a more intuitive meaning. The transformations from (α,β) to (σ,K) are

$$\sigma = \frac{1 - \beta}{2 - \alpha - \beta}, \tag{1}$$

$$K = \frac{1}{2 - \alpha - \beta} \tag{2}$$

and from (σ,K) to (α,β)

$$\alpha = 1 - \frac{1-\sigma}{K}, \tag{3}$$

$$\beta = 1 - \frac{\sigma}{K}. \tag{4}$$

The intuitive meaning behind the parameter σ is that it represents the relative frequency of the type 1 customers, i.e., the fraction of customers that are of type 1 (2) is σ (1 − σ respectively). The parameter K on the other hand gives a clear indication about the correlation. The parameter is directly proportional to one minus the mean number of customers of the same type that arrive back-to-back. More specifically, we have

$$E\left[\text{number of customers of type 1 arriving back-to-back}\right] = \tfrac{1}{1-\alpha} = \frac{K}{1-\sigma}, \tag{5}$$

$$E\left[\text{number of customers of type 2 arriving back-to-back}\right] = \tfrac{1}{1-\beta} = \frac{K}{\sigma}, \tag{6}$$

where $E\left[\cdots\right]$ represents the expected value of the quantity between brackets. Notice here that when K equals 1, the types of consecutive customers in the arrival stream are uncorrelated.

3 Stability Condition

When deriving the stability condition, we can presume that the system is constantly provided with new customers and the system will therefore be filled with at least 2 customers all the time. Note that we are only interested in the class of the customers in the set of leading customers, i.e., the first 2 customers of the system (possibly being served). These observations lead to the 4-state Markov chain depicted in Fig. 2. In state (m, t), m customers are of type 2 (and thus $2-m$ are of type 1) and the last customer in the set of leading customers has type

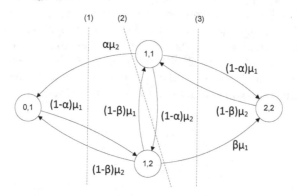

Fig. 2. 4-state Markov chain to determine the stability condition

t. Notice that we do not have states $(0, 2)$ and $(2, 1)$ since the "last" customer cannot be of type 2 (1) if all leading customers are of type 1 (2).

If we define $p(m, t)$ as the steady-state probably to be in state (m, t), then we end up with the following balance equations (corresponding to the dotted lines (1) to (3) in Fig. 2):

$$(1 - \alpha)\mu_1 p(0, 1) = \alpha\mu_2 p(1, 1) + (1 - \beta)\mu_2 p(1, 2), \tag{7}$$

$$\mu_2 p(1, 1) = \mu_1 p(1, 2), \tag{8}$$

$$(1 - \beta)\mu_2 p(2, 2) = (1 - \alpha)\mu_1 p(1, 1) + \beta\mu_1 p(1, 2). \tag{9}$$

These balance equations combined with the normalization condition

$$\sum_{m=0}^{2} (p(m, 1) + p(m, 2)) = 1, \tag{10}$$

where $p(0, 2) = p(2, 1) = 0$ by definition, yields

$$p(0, 1) = \frac{\mu_2^2(1 - \beta)(\alpha\mu_1 + (1 - \beta)\mu_2)}{(1 - \alpha)^2\mu_1^3 + (1 - \alpha)\mu_1^2\mu_2 + (1 - \beta)\mu_1\mu_2^2 + (1 - \beta)\mu_2^3}, \tag{11}$$

$$p(1, 1) = \frac{\mu_1^2\mu_2(1 - \beta)(1 - \alpha)}{(1 - \alpha)^2\mu_1^3 + (1 - \alpha)\mu_1^2\mu_2 + (1 - \beta)\mu_1\mu_2^2 + (1 - \beta)\mu_2^3}, \tag{12}$$

$$p(1, 2) = \frac{\mu_1\mu_2^2(1 - \beta)(1 - \alpha)}{(1 - \alpha)^2\mu_1^3 + (1 - \alpha)\mu_1^2\mu_2 + (1 - \beta)\mu_1\mu_2^2 + (1 - \beta)\mu_2^3}, \tag{13}$$

$$p(2, 2) = \frac{\mu_1^2(1 - \alpha)((1 - \alpha)\mu_1 + \beta\mu_2)}{(1 - \alpha)^2\mu_1^3 + (1 - \alpha)\mu_1^2\mu_2 + (1 - \beta)\mu_1\mu_2^2 + (1 - \beta)\mu_2^3}. \tag{14}$$

Having obtained the $p(m, t)$'s, we can now move on to the stability condition. Therefore, we postulate that the average amount of work per unit time that enters the system (ρ) is smaller than the average amount of work the system can execute per unit time, i.e., the average amount of work the system would execute per unit time when it would be constantly provided with new customers. Here, the system is able to execute 2 units of work per unit of time when both servers are able to work (when the system is in the state (1,1) or (1,2)). The system is able to execute 1 unit of work per unit of time when only one server is able to work (when the system is in state (0,1) or (2,2)). The stability condition is thus

$$\rho < p(0, 1) + 2(p(1, 1) + p(1, 2)) + p(2, 2), \tag{15}$$

or after using expressions (11) to (14)

$$\rho < \frac{\left(1 + \frac{(1-\alpha)\mu_1}{(1-\beta)\mu_2}\right)\left((1 - \alpha)\frac{\mu_1}{\mu_2} + (1 - \beta)\frac{\mu_2}{\mu_1} + 1\right)}{\frac{(1-\alpha)\mu_1}{(1-\beta)\mu_2}\left((1 - \alpha)\frac{\mu_1}{\mu_2} + 1\right) + (1 - \beta)\frac{\mu_2}{\mu_1} + 1}, \tag{16}$$

where ρ (average amount of work that enters the system) is defined as follows

$$\rho = \rho_1 + \rho_2 \triangleq \frac{\sigma\lambda}{\mu_1} + \frac{(1 - \sigma)\lambda}{\mu_2}. \tag{17}$$

To make the numerical results in section 5 more intuitive we use the transformations from equations (3) and (4). Here, we also already see that not the exact values of μ_1 and μ_2 are of importance but only the ratio. The stability condition becomes

$$\rho < \frac{\left(1 + \frac{1-\sigma}{\sigma}c\right)\left(\frac{1-\sigma}{K}c + \frac{\sigma}{K}\frac{1}{c} + 1\right)}{\frac{1-\sigma}{\sigma}c\left(\frac{1-\sigma}{K}c + 1\right) + \frac{\sigma}{K}\frac{1}{c} + 1}, \tag{18}$$

where

$$c = \frac{\mu_1}{\mu_2}. \tag{19}$$

4   System Occupancy

4.1   System State Diagram and Balance Equations

The system can be described by a continuous-time Markov chain where the state of the system is described by the triple (n, m, t). Here, n represents the number of customers in the system, m represents the number of customers of type 2 in the set of leading customers and t represents the type of the last customer in this set of customers (1 or 2). Notice that we do not have states $(n, 0, 2)$ and $(n, 2, 1)$ for $n > 1$ since the "last" customer cannot be of type 2 (1) if all leading customers are of type 1 (2). Notice, we do have states $(1, 0, 2)$ and $(1, 1, 1)$. It is possible that the last customer has already left the system and thus has overtaken the customer still in the system (by having a shorter service time). The remaining customer is not necessarily the last customer. State $(0, t)$ represents the empty system where the last customer that arrived is of type t. This is thus a Quasi-birth-and-death (QBD) process (see also [15]) with four phases and the levels are represented by the number of customers in the system.

If we define $p(n, m, t)$ as the steady-state probability to be in state (n, m, t) (and $p(0, t)$ to be in state $(0, t)$), we end up with the following balance and boundary equations (observe transitions to and from states (1)-(14) in Fig. 3):

$$\lambda p(0, 1) = \mu_1 p(1, 0, 1) + \mu_2 p(1, 1, 1) \tag{20}$$

$$\lambda p(0, 2) = \mu_1 p(1, 0, 2) + \mu_2 p(1, 1, 2) \tag{21}$$

$$(\lambda + \mu_1)p(1, 0, 1) = \mu_1 p(2, 0, 1) + \mu_2 p(2, 1, 1)$$
$$+ \lambda(\alpha p(0, 1) + (1 - \beta)p(0, 2)) \tag{22}$$

$$(\lambda + \mu_1)p(1, 0, 2) = \mu_2 p(2, 1, 2) \tag{23}$$

$$(\lambda + \mu_2)p(1, 1, 1) = \mu_1 p(2, 1, 1) \tag{24}$$

$$(\lambda + \mu_2)p(1, 1, 2) = \mu_1 p(2, 1, 2) + \mu_2 p(2, 2, 2)$$
$$+ \lambda((1 - \alpha)p(0, 1) + \beta p(0, 2)) \tag{25}$$

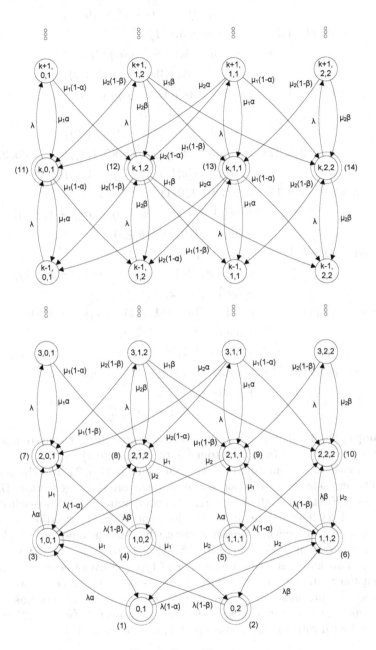

Fig. 3. State Diagram

$$(\lambda + \mu_1)p(2,0,1) = \mu_1\alpha p(3,0,1) + \mu_2((1-\beta)p(3,1,2) + \alpha p(3,1,1))$$
$$+ \lambda(\alpha p(1,0,1) + (1-\beta)p(1,0,2)) \tag{26}$$
$$(\lambda + \mu_1 + \mu_2)p(2,1,2) = \mu_1(1-\alpha)p(3,0,1)$$
$$+ \mu_2(\beta p(3,1,2) + (1-\alpha)p(3,1,1))$$
$$+ \lambda((1-\alpha)p(1,0,1) + \beta p(1,0,2)) \tag{27}$$
$$(\lambda + \mu_1 + \mu_2)p(2,1,1) = \mu_1((1-\beta)p(3,1,2) + \alpha p(3,1,1))$$
$$+ \mu_2(1-\beta)p(3,2,2)$$
$$+ \lambda(\alpha p(1,1,1) + (1-\beta)p(1,1,2)) \tag{28}$$
$$(\lambda + \mu_2)p(2,2,2) = \mu_1(\beta p(3,1,2) + (1-\alpha)p(3,1,1)) + \mu_2\beta p(3,2,2)$$
$$+ \lambda((1-\alpha)p(1,1,1) + \beta p(1,1,2)) \tag{29}$$
$$(\lambda + \mu_1)p(k,0,1) = \mu_1\alpha p(k+1,0,1)$$
$$+ \mu_2((1-\beta)p(k+1,1,2) + \alpha p(k+1,1,1))$$
$$+ \lambda p(k-1,0,1) \quad, k \geq 3 \tag{30}$$
$$(\lambda + \mu_1 + \mu_2)p(k,1,2) = \mu_1(1-\alpha)p(k+1,0,1)$$
$$+ \mu_2(\beta p(k+1,1,2) + (1-\alpha)p(k+1,1,1))$$
$$+ \lambda p(k-1,1,2) \quad, k \geq 3 \tag{31}$$
$$(\lambda + \mu_1 + \mu_2)p(k,1,1) = \mu_1((1-\beta)p(k+1,1,2) + \alpha p(k+1,1,1))$$
$$+ \mu_2(1-\beta)p(k+1,2,2)$$
$$+ \lambda p(k-1,1,1) \quad, k \geq 3 \tag{32}$$
$$(\lambda + \mu_2)p(k,2,2) = \mu_1(\beta p(k+1,1,2) + (1-\alpha)p(k+1,1,1))$$
$$+ \mu_2\beta p(k+1,2,2)$$
$$+ \lambda p(k-1,2,2)) \quad, k \geq 3 \tag{33}$$

For example, the left-hand side of equation (31) represents the system leaving state $(k,1,2)$ with rate λ (a new customer enters the system), rate μ_1 (a customer of type 1 leaves the system) and rate μ_2 (a customer of type 2 leaves the system). The right-hand side of the equation is a bit more involved. We go to state $(k,1,2)$ in four cases. First, with rate λ (an arrival occurs) state $(k,1,2)$ is reached from state $(k-1,1,2)$. The arriving customer does not change the leading customers since at least 2 customers are already present when the customer arrives ($k \geq 3$). Secondly, the system goes from state $(k+1,0,1)$ to state $(k,1,2)$ with rate $\mu_1(1-\alpha)$. This happens when a customer of type 1 leaves the system and the "new" customer in the set of leading customers is of type 2 (with probability $1-\alpha$ since the previous "last" customer of the leading customers was of type 1). Analogously, the system can go with rate $\mu_2\beta$ from state $(k+1,1,2)$ to state $(k,1,2)$ and with rate $\mu_2\beta$ from state $(k+1,1,1)$ to state $(k,1,2)$.

4.2   Distribution and Moments of System Occupancy

To tackle this problem, generating functions are used. We first introduce the three following partial probability generating functions (pgf's)

$$P_0(z) \triangleq \sum_{k=2}^{\infty} p(k,0,1)z^k \ , \tag{34}$$

$$P_1(z) \triangleq \sum_{k=2}^{\infty} p(k,1,2)z^k \ , \tag{35}$$

$$P_2(z) \triangleq \sum_{k=2}^{\infty} p(k,1,1)z^k \ , \tag{36}$$

$$P_3(z) \triangleq \sum_{k=2}^{\infty} p(k,2,2)z^k. \tag{37}$$

Equations (30) to (33) are multiplied by z^k and summed over all $k \geqslant 3$. We find

$$(\lambda + \mu_1)(P_0(z) - z^2 p(2,0,1)) =$$
$$\frac{1}{z} \left[\mu_1 \alpha(P_0(z) - z^3 p(3,0,1) - z^2 p(2,0,1)) \right.$$
$$+ \mu_2((1-\beta)(P_1(z) - z^3 p(3,1,2) - z^2 p(2,1,2))$$
$$\left. + \alpha(P_2(z) - z^3 p(3,1,1) - z^2 p(2,1,1))) \right]$$
$$+ \lambda z P_0(z), \tag{38}$$

$$(\lambda + \mu_1 + \mu_2)(P_1(z) - z^2 p(2,1,2)) =$$
$$\frac{1}{z} \left[\mu_1(1-\alpha)(P_0(z) - z^3 p(3,0,1) - z^2 p(2,0,1)) \right.$$
$$+ \mu_2(\beta(P_1(z) - z^3 p(3,1,2) - z^2 p(2,1,2))$$
$$\left. + (1-\alpha)(P_2(z) - z^3 p(3,1,1) - z^2 p(2,1,1))) \right]$$
$$+ \lambda z P_1(z), \tag{39}$$

$$(\lambda + \mu_1 + \mu_2)(P_2(z) - z^2 p(2,1,1)) =$$
$$\frac{1}{z} \left[\mu_1((1-\beta)(P_1(z) - z^3 p(3,1,2) - z^2 p(2,1,2)) \right.$$
$$+ \alpha(P_2(z) - z^3 p(3,1,1) - z^2 p(2,1,1)))$$
$$\left. + \mu_2(1-\beta)(P_3(z) - z^3 p(3,2,2) - z^2 p(2,2,2)) \right]$$
$$+ \lambda z P_2(z), \tag{40}$$

$$(\lambda + \mu_2)(P_3(z) - z^2 p(2,2,2)) =$$
$$\frac{1}{z} \left[\mu_1(\beta(P_1(z) - z^3 p(3,1,2) - z^2 p(2,1,2)) \right.$$
$$+ (1-\alpha)(P_2(z) - z^3 p(3,1,1) - z^2 p(2,1,1)))$$
$$\left. + \mu_2\beta(P_3(z) - z^3 p(3,2,2) - z^2 p(2,2,2)) \right]$$
$$+ \lambda z P_3(z). \tag{41}$$

The common probability generating function of the (total) number of customers in the system is given by

$$P(z) = p(0) + z \cdot (p(1,0,1) + p(1,0,2) + p(1,1,1) + p(1,1,2))$$
$$+ P_0(z) + P_1(z) + P_2(z) + P_3(z). \tag{42}$$

If we solve equations (20) to (29) and (38) to (41) and insert the solutions in (42) this equation translates into a equation that only contains known quantities, except for four unknown probabilities in the numerator. These can be determined, in general, by invoking the well-known property that pgfs such as $P(z)$ are bounded inside the closed unit disk $\{z : |z| \leq 1\}$ of the complex z-plane, at least when the stability condition (15) of the queueing system is met (only in such a case our analysis was justified and $P(z)$ can be viewed as a legitimate pgf). Now, it can be shown by means of Rouché's theorem from complex analysis [16,17] that the denominator of (42) has exactly four zeroes inside the closed unit disk of the complex z-plane, one of which is equal to 1, as soon as the stability condition (15) is fulfilled. It is clear that these four zeroes should also be zeroes of the numerator of (42), as $P(z)$ must remain bounded in those points. We conclude with the calculation of the three remaining zeroes that are inside the closed unit disk, using numerical methods. For the zeroes inside the closed unit disk (z_0, z_1, z_2), the requirement that the numerator should vanish yields three linear equations for the four unknowns. For the zero $z = 1$, this condition is fulfilled regardless of the values of the unknowns, since the numerator of (42) contains a factor $z - 1$. A fourth linear equation can however be obtained by invoking the normalizing condition of the pgf $P(z)$, i.e., the condition $P(1) = 1$. In general, the four unknown probabilities can be found as the solutions of the four established linear equations. Substitution of the obtained values in (42) then leads to a fully determined expression for the steady-state pgf $P(z)$ of the system occupancy. $P(z)$ is a rational function (the quotient of two polynomials of degree 4).

From this result, various performance measures of practical importance can then be derived. For instance, the mean system occupancy can be found as $\overline{N} = P'(1)$. The mean system delay T can then be calculated using Little's Law [18].

$$T = \frac{\overline{N}}{\lambda}. \tag{43}$$

5   Discussion of Results and Numerical Examples

In this section, we discuss the results obtained in the previous sections, both from a qualitative perspective and by means of some numerical examples. Before discussing the results, we introduce two new parameters

$$\omega \triangleq \frac{\frac{\sigma}{\mu_1}}{\frac{\sigma}{\mu_1} + \frac{1-\sigma}{\mu_2}} = \frac{\rho_1}{\rho_1 + \rho_2}, \tag{44}$$

$$d \triangleq \frac{\mu_1}{\mu_1 + \mu_2} = \frac{c}{c + 1} \tag{45}$$

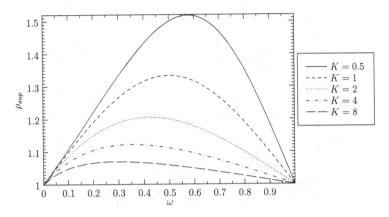

Fig. 4. ρ_{sup}, the least upper bound of the set of values ρ where the system is stable versus parameter ω, with $\mu_1 = 20$ and $\mu_2 = 1$ $(d = \frac{20}{21})$

These parameters will allow us to interpret the results more intuitively. The parameter ω represents the relative load of customers of type 1 and d represents the relative service rate of type 1.

Impact of Class Clustering (Parameter K). Fig. 4 shows ρ_{sup}, least upper bound of the set of values ρ where the system is stable versus parameter ω, with $\mu_1 = 20$ and $\mu_2 = 1$ $(d = \frac{20}{21})$. It is clear that the system where customers have the tendency to arrive back-to-back (higher K) performs "worse" than the system where customers have the tendency to arrive more alternatingly (smaller K). An observation that is also confirmed in Fig. 5 where the mean system occupancy versus parameter ρ with $\sigma = \frac{2}{5}$, $\mu_1 = 1$ and $\mu_2 = 20$ (and thus $\omega = \frac{40}{43}$ and $d = \frac{1}{21}$) is shown. Those figures illustrate that it is not possible to ignore the concept of class clustering for our system.

Impact of the Load and Service Rate Balance between Customers of Type 1 and Customers of Type 2 (Parameters ω and d). In Fig. 4, we notice that the maximum achievable throughput when $K = 1$ is obtained for a perfectly balanced system $(\omega = \frac{1}{2})$. However, the more class clustering (K increases), the more this maximum will move towards a situation where the fastest server gets a higher relative load. This might be a little contra-intuitive. In a system without blocking, the maximum achievable throughput is achieved when our system is perfectly balanced (both servers get a load of 1, or a total load of 2) irrespective of K. In the system with blocking, this maximum achievable throughput lies between 1 (the load one server can process) and 2 (the load two servers can process), as can also be seen from Fig. 4. The exact maximum is actually determined by maximizing the fraction of time both servers work simultaneously, i.e., when the two leading customers are of opposite type. This observation that is also confirmed by Fig. 6, which represents ρ_{sup}, the least

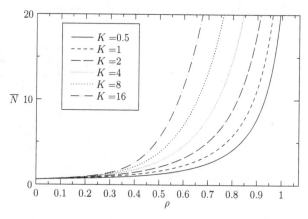

Fig. 5. Mean system occupancy versus parameter ρ, with $\sigma = \frac{2}{5}$, $\mu_1 = 1$ and $\mu_2 = 20$ ($\omega = \frac{40}{43}$ and $d = \frac{1}{21}$)

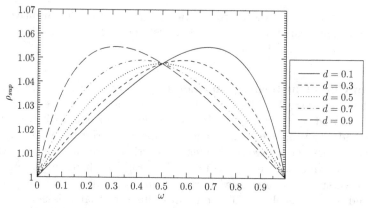

Fig. 6. ρ_{sup}, the least upper bound of the set of values ρ where the system is stable versus parameter ω, with $K = 10$

upper bound of the set of values ρ where the system is stable versus parameter ω, with $K = 10$. We notice here that it not always ideal to have a symmetric system (where the workload is equally balanced and both servers have the same service rate). Even more surprising is that when there is negative correlation in the types of consecutively arriving customers ($K < 1$), the slowest server should get a higher relative load.

In Fig. 7, the mean system time versus parameter ω, with $K = 5$, $\mu_1 = 1$ and $\mu_2 = 2$ ($d = \frac{1}{3}$) is shown. We see that the mean system time is rather independent of the load balance (ω), for small total load (ρ). The load balance (whether or not the slowest server is also working) becomes only of importance when the total load becomes too much for one server to handle (when ρ approaches 1). In Fig. 7, it is even better for our system to have only one type of customer (of the

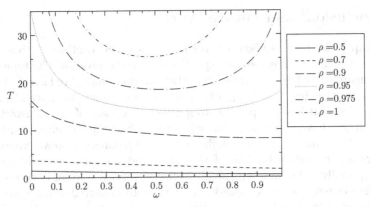

Fig. 7. Mean system time versus parameter ω, with $K = 5$, $\mu_1 = 1$ and $\mu_2 = 2$ $(d = \frac{1}{3})$

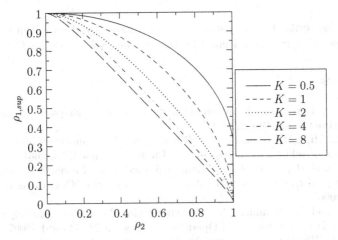

Fig. 8. $\rho_{1,sup}$, the least upper bound of the set of values ρ_1 where the system is stable versus parameter ρ_2, with $\sigma = \frac{1}{4}$

fastest server) for a small load. This, however, is mainly due to the fact that the service time is included in the system time.

Impact of Customers of One Type on Customers of the Other Type.
Fig. 8 represents $\rho_{1,sup}$, the least upper bound of the set of values ρ_1 where the system is stable versus parameter ρ_2, with $\sigma = \frac{1}{4}$. The more class clustering (higher K), the more the customers of different types have an influence on each other. This is intuitively also clear since the more class clustering (or more customers of the same type arriving consecutively), the more a customer of a different type will be blocked behind the group of customers of the same type.

6   Conclusions and Future Work

In this paper, we have studied a two-class, two-server, continuous-time queue with class-dedicated servers assuming a Poisson arrival process. The type of the arriving customer is determined by a two-state Markov chain. We have derived an expression for the steady-state pgf of the system occupancy. We have illustrated and quantified that it is not possible to ignore the concept of class clustering for our system. We have also shown that when we look at the stability condition or when the system has to handle a high load, a well balanced system (where both customers accommodate for half of the total load) performs not always best. This is especially the case when the difference in service rates of both servers is large. The system where the fastest server is more preferred, often gives more performant results. The bigger the difference between the service times of the servers, the more the fastest server should be preferred.

Acknowledgment. This research has been funded by the Interuniversity Attraction Poles Programme initiated by the Belgian Science Policy Office.

References

1. Ngoduy, D.: Derivation of continuum traffic model for weaving sections on freeways. Transportmetrica 2, 199–222 (2006)
2. Nishi, R., Miki, H., Tomoeda, A., Nishinari, K.: Achievement of alternative configurations of vehicles on multiple lanes. Physical Review E 79, 066119 (2009)
3. Van Woensel, T., Vandaele, N.: Empirical validation of a queueing approach to uninterrupted traffic flows. 4OR, A Quarterly Journal of Operations Research 4, 59–72 (2006)
4. Van Woensel, T., Vandaele, N.: Modeling traffic flows with queueing models: A review. Asia-Pacific Journal of Operational Research 24, 435–461 (2007)
5. Karol, M., Hluchyj, M., Morgan, S.: Input versus output queueing on a space-division packet switch. IEEE Transactions on Communications 35, 1347–1356 (1987)
6. Liew, S.: Performance of various input-buffered and output-buffered ATM switch design principles under bursty traffic: simulation study. IEEE Transactions on Communications 42, 1371–1379 (1994)
7. Laevens, K.: A processor-sharing model for input-buffered ATM-switches in a correlated traffic environment. Microprocessors and Microsystems 22, 589–596 (1999)
8. Stolyar, A.: MaxWeight scheduling in a generalized switch: state space collapse and workload minimization in heavy traffic. Annals of Applied Probability 14, 1–53 (2004)
9. Beekhuizen, P., Resing, J.: Performance analysis of small non-uniform packet switches. Performance Evaluation 66, 640–659 (2009)
10. He, Q.: Analysis of a continuous time sm[k]/ph[k]/1/fcfs queue: Age process, sojourn times, and queue lengths. Journal of Systems Science and Complexity 25(1), 133–155 (2012)
11. Houdt, B.V.: Analysis of the adaptive mmap[k]/ph[k]/1 queue: A multi-type queue with adaptive arrivals and general impatience. European Journal of Operational Research 220(3), 695–704 (2012)

12. Houdt, B.V.: A matrix geometric representation for the queue length distribution of multitype semi-markovian queues. Performance Evaluation 69(7-8), 299–314 (2012)
13. Mélange, W., Bruneel, H., Steyaert, B., Walraevens, J.: A two-class continuous-time queueing model with dedicated servers and global fcfs service discipline. In: Al-Begain, K., Balsamo, S., Fiems, D., Marin, A. (eds.) ASMTA 2011. LNCS, vol. 6751, pp. 14–27. Springer, Heidelberg (2011)
14. Mélange, W., Bruneel, H., Steyaert, B., Claeys, D., Walraevens, J.: A continuous-time queueing model with class clustering and global fcfs service discipline. Journal of Industrial and Management Optimization 10, 193–206 (2014)
15. Neuts, M.: Matrix-Geometric Solutions in Stochastic Models: An Algorithmic Approach. The John Hopkins University Press (1981)
16. Gonzáles, M.: Classical complex analysis. Marcel Dekker, New York (1992)
17. Bruneel, H., Kim, B.: Discrete-time models for communication systems including ATM. Kluwer Academic, Boston (1993)
18. Kleinrock, L.: Queueing Systems, Theory, vol. 1. Wiley-Interscience (1975)

Time-Parallel Simulation for Stochastic Automata Networks and Stochastic Process Algebra

Thu Ha Dao Thi, Jean-Michel Fourneau, and Franck Quessette

PRiSM, Univ. Versailles St Quentin, UMR CNRS 8144, Versailles France
{jmf,qst}@prism.uvsq.fr

Abstract. Time Parallel Simulation (TPS) is the construction of the time-slices of a sample-path on a set of parallel processors (see [11] chap. 6 and references therein). TPS has a potential to massive parallelism as the number of logical processes is only limited by the number of time intervals which is a direct consequence of the time granularity and the simulation length. Stochastic Automata Networks (SAN in the following) and some stochastic process algebra (like PEPA) allow the construction of extremely large Markov chains which are difficult to analyze due to their size. Here, we show how we can use TPS to solve efficiently some models based on SAN or PEPA. The approach uses some graph theoretical properties which can be checked easily on a SAN or a PEPA model. The quantitative results are obtained by a TPS based on linear recurrence equations of the daters with associative operators.

1 Introduction

Stochastic Automata Networks (SAN in the following) have been introduced by Plateau in a seminal paper [20]. They offer many interesting properties for the modeling of systems and their analysis. First, as a component based modeling paradigm, they allow to design complex interactions between sub-models. An automaton consists of states and transitions which represent the effects of events. These events are classified into two categories: local events and synchronizing events. A local event affects a single automaton and is modeled by one local transition. On the opposite, a synchronizing event modifies the state of more than one automaton. The key property is the representation of the transition rate matrix M of the Continuous Time Markov Chain (CTMC) associated with the SAN as a sum of tensor product of smaller matrices used to describe the behavior of the components and the ways they interact. Therefore, it is possible to design chains with extremely large state space.

Many algorithms have been developed to use this tensor representation of M to numerically solve the steady-state distribution of the chain (see for instance [5]). Further properties such as lumpability or existence of a product form solution can be checked on the automaton level [13,8,4], with a smaller complexity

B. Sericola, M. Telek, and G. Horváth (Eds.): ASMTA 2014, LNCS 8499, pp. 140–154, 2014.
© Springer International Publishing Switzerland 2014

than at the global level. SANs also have strong relations with Stochastic Process Algebra with exponential durations of transition. This is typically the case for PEPA (Performance Evaluation Process Algebra) introduced by Hillston [14]. Thus the approach we present here can be applied to PEPA models as well. Formally, the approach is based on a graph representation and this representation can be obtained in both frameworks. Here we present a new method to compute the asymptotic throughput of some SAN and PEPA models using time parallel simulation. The presentation put more emphasis on SANs to be simpler but it must be clear that the results apply to PEPA models.

Time Parallel Simulation (TPS in the following) is based on a decomposition of the time axis. One performs the simulations on time intervals in parallel on adjacent time intervals (see [11] chap. 6, and references therein). Afterwards the simulation results are combined to build the overall sample-path. TPS has a potential to massive parallelism [19] as the number of logical processes is only limited by the number of time intervals which is a direct consequence of the time granularity and the simulation length. But the final and initial states of adjacent time intervals do not necessarily coincide at interval boundaries, possibly resulting in incorrect state changes. The efficiency of TPS depends on our ability to guess the state of the system at the beginning of the simulation intervals or to efficiently correct the guessed states to finally obtain a consistent sample-path after a small number of trials. Several properties had already been studied: regeneration [17], efficient forecasting of some points of the sample-path [10], parallel prefix computation [12], a guessed initial state followed up by some fix-up computations when the first state has a weak influence on the sample-path [19]. Relaxing these assumptions, one may obtain an approximation of the results [15] or some bounds on the sample-path [6,9].

Here we consider some families of SANs and we show that we can obtain some linear recurrence equations using operators "max" and "+" on the daters (i.e. the time instants when transitions occur). Such a property has already been proved for some Stochastic Event Graph in [1]. It is also known, following [2] that, in that case, one can develop a parallel prefix technique to obtain a time parallel simulation of the recurrence equation and estimate the asymptotic throughput and the steady-state distribution.

The technical part of the paper is organized as follows. In the next section, we give a brief introduction of SANs, we introduce the Synchronized Product of Directed Cycles (SPDC) subset of SANs and we prove that these models exhibit a linear recurrence equation on daters based on associative operators. Due to this associativity, following [2], we derive in Section 3 a time parallel simulation approach based on the parallel prefix technique. Then we show how to obtain the throughput and how to compute stochastic bounds on the instant of transitions with a faster simulation. It is important to note that we never generate the whole state space and we study the dater (i.e. the times of transition) rather than the states.

2   Stochastic Automata Networks Associated with a (max,+) Evolution of Daters

We briefly introduce SANs. We refer to the literature [5,4,21] to obtain more details and results.

Definition 1. *A SAN is defined by*

1. *N automata denoted as $\mathcal{A}_n$ for $n = 1$ to N with finite state space.*
2. *The local transitions are defined by matrix L_n for automaton $\mathcal{A}_n$. L_n is an infinitesimal generator.*
3. *K synchronizations between automata. Synchronization i on Automaton $\mathcal{A}_n$ is described by stochastic matrix S_i^n. Synchronization i has rate δ_i.*
4. *A reachability function which describes which states of the Cartesian product of the automata state spaces are reachable.*
5. *A set of functions which describe how a transition rate in one automaton may depend on the state of the whole set of automata.*

Property 1. (see [21] for a proof). The continuous time Markov Chain associated to a SAN has a transition rate matrix M built with the tensor products and sums of the matrices used to describe the local transitions and the synchronizations:

$$M = \bigoplus_{n=1}^{N} L_n + \sum_{i=1}^{K} \delta_i \bigotimes_{n=1}^{N} S_i^n - \sum_{i=1}^{K} \delta_i \bigotimes_{n=1}^{N} N_i^n,$$

where $\bigoplus$ and $\bigotimes$ are generalized tensor sum and product defined in [21] and matrices N_i^n are normalization matrices associated with S_i^n (see the references for a more detailed discussion).

This representation of the transition rate matrix allows to consider extremely large Markov chains to study protocols or systems [3] as one only stores matrices L_n and S_i^n and matrix M is neither explicitly stored in memory. Structural properties can also be checked at the automaton level in a more efficient way. These properties allow to chose a solver well-suited for this type of models. It is typically the approach we have used here. We show how the Time Parallel Simulation based of Parallel Prefix computation of the daters can be used to compute in parallel a sample-path for some easy to identify SAN models.

2.1   PEPA

In PEPA, a system is viewed as a set of *components* which carry out *activities*. Each activity is characterized by an *action type* and a duration which is exponentially distributed. PEPA formalism provides a set of combinators which allows expressions to be built, defining the behavior of components, via the activities they engage in. One may refer to [14] for a more formal and detailed description of PEPA. We just give here the necessary information to understand the syntax of the example and to show how to transform the model into a labelled multigraph.

- Prefix: noted $(\alpha, r).P$, this combinator is the basic mechanism by which the behavior of the components are constructed. The component carries out activity (α, r) (where α is an action and r is its rate) and subsequently behaves as component P.
- Choice: noted $P_1 + P_2$, this combinator represents competition between components. The system may behave either as component P_1 or as P_2. All current activities of the components are enabled. The first activity to complete, determined by the race condition, distinguishes one of these components, the other is discarded.
- Cooperation: noted $P_1 \bowtie_L P_2$, it allows the synchronization of components P_1 and P_2 over the activities in the cooperation set L. Components may proceed independently with activities whose types do not belong to this set. A particular case of the cooperation is when $L = \emptyset$. In this case, components proceed with activities independently and are noted $P_1 \| P_2$.

Necessary (but not sufficient) conditions for the ergodicity of the Markov process in terms of the structure of the PEPA model have been identified and can be readily checked [14]. These conditions imply that the model must be a *cyclic* PEPA component. The model should be constructed as a cooperation of *sequential* components, i.e. components constructed using only prefix, choice and constants. This leads to formally define the syntax of PEPA expressions in terms of *model components P* and *sequential components S*:

$$P ::= A \mid P \bowtie_L P \mid P/L \qquad\qquad S ::= (\alpha, r).S \mid S + S \mid A_s$$

where A denotes a constant which is either a model or a sequential component and A_s denotes a constant which is a sequential component. In our graphical approach detailed in the next section, the model components will be defined by labelled multi-graphs. The labels will be the rates, the action names and the synchronization list. An example is given in the next section.

2.2 Synchronized Product of Directed Cycles

The approach is based on the directed graphs of the automata. Therefore we just have to translate the SAN or the PEPA model into a set of labelled directed graphs. The conditions to apply the method only depend on local conditions on the graphs. It is therefore easy to check them with a small complexity. We never build the global state space.

Typically at the first step of the model, we can transform the stochastic automata or the PEPA model of one component into a directed and labelled multi-graph . The nodes of the multigraph are the states of the component. Each edge represents a transition which carry only one label. As some transitions between states may occur due to several events or actions and therefore carry multiple labels, we use several edges, each of them carrying only one label. Therefore, we obtain a directed multigraph with simple labels while the usual representation of a SAN is a directed graph with multiple labels. This multigraph representation allows to describe in an easier way the deletion of arcs associated to a label.

Definition 2 (Subgraph). *Let $G = (V, E)$ be a labelled directed graph with vertices set V and set of directed edges with labels E, H is a subgraph of G if $V(H) \subset V(G)$ and $E(H) \subset E(G)$. Furthermore, if (x, y) is an arc from x to y with label l in H, it also exists in G with the same label.*

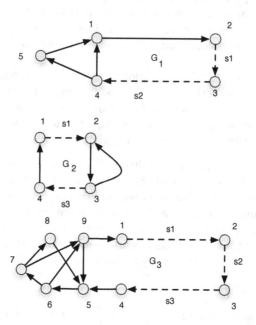

Fig. 1. A SPDC composed of three automata, the synchronization arcs are depicted as dotted lines and the local transitions by black lines. The rates are omitted to simplify the presentation.

Definition 3 (Minor). *Let $G = (V, E)$ be a labelled directed graph with vertices set V and set of directed edges with labels E, H is a minor of G if H can be obtained from G by deleting edges and vertices and contracting edges. An edge contraction consists in deleting the edge from the graph while simultaneously merging the two vertices previously connected by the edge. Of course the labels carried by the arcs are kept unchanged when the arcs are not deleted.*

Definition 4. *A SAN is a Synchronized Product of Directed Cycles (SPDC in the following) if the multi-graphs of the automata satisfy the following condition.*

A1 Let G_i the multigraph of Automaton A_i. We assume that G_i is strongly connected. The directed edges of G_i carry a transition rate when the transition is local or a synchronization label. A synchronization label is the name of the synchronization and a rate. The syntax for the rate description for a local transition and a synchronization are not the same for SAN and PEPA models but we do not detailed them here. The time description is given in Property 4.

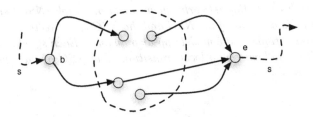

Fig. 2. An abstracted representation of a connected component of $\mathcal{A}_i$ after edge-extraction of the synchronizations. The edges inside the connected components are omitted.

A2 We delete in G_i the edges carrying a synchronization label. We do not delete the nodes even if their degree is 0. Let H_i be this graph. We decompose H_i into connected components. Let cc_i be the number of connected components in H_i. We assume that for each connected component (say component C_i^j for the j-th connected component of H_i), there exist two unique nodes b_i^j and e_i^j such that (the assumptions are illustrated in Fig. 2):

    1. b_i^j and e_i^j are distinct,

    2. for all nodes u in C_i^j, there exists a directed path from u to e_i^j,

    3. for all nodes v in C_i^j, there exists a directed path from b_i^j to v.

    4. in G_i the synchronization arcs between C_i^k and C_i^j begin in node e_i^k and finish at node b_i^j.

    Clearly, all the paths crossing component C_i^j begin in node b_i^j and finish at node e_i^j.

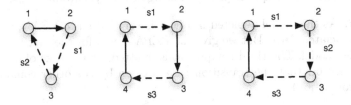

Fig. 3. The reduction of the SPDC detailed in assumption A3

A3 Let us consider G_i again. We replace each of the connected components by a simple directed graph with the two distinguished nodes b_i^j and e_i^j and a directed edge from b_i^j to e_i^j. The synchronization arcs are kept unchanged. Let F_i be this new graph. Graphs F_i will be called the reduced version of the SPDC. We have depicted in Fig. 3 the reduced version of the SPDC presented in Fig. 1. We assume that for all i, F_i is a directed cycle, eventually an isolated node.

A4 Now we consider the whole model (the network of stochastic automata or the complete specification in PEPA). We assume that there exists a labelled *directed graph called the network synchronization graph (NSG) such that for all i, each F_i is a minor of the NSG and the NSG is a directed cycle where all the synchronizations of the model appear only once. Each synchronization arc in the NSG is followed by a local transition. The NSG of the small example is depicted in Fig. 4.*

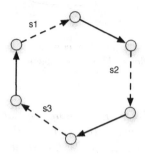

Fig. 4. The network synchronization graph

Note that assumption A4 implies that the graphs F_i are directed cycles (i.e. assumption A3) but we keep A3 to be clearer.

Note that, with this definition, the transitions between two distinct states of an automaton cannot be both local and synchronized. In a general SAN, the graph of an automaton is a multigraph, thus we may have a synchronization and a local transition between two states. This is strictly forbidden here.

Property 2. As the NSG is a directed graph, it defines, once the initial state of the model is fixed, a unique sequence of firing of the synchronizations.

Example 1. As the model depicted in Fig. 1 is very closed to a SAN, we do not give the associated SAN. But we give the PEPA model for the first two components of the model. The third component is omitted for the sake of conciseness (symbol $\top$ means that the transition rate is given by the other component of the synchronization):

$$P \quad ::= G_1.1 \underset{s1,s2}{\bowtie} G_3.1 \underset{s1,s3}{\bowtie} G_2.1$$

$$G_1.1 ::= (local1, \alpha_1)G_1.2$$

$$G_1.2 ::= (s1, 1/B_1)G_1.3$$

$$G_1.3 ::= (s2, 1/B_2)G_1.4$$

$$G_1.4 ::= (local2, \alpha_2)G_1.1 + (local3, \alpha_3)G_1.5$$

$$G_1.5 ::= (local4, \alpha_4)G_1.1$$

$$G_2.1 ::= (s1, \top)G_2.2$$

$$G_2.2 ::= (local5, \alpha_5)G_2.3$$

$$G_2.3 ::= (s1, 1/B_3)G_2.4$$

$$G_2.4 ::= (local6, \alpha_6)G2.1$$

The state space of a SPDC is included into the Cartesian product of the states of the automata. Thus, a state will be defined by $(x_1, \ldots, x_N)$ where x_n is the state of automaton $\mathcal{A}_n$. For the analysis, one must have a description of the reachable states space to correctly initialize the transient distribution. We now define globally consistent states. Starting the simulation in a globally consistent state ensures that the simulation will never deadlock.

Definition 5. *A synchronization is fireable if and only if all the automata involved in the synchronization are in the initial state of the synchronization.*

Note that a synchronization may synchronize only a subset of the automata and not necessarily all the automata. In the example of Fig. 1, $S1$ synchronizes the three automata while $S2$ synchronizes only two. From a global state, a transition may change only one component if it is a local transition, or it may change some (resp. all) of the components for a synchronization that acts on some (resp. all) automata.

Definition 6. *Let x_i be a state in G_i, we associate a path P_i to x_i as follows:*

1. *If x_i is in a connected component C_i^j, we replace x_i by j (i.e. the connected component index). Otherwise we keep x_j.*
2. *After step 1, we obtain a node in F_i. As F_i is a directed cycle, we find the last synchronization before the node (say s_a) and the next synchronization after the node (say s_b).*
3. *As F_i is a minor of the NSG, synchronizations s_a and s_b are also labels carried by some edges of the NSG. Let a the finishing node of s_a and b the starting node of s_b in the NSG.*
4. *P_i is a subgraph of the NSG consisting in the path from a to b.*

Definition 7. *A state $X = (x_1, \ldots, x_N)$ is globally consistent if and only if the intersection of paths P_i associated to state x_i and graph G_i is not empty. For instance node $(5, 1, 1)$ is consistent while node $(4, 2, 2)$ is not.*

Property 3. If X is a globally consistent state, then after any transitions, local or synchronized when they are fireable, the state reached is also globally consistent.

The proof is omitted for the sake of conciseness.

Definition 8. *We consider the daters of the sample path of automaton $\mathcal{A}_n$ in its reduced representation (with graph F_n). In the following, $t^n(u, k)$ will denote the time instant where reduced automaton $\mathcal{A}_n$ reaches state u for the k-th time.*

Property 4. By construction, all the transitions in G_i have an exponential duration. These delays are also assumed to be independent. However this is not mandatory for the structural properties we will prove now. These assumptions are inherited from the SAN (or PEPA) methodology.

Let us now consider the timing of the transitions in F_i (i.e. after transformation of the connected components into the arc b_i^j to e_i^j). We have two types of transitions:

- local: The local transition out of state u in Automaton F_i has a duration which is denoted by D_u^n. The durations follow a Phase type distribution. Indeed, the transition between b_i^j and e_i^j represents an absorbing Markov chain beginning in b_i^j and finishing at e_i^j. This is the definition of a PH distribution.
- synchronization: Denote by $On(i)$ the set of all automata involved in synchronization i. When all the automata in $On(i)$ are ready to trigger the transition, synchronization i takes place and its duration is denoted by B_i.

Like a Stochastic Event graph [1], a SPDC with a globally consistent initial state is associated with equations on the daters which have a linear representation with (max,+) operations when we model it at the aggregated level (i.e. when the connected components are replaced by transitions from b_i^j to e_i^j). Note that as F_i is a directed cycle, we define $u - 1$ as the node which is the origin of the arc finishing at node u.

Theorem 1. *If a SPDC is associated with a globally consistent initial state, then the equations on the daters for the aggregated model have a linear representation with (max,+) operations. More precisely, we have:*

- *If u is not the initial state of $\mathcal{A}_n$, then*
  - *if the arc arriving at u is local,*

$$t^n(u, k) = t^n(u - 1, k) + D_{u-1}^n$$

  - *if the arc arriving at u is a synchronization,*

$$t^n(u, k) = \max_{m \in On(r)} (t^m(d_r^m, k)) + B_r$$

- *If u is the initial state of $\mathcal{A}_n$, then:*
  - *if the arc arriving at u is local,*

$$t^n(u, k) = t^n(u - 1, k - 1) + D_{u-1}^n$$

  - *if the arc arriving at u is a synchronization,*

$$t^n(u, k) = \max_{m \in On(r)} (t^m(d_r^m, k - 1)) + B_r$$

where d_r^m is the initial state of synchronization r in F_m. As F_m is a minor of the NSG, synchronization r appears at most once and as $m \in On$, d_r^m is defined without ambiguity.

Proof. The proof is simply based on the fact that there is only one transition for each node. This transition may be local or a synchronization. The timing of local transitions simply requires that we add the delay. As usual, synchronizations implies that all the components are ready to synchronize, therefore we must wait $\max_{m \in On(r)}(t^m(d_r^m, k))$ before firing the transition and this transition has a duration equal to B^r.

Example 2. Let consider again the small example presented in Fig. 3. Assume that the initial state is $(2, 1, 4)$. We now give the equations on the daters.

$$\begin{bmatrix} t^1(1, k) = max(t^1(3, k), t^3(2, k)) + B_2 \\ t^1(2, k) = t^1(1, k-1) + D_1^1 \\ t^1(3, k) = max(t^1(2, k), t^2(1, k), t^3(1, k)) + B_1 \\ t^2(1, k) = t^2(4, k-1) + D_4^2 \\ t^2(2, k) = max(t^1(2, k), t^2(1, k), t^3(1, k)) + B_1 \\ t^2(3, k) = t^2(2, k) + D_2^2 \\ t^2(4, k) = max(t^2(3, k), t^3(3, k)) + B_3 \\ t^3(1, k) = t^3(4, k) + D_4^3 \\ t^3(2, k) = max(t^1(2, k), t^2(1, k), t^3(1, k)) + B_1 \\ t^3(3, k) = max(t^1(3, k), t^3(2, k)) + B_2 \\ t^3(4, k) = max(t^2(3, k-1), t^3(3, k-1)) + B_3 \end{bmatrix}$$

3   Quantitative Analysis of SPDC

We proceed in a hierarchical way. First we analyze the network at the F_i level (i.e. the connected components C_i^j have been replaces by an arc from b_i^j to e_i^j). Then we combine this analysis with the model of a connected components to obtain more detailed results. Note that we obtain the marginal distributions rather than the distribution on the global state space (i.e. the Cartesian product of the states of the component). Finally we show how we can simplify the computation using stochastic bounds to obtain stochastic guarantees with a faster simulation.

3.1   Time Parallel Simulation of Reduced SPDCs based on the Parallel Prefix Algorithm

We briefly present the Parallel Prefix approach for the Time Parallel Simulation. It is known for a long time that if the simulation consists in the computation of associative operators on the input sequences of random variables, one can use the parallel prefix method with parallel processors (see [11] chap. 6). Assume that we have a sequence y_i of M values and we want to compute all the values of $\odot_{i=1}^T y_i$ for all T. This is the computation of the sample-path of the simulation built from the input traces or the realization of the random variables (i.e. the sequence y_i). We assume that $\odot$ is an associative operator which represents the simulation models. We have at our disposal M synchronous processors. The algorithm will need $log_2(M)$ steps. At the end of these steps, processor J has computed $\odot_{i=1}^{J} y_i$. The processors proceed as follows:

- initially (i.e. at step 0), processor J receives y_J and stores it in $Y^{(0)}$.
- At step $M+1$, it has a current value $Y^{(M)}$. It receives from processor $J-2^M$ a value $X^{(M)}$, it computes $X^{(M)} \odot Y^{(M)}$ and stores it in $Y^{(M+1)}$.
- To complete this step, it sends $Y^{(M+1)}$ to processor $J+2^{M+1}$.

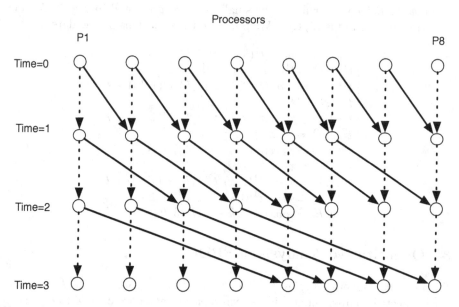

Fig. 5. Computations and exchanges of message for a Time parallel simulation on 8 processors

The global computation is depicted in Fig. 5 for 8 processors. After $log_2(M)$ steps, the sample-path (i.e. $\odot_{i=1}^{T} y_i$ for all T) has been computed on the M processors and it is locally stored as $Y^{(M)}$. The only property we need here is the possibility to build the sample-path of the simulation with an associative operator applied on the input sequence. Following the approach developed by Baccelli and his colleagues [1], we use a matrix and vector representation of the previous set of equations on $R \cup \{-\infty\}$ semi-ring with (max,+) operations where $-\infty$ is used as an absorbing element for the max operation.

Example 3. This is illustrated by two equations for the former example (we cannot give all of them because of the lack of space).

$$t^1(1, k+1) = max(t^1(1, k) + B_1 + B_2 + D_1^1, t^1(2, k) - \infty, t^1(3, k) - \infty,$$
$$t^2(1, k) - \infty, t^2(2, k) - \infty, t^2(3, k) + B_1 + B_2 + B_3 + D_4^3,$$
$$t^2(4, k) + B_1 + B_2 + D_4^2, t^3(1, k) - \infty, t^3(2, k) - \infty,$$
$$t^3(3, k) + B_1 + B_2 + B_3 + D_4^3, t^3(4, k) - \infty),$$

$$t^1(2, k+1) = max(t^1(1, k) + D_1^1, t^1(2, k) - \infty, t^1(3, k) - \infty,$$
$$t^2(1, k) - \infty, t^2(2k) - \infty, t^2(3k) - \infty, t^2(4, k) - \infty,$$
$$t^3(1, k) - \infty, t^3(2, k) - \infty, t^3(3, k) - \infty, t^3(4, k) - \infty).$$

The set of equations is a vector matrix product on the $R \cup \{-\infty\}$ semi-ring with (max,+) operations [1]. It is well known that this representation is associative. Thus, one can design the same parallel prefix techniques to get all the daters for the simulation. The operation $\odot$ mentioned earlier is this product of the vector of daters by the matrix of the model in the considered semi-ring.

3.2   Analysis of SPDCs

Let us now turn to the steady-state. Under the assumptions about the directions of the transitions, the system is ergodic [1]. And we obtain the cycle time and the throughput of the model as a function of the limit.

Property 5. The cycle time c is estimated as $t^n(u, k)/k$ for a large value for k. The value of the limit when k lead to infinity does not depend on the state u and the automaton n. However as the simulation is finite, the estimates may depend on n and u. The throughput γ of the system is $1/c$.

Property 6. The probability of state u in F_i when the transition out of u is local is estimated by :
$$\pi_F^i(u) = \gamma D_u^i.$$

Consider now the original problem with the initial description of the automata (i.e. graphs G_i for all $i = 1..N$). We want to estimate the probability that the state is x where x is a node of C_i^u. After conditioning on the component we obtain

$$\pi_G^i(x) = Pr(state = x \mid component = u)\pi_F^i(u).$$

Therefore we have to estimate the conditional probability that the state is x knowing the component is u. Remember that the component is an absorbing Markov chain. We propose to make a simulation of this Markov chain to obtain such an estimate. We can take advantage of the parallel processors to obtain several samples more efficiently. The algorithm proceeds as follows for each processor creating samples:

1. the processor begins in state b_i^j
2. it simulates the continuous time Markov chain counting the time spent at each state and the current time.
3. until it reaches e_i^j

The conditional probability is estimated by the ratio of the time spend at state x before being absorbed.

It is also possible to make some numerical analysis of the absorbing Markov chain to obtain the conditional probability but we prefer using a simulation approach to be consistent with the analysis at the reduced level which is also performed by a simulation approach.

3.3   Stochastic Bound and Simplification

Note that we have to compute k samples of the durations D_i^j which are PH distributions. Such a task may be extremely time consuming. Note that it can be done in parallel using the M processors before computing the sample path. Here we show how we can use the stochastic ordering methodology to simplify the computation of the sample-path while obtaining bounds on the daters. For more details on stochastic ordering, one should refer to [16,18].

Definition 9 (Stochastic Ordering). *Let x and y be R^m-valued random variables.*

We say that x is smaller than y with the strong stochastic ordering (st ordering for short) and we note $x \preceq_{st} y$ if $E[f(x)] \geq E[f(y)]$ for all non decreasing functions f from R^m to R whenever the expectations exist.

Similarly we say that x is smaller than y with the increasing convex stochastic ordering and we note $x \preceq_{icx} y$ if $E[f(x)] \geq E[f(y)]$ for all functions f from R^m to R which are increasing and convex (whenever the expectations exist).

Baccelli and his coauthors [1] had proved many results on the stochastic monotonicity property for systems of linear equations on $(max, +)$ semi-ring. This properties still hold as the equations we obtain are similar to the equations obtained in [1] for Stochastic Event Graphs, a subset of stochastic Petri nets. They are all due to the following property: the Max and Plus operators are non decreasing and convex functions. Note that if we compute bounds of the cycle time, we are not able anymore to perform the Monte Carlo simulation described in section 3.2 to analyze the states in a connected component.

Property 7 (Bounds for st-ordering). if we build L_i^j and U_i^j such that $L_i^j \preceq_{st} D_i^j \preceq_{st} U_i^j$, we clearly obtain a lower and un upper stochastic bound of the sequence $t^i()$.

For instance the truncation algorithm given in the following provides a sample of a distribution which is a st-lower bound of D_i^j. Clearly this algorithm has a smaller average complexity. For many algorithms to find stochastic bounds on Markov chains, one can refer to [7].

1. the processor begins in state b_i^j
2. it simulates the continuous time Markov chain counting the time spent at each state.
3. until it reaches e_i^j or the current time equal to $T0$.

Property 8 (Lower bound for icx-ordering). If we replace D_i^j by its expectation, we get an icx bound for the sequence $t^i(,)$.

Proof. Due to Jensen's inequality, we have $E[D_i^j] \preceq_{icx} D_i^j$. As the system of is modeled by a linear system of equation in (max,+) semi-ring and as the operators Max and "+" are convex and non decreasing, we get the result. Note that using $E[D_i^j]$ instead of generating a sample has a large influence of the complexity. The expectation is computed once, before the simulation starts. Using the constant during the simulation now has a complexity of $O(1)$ while the generation of a sample of the PH distribution may be much more time consuming.

4   Conclusion

We have proposed a method based of time parallel simulation to estimate the throughput of a system modeled by some family of SANs or PEPA specifications. We combine two approaches: a time parallel simulation for the analysis at the reduced level and a Monte Carlo simulation of the component for the results at the initial level. We are now investigating the implementation of the Time Parallel Simulation approach based on Parallel prefix on a GPU with a large number of cores.

Acknowledgement. This work was partially supported by a PEPS 2011 grant from CNRS (MONOSIMPA) and ANR MARMOTE (ANR-12-MONU-0019).

References

1. Baccelli, F., Cohen, G., Olsder, G.J., Quadrat, J.-P.: Synchronization and Linearity: An Algebra for Discrete Event Systems. Willey, New York (1992)
2. Baccelli, F., Gaujal, B., Furmento, N.: Parallel and Distributed Simulation of Free Choice Petri Nets. In: Parallel and Distributed Simulation, Lake Placid, USA (1995)
3. Ben-Othman, J., Mokdad, L., Cheikh, M.O., Sene, M.: Performance analysis of composite web services using stochastic automata networks over ip network. In: Proceedings of the 14th IEEE Symposium on Computers and Communications (ISCC 2009), Sousse, Tunisia, pp. 92–97. IEEE (2009)
4. Dao Thi, T.H., Fourneau, J.M.: Stochastic automata networks with master/slave synchronization: Product form and tensor. In: Al-Begain, K., Fiems, D., Horváth, G. (eds.) ASMTA 2009. LNCS, vol. 5513, pp. 279–293. Springer, Heidelberg (2009)
5. Fernandes, P., Plateau, B., Stewart, W.J.: Efficient descriptor-vector multiplications in Stochastic Automata Networks. J. ACM 45(3), 381–414 (1998)
6. Fourneau, J.-M., Kadi, I., Pekergin, N.: Improving time parallel simulation for monotone systems. In: Turner, S.J., Roberts, D., Cai, W., El-Saddik, A.B. (eds.) 13th IEEE/ACM International Symposium on Distributed Simulation and Real Time Applications, pp. 231–234. IEEE Computer Society Press, Singapore (2009)
7. Fourneau, J.-M., Pekergin, N.: An algorithmic approach to stochastic bounds. In: Calzarossa, M.C., Tucci, S. (eds.) Performance 2002. LNCS, vol. 2459, pp. 64–88. Springer, Heidelberg (2002)

8. Fourneau, J.-M., Plateau, B., Stewart, W.: An algebraic condition for product form in Stochastic Automata Networks without synchronizations. Performance Evaluation 85, 854–868 (2008)
9. Fourneau, J.-M., Quessette, F.: Monotonicity and efficient computation of bounds with time parallel simulation. In: Thomas, N. (ed.) EPEW 2011. LNCS, vol. 6977, pp. 57–71. Springer, Heidelberg (2011)
10. Fujimoto, R.M., Cooper, C.A., Nikolaidis, I.: Parallel simulation of statistical multiplexers. J. of Discrete Event Dynamic Systems 5, 115–140 (1994)
11. Fujimoto, R.M.: Parallel and Distributed Simulation Systems. Wiley Series on Parallel and Distributed Computing (2000)
12. Greenberg, A.G., Lubachevsky, B.D., Mitrani, I.: Algorithms for unboundedly parallel simulations. ACM Trans. Comput. Syst. 9(3), 201–221 (1991)
13. Gusak, O., Dayar, T., Fourneau, J.-M.: Lumpable continuous-time stochastic automata networks. European Journal of Operational Research 148(2), 436–451 (2003)
14. Hillston, J.: A compositional approach to Performance Modeling. PhD thesis. University of Edinburgh (1994)
15. Kiesling, T.: Using approximation with time-parallel simulation. Simulation 81, 255–266 (2005)
16. Kijima, M.: Markov Processes for Stochastic Modeling. Chapman & Hall, London (1997)
17. Lin, Y., Lazowska, E.: A time-division algorithm for parallel simulation. ACM Transactions on Modeling and Computer Simulation 1(1), 73–83 (1991)
18. Muller, A., Stoyan, D.: Comparison Methods for Stochastic Models and Risks. Wiley, New York (2002)
19. Nicol, D.M., Greenberg, A.G., Lubachevsky, B.D.: Massively parallel algorithms for trace-driven cache simulations. IEEE Trans. Parallel Distrib. Syst. 5(8), 849–859 (1994)
20. Plateau, B.: On the stochastic structure of parallelism and synchronization models for distributed algorithms. In: Proc. of the SIGMETRICS Conference, Texas, pp. 147–154 (1985)
21. Plateau, B., Fourneau, J.M., Lee, K.H.: PEPS: A package for solving complex Markov models of parallel systems. In: Proceedings of the 4th Int. Conf. on Modeling Techniques and Tools for Computer Performance Evaluation, Majorca, Spain, pp. 341–360 (1988)

Statistical Techniques for Validation of Simulation and Analytic Stochastic Models

Nicholas Nechval[1], Konstantin Nechval[2], Vadim Danovich[1],
and Natalija Ribakova[1]

[1] University of Latvia, EVF Research Institute, Statistics Department,
Raina Blvd 19, LV-1050 Riga, Latvia
Nicholas Nechval, Vadim Danovich, Natalija Ribakova
nechval@junik.lv
[2] Transport and Telecommunication Institute, Applied Mathematics Department,
Lomonosov Street 1, LV-1019 Riga, Latvia
konstan@tsi.lv

Abstract. In this paper, we consider the problem of statistical validation of multivariate stationary response simulation and analytic stochastic models of observed systems (say, transportation or service systems), which have p response variables. The problem is reduced to testing the equality of the mean vectors for two multivariate normal populations. Without assuming equality of the covariance matrices, it is referred to as the Behrens–Fisher problem. The main purpose of this paper is to bring to the attention of applied researchers the satisfactory tests that can be used for testing the equality of two normal mean vectors when the population covariance matrices are unknown and arbitrary. To illustrate the proposed statistical techniques, application examples are given.

Keywords: Simulation model, stochastic model, validation, hypothesis tests.

1 Introduction

Validation is a central aspect to the responsible application of models to scientific and managerial problems. The importance of validation to those who construct and use models is well recognized [1-7]. However, there is little consensus on what is the best way to proceed. This is at least in part due to the variety of models, model applications, and potential tests. The options are manifold, but the guidelines are few.

It is generally preferable to use some form of objective analysis to perform model validation. A common form of objective analysis for validating simulation models is statistical hypothesis testing. Statistical hypothesis testing, as distinguished from graphical or descriptive techniques, offers a framework that is particularly attractive for model validation. A test would compare a sample of observations taken from the target population against a sample of predictions taken from the model. The validity of the model is then assessed by examining the accuracy of model predictions. Such tests have numerous advantages: they provide an objective and quantifiable metric, they are amenable to reduction to a binary outcome, and therefore permit computation of error probability rates, and they accommodate sample-based uncertainty into the test result.

B. Sericola, M. Telek, and G. Horváth (Eds.): ASMTA 2014, LNCS 8499, pp. 155–169, 2014.
© Springer International Publishing Switzerland 2014

Not surprisingly, a number of statistical tools have been applied to validation problems. For example, Freese [8] introduced an accuracy test based on the standard χ^2 test. Ottosson and Håkanson [9] used R^2 and compared with so-called highest-possible R^2, which are predictions from common units (parallel time-compatible sets). Jans-Hammermeister and McGill [10] used an F-statistic-based lack of fit test. Landsberg et al. [11] used R^2 and relative mean bias. Bartelink [12] graphed field data and predictions with confidence intervals. Finally, Alewell and Manderscheid [13] used R^2 and normalized mean absolute error (NMAE).

The purpose of this paper is to give a methodology based on pivotal quantities for constructing statistical hypothesis tests, which are used for validating a simulation or analytic stochastic model of a real, observable system.

2   Preliminaries

Theorem 1 (Characterization of the multivariate normality). Let $\mathbf{Z}_i$, $i=1(1)n$, be n independent p-multivariate random variables ($n \geq p+2$) with common mean $\boldsymbol{\mu}$ and co-variance matrix (positive definite) $\mathbf{Q}$. Let W_r, $r=p+2, \ldots, n$, be defined by

$$W_r = [r-(p+1)](r-1)(pr)^{-1}(\mathbf{Z}_r - \overline{\mathbf{Z}}_{r-1})'\mathbf{S}_{r-1}^{-1}(\mathbf{Z}_r - \overline{\mathbf{Z}}_{r-1})$$

$$= \frac{r-(p+1)}{p}\left(\frac{|\mathbf{S}_r|}{|\mathbf{S}_{r-1}|} - 1 \right), \quad r = p+2, \ldots, n, \tag{1}$$

where

$$\overline{\mathbf{Z}}_{r-1} = \sum_{i=1}^{r-1} \mathbf{Z}_i / (r-1), \tag{2}$$

$$\mathbf{S}_{r-1} = \sum_{i=1}^{r-1} (\mathbf{Z}_i - \overline{\mathbf{Z}}_{r-1})(\mathbf{Z}_i - \overline{\mathbf{Z}}_{r-1})', \tag{3}$$

then the $\mathbf{Z}_i$ ($i=1, \ldots, n$) are $N_p(\boldsymbol{\mu}, \mathbf{Q})$ if and only if $W_{p+2}, \ldots, W_n$ are independently distributed according to the central F distribution with p and $1, 2, \ldots, n-(p+1)$ degrees of freedom, respectively.

Proof. The proof is similar to that of the theorems of characterization (Nechval et al. [14, 15]) and so it is omitted here.

Goodness-of-Fit Testing for the Multivariate Normality. The results of Theorem 1 can be used to obtain test for the hypothesis of the form H_0: $\mathbf{Z}_i$ follows $N_p(\boldsymbol{\mu}, \mathbf{Q})$ versus H_a: $\mathbf{Z}_i$ does not follow $N_p(\boldsymbol{\mu}, \mathbf{Q})$, $\forall i=1(1)n$. The general strategy is to apply the probability integral transforms of W_r, $\forall r=p+2(1)n$, to obtain a set of i.i.d. $U(0,1)$ random variables under H_0 [16]. Under H_a this set of random variables will, in general, not be i.i.d. $U(0,1)$. Any statistic, which measures a distance from uniformity in the transformed sample (say, a Kolmogorov-Smirnov statistic), can be used as a test statistic.

3   Statistical Validation of Simulation and Analytic Models

Suppose that we desire to validate a multivariate stationary response simulation model having p normally distributed response variables. In using statistical hypothesis testing to test the validity of a simulation model under a given experimental frame and for an acceptable range of accuracy consistent with the intended application of the model, we have the following hypotheses:

H_0: Model is valid for the acceptable range of accuracy under the experimental frame.

H_1: Model is invalid for the acceptable range of accuracy under the experimental frame.

Let μ_m and μ_s be the population means of the simulation model (subscript m) and system (subscript s) response variables, and let Q_m and Q_s be the model and system covariance matrices. We are interested in the testing problem

$$H_0: \mu_m = \mu_s \text{ vs. } H_1: \mu_m \neq \mu_s. \tag{4}$$

Furthermore, let us assume that for the purpose, for which the simulation model is intended, the validity of the model can be determined with respect to its mean response, and the acceptable range of accuracy can be expressed as the difference between the means of the model and the system responses and can be stated as

$$\left| \mu_m - \mu_s \right| \leq \Delta, \tag{5}$$

where Δ is a vector of the largest acceptable differences. Thus, we deal with the multivariate Behrens-Fisher problem, where it is assumed that μ_m and μ_s are unknown $p \times 1$ vectors and Q_m and Q_s are unknown $p \times p$ positive definite covariance matrices.

3.1   Validation via Testing Two Mean Vectors with Equal Covariance Matrices

Let the p-vectors $X_1, X_2, \ldots, X_{n_m}$ and $Y_1, Y_2, \ldots, Y_{n_s}$ be independent (potential) random samples from X (model) and Y (system), respectively, where $X \sim N_p(\mu_m, Q_m)$ and $Y \sim N_p(\mu_s, Q_s)$. The problem is to compare the mean vectors μ_m and μ_s.

Here we consider the case when the covariance matrices of the two normal distributions are equal (say $Q_m = Q_s = Q$). In this case, Hotelling's T^2 statistic is used for testing the hypothesis $H_0: \mu_m = \mu_s$ vs. $H_1: \mu_m \neq \mu_s$.

Define

$$\overline{X} = \frac{1}{n_m} \sum_{i=1}^{n_m} X_i, \quad \overline{Y} = \frac{1}{n_s} \sum_{j=1}^{n_s} Y_j, \tag{6}$$

$$S_m = \frac{1}{n_m - 1} \sum_{i=1}^{n_m} (X_i - \overline{X})(X_i - \overline{X})', \tag{7}$$

and

$$S_s = \frac{1}{n_s - 1} \sum_{j=1}^{n_s} (Y_j - \overline{Y})(Y_j - \overline{Y})'. \tag{8}$$

Then

$$\overline{\mathbf{X}} \sim N_p(\boldsymbol{\mu}_m, (1/n_m)\mathbf{Q}_m), \tag{9}$$

$$\overline{\mathbf{Y}} \sim N_p(\boldsymbol{\mu}_s, (1/n_s)\mathbf{Q}_s), \tag{10}$$

$$(n_m - 1)\mathbf{S}_m \sim W_p(n_m - 1, \mathbf{Q}_m), \tag{11}$$

$$(n_s - 1)\mathbf{S}_s \sim W_p(n_s - 1, \mathbf{Q}_s), \tag{12}$$

and these are mutually independent, where $W_p(n,\mathbf{Q})$ denotes the Wishart distribution with n degrees of freedom and corresponding covariance matrix $\mathbf{Q}$.

If $\mathbf{Q}_m = \mathbf{Q}_s = \mathbf{Q}$, then

$$\overline{\mathbf{X}} - \overline{\mathbf{Y}} \sim N_p(\boldsymbol{\mu}_m - \boldsymbol{\mu}_s, (1/n_m + 1/n_s)\mathbf{Q}), \tag{13}$$

$$(n_m - 1)\mathbf{S}_m + (n_s - 1)\mathbf{S}_s = (n_m + n_s - 2)\mathbf{S} \sim W_p(n_m + n_s - 2, \mathbf{Q}), \tag{14}$$

where " $\sim$ " denotes "is distributed as". Thus the test statistic is

$$T^2 = (\overline{\mathbf{X}} - \overline{\mathbf{Y}})'[(1/n_m + 1/n_s)\mathbf{S}]^{-1}(\overline{\mathbf{X}} - \overline{\mathbf{Y}}) = \frac{n_m n_s}{n_m + n_s}(\overline{\mathbf{X}} - \overline{\mathbf{Y}})'\mathbf{S}^{-1}(\overline{\mathbf{X}} - \overline{\mathbf{Y}}). \tag{15}$$

The T^2 statistic has the central T^2 distribution when $\boldsymbol{\mu}_m = \boldsymbol{\mu}_s$ is true. For T^2 to have an F distribution, the expression for T^2 must be weighted by the factor $(n_m + n_s - p - 1)/[p(n_m + n_s - 2)]$ so that

$$F = \frac{n_m + n_s - p - 1}{p(n_m + n_s - 2)}T^2 \sim F_{p, n_m + n_s - p - 1}, \tag{16}$$

where $F_{p, n_m + n_s - p - 1}$ is the F distribution with degrees of freedom p and $n_m + n_s - p - 1$. If

$$F \le F_{p, n_m + n_s - p - 1; 1 - \alpha} \tag{17}$$

or

$$T^2 \le \frac{p(n_m + n_s - 2)}{n_m + n_s - p - 1}F_{p, n_m + n_s - p - 1; 1 - \alpha}, \tag{18}$$

where

$$\Pr\{F \le F_{p, n_m + n_s - p - 1; 1 - \alpha}\} = 1 - \alpha, \tag{19}$$

the null hypothesis $H_0 : \boldsymbol{\mu}_m = \boldsymbol{\mu}_s$ is accepted and rejected otherwise.

A $100(1 - \alpha)\%$ confidence region for $\boldsymbol{\mu}_m - \boldsymbol{\mu}_s$ is given by

$$C_{\boldsymbol{\mu}_m - \boldsymbol{\mu}_s} = \left\{ \begin{matrix} \boldsymbol{\mu}_m - \boldsymbol{\mu}_s : \dfrac{n_m n_s}{n_m + n_s}[(\overline{\mathbf{X}} - \overline{\mathbf{Y}}) - (\boldsymbol{\mu}_m - \boldsymbol{\mu}_s)]'\mathbf{S}^{-1}[(\overline{\mathbf{X}} - \overline{\mathbf{Y}}) - (\boldsymbol{\mu}_m - \boldsymbol{\mu}_s)] \\ \le \dfrac{p(n_m + n_s - 2)}{n_m + n_s - p - 1}F_{p, n_m + n_s - p - 1; 1 - \alpha} \end{matrix} \right\}, \tag{20}$$

where

$$V = \frac{n_m + n_s - p - 1}{p(n_m + n_s - 2)} \frac{n_m n_s}{n_m + n_s} [(\overline{X} - \overline{Y}) - (\mu_m - \mu_s)]'$$

$$\times S^{-1}[(\overline{X} - \overline{Y}) - (\mu_m - \mu_s)] \sim F_{p, n_m + n_s - p - 1} \tag{21}$$

represents a pivotal quantity whose distribution does not depend on unknown parameters.

Thus, the decision rule for testing the validity of the model with specified model user's risk (α) is the following: Accept the validity of the model for the acceptable range of accuracy (5) under the given experimental frame if (18) takes place and

$$|\mu_m - \mu_s| \le \Delta \text{ for all } (\mu_m - \mu_s) \in C_{\mu_m - \mu_s}. \tag{22}$$

With

$$\mu'_m - \mu'_s = [\mu_{m1} - \mu_{s1}, ..., \mu_{mk} - \mu_{sk}, ..., \mu_{mp} - \mu_{sp}], \tag{23}$$

the $100(1-\alpha)\%$ simultaneous confidence intervals for the p population mean differences (components of the vector $\mu_m - \mu_s$) are

$$\mu_{mk} - \mu_{sk} : \overline{X}_k - \overline{Y}_k \pm \sqrt{\frac{p(n_m + n_s - 2)}{n_m + n_s - p - 1} F_{p, n_m + n_s - p - 1, 1 - \alpha}} \sqrt{\left(\frac{1}{n_m} + \frac{1}{n_s}\right) S_k^2}, \quad k = 1, ..., p, \tag{24}$$

where S_k^2 is the pooled variance for variables X_k and Y_k.

3.2   Testing the Equality of Two Covariance Matrices

For testing the hypothesis H_0: $Q_m = Q_s$, for two multivariate normal populations, the statistic

$$M = (n_m + n_s - 2)\ln|S| - [(n_m - 1)\ln|S_m| + (n_s - 1)\ln|S_s|] \tag{25}$$

is used. χ^2 approximation for the probability distribution of M is given by Box [17] as

$$M(1 - A) \sim \chi_f^2, \tag{26}$$

where

$$f = \frac{p(p + 1)}{2} \tag{27}$$

is the number of degrees of freedom,

$$A = \frac{2p^2 + 3p - 1}{6(p + 1)} \left(\frac{1}{n_m - 1} + \frac{1}{n_s - 1} - \frac{1}{n_m + n_s - 2}\right). \tag{28}$$

The hypothesis H_0 is rejected if

$$M(1 - A) > \chi_{f; 1 - \alpha}^2, \tag{29}$$

where α is the significance level.

3.3   Validation via Testing Two Mean Vectors with Unequal Covariance Matrices

Let the p-vectors $\mathbf{X}_1, \mathbf{X}_2, \ldots, \mathbf{X}_{n_m}$, and $\mathbf{Y}_1, \mathbf{Y}_2, \ldots, \mathbf{Y}_{n_s}$ be independent (potential) random samples from $\mathbf{X}$ (model) and $\mathbf{Y}$ (system), respectively, where $\mathbf{X} \sim N_p(\boldsymbol{\mu}_m, \mathbf{Q}_m)$ and $\mathbf{Y} \sim N_p(\boldsymbol{\mu}_s, \mathbf{Q}_s)$. The problem is to compare the mean vectors $\boldsymbol{\mu}_m$ and $\boldsymbol{\mu}_s$. In this section, we consider the case when the covariance matrices of the two normal distributions are unequal (i.e., $\mathbf{Q}_m \neq \mathbf{Q}_s$).

Version 1. In this case, for testing the hypothesis $H_0: \boldsymbol{\mu}_m = \boldsymbol{\mu}_s$ vs. $H_1: \boldsymbol{\mu}_m \neq \boldsymbol{\mu}_s$, a very natural statistic

$$T_\bullet^2 = (\overline{\mathbf{X}} - \overline{\mathbf{Y}})'[\mathbf{S}_m / n_m + \mathbf{S}_s / n_s]^{-1}(\overline{\mathbf{X}} - \overline{\mathbf{Y}}) \tag{30}$$

can be used. This test statistic has approximately Hotelling's distribution, as it is shown in Krisnamoorthy and Yu [18], i.e.,

$$T_\bullet^2 \sim [p\nu/(\nu - p + 1)]F_{p,\nu-p+1} \text{ approximately}, \tag{31}$$

where

$$\nu = p(p+1)\left(\frac{\text{tr}(\mathbf{S}_m \mathbf{S}^{-1})^2 + [\text{tr}(\mathbf{S}_m \mathbf{S}^{-1})]^2}{n_m^2(n_m - 1)} + \frac{\text{tr}(\mathbf{S}_s \mathbf{S}^{-1})^2 + [\text{tr}(\mathbf{S}_s \mathbf{S}^{-1})]^2}{n_s^2(n_s - 1)} \right)^{-1}. \tag{32}$$

Krishnamoorthy and Yu [18] showed that ν is bound in the same way as in the one-dimensional case,

$$\min(n_m - 1, n_s - 1) \leq \nu \leq n_m + n_s - 2, \tag{33}$$

ν being close to the upper bound tells us that the two variance matrices are (almost) equal. The closer ν is to the lower bound, the bigger the discrepancy is between them. The lower bound is attained only if one of $\mathbf{S}_m, \mathbf{S}_s$ is a zero matrix.

A $100(1-\alpha)\%$ confidence region for $\boldsymbol{\mu}_m - \boldsymbol{\mu}_s$ is given by

$$C_{\boldsymbol{\mu}_m - \boldsymbol{\mu}_s}^\bullet = \left\{ \begin{array}{l} \boldsymbol{\mu}_m - \boldsymbol{\mu}_s : [(\overline{\mathbf{X}} - \overline{\mathbf{Y}}) - (\boldsymbol{\mu}_m - \boldsymbol{\mu}_s)]'\left(\dfrac{\mathbf{S}_m}{n_m} + \dfrac{\mathbf{S}_s}{n_s} \right)^{-1} \\ \times [(\overline{\mathbf{X}} - \overline{\mathbf{Y}}) - (\boldsymbol{\mu}_m - \boldsymbol{\mu}_s)] \leq \dfrac{p\nu}{\nu - p + 1} F_{p,\nu-p+1;1-\alpha} \end{array} \right\}. \tag{34}$$

Thus, in this case, the decision rule for testing the validity of the model with specified model builder's risk is the following: Accept the validity of the model for the acceptable range of accuracy (5) under the given experimental frame if

$$T_\bullet^2 \leq [p\nu/(\nu - p + 1)]F_{p,\nu-p+1;1-\alpha} \tag{35}$$

takes place and

$$|\boldsymbol{\mu}_m - \boldsymbol{\mu}_s| \leq \Delta \text{ for all } (\boldsymbol{\mu}_m - \boldsymbol{\mu}_s) \in C_{\boldsymbol{\mu}_m - \boldsymbol{\mu}_s}^\bullet. \tag{36}$$

The $100(1-\alpha)\%$ simultaneous confidence intervals for the p components of the vector $\boldsymbol{\mu}_m - \boldsymbol{\mu}_s$ are

$$\mu_{mk} - \mu_{sk} : \overline{X}_k - \overline{Y}_k \pm \sqrt{\frac{p\nu}{\nu - p + 1}} F_{p,\nu-p+1;1-\alpha} \sqrt{\frac{S_{mk}^2}{n_m} + \frac{S_{sk}^2}{n_s}}, \quad k = 1, ..., p, \qquad (37)$$

where S_{mk}^2 and S_{sk}^2 are the variances for variables X_k and Y_k, respectively.

Version 2. In this case, for testing the hypothesis H_0: $\mu_m = \mu_s$ vs. H_1: $\mu_m \neq \mu_s$, the statistic (30) is used. This test statistic has approximately Hotelling's distribution, as it is shown in Seber [19], i.e.,

$$T_\bullet^2 \sim \frac{p(n_m + n_s - 2)}{n_m + n_s - p - 1} F_{p,\vartheta} \text{ approximately}, \qquad (38)$$

where

$$\frac{1}{\vartheta} = \frac{1}{n_m - 1} \frac{(\overline{X} - \overline{Y})'[S_m / n_m + S_s / n_s]^{-1}[S_m / n_m][S_m / n_m + S_s / n_s]^{-1}(\overline{X} - \overline{Y})}{T_\bullet^2}$$

$$+ \frac{1}{n_s - 1} \frac{(\overline{X} - \overline{Y})'[S_m / n_m + S_s / n_s]^{-1}[S_s / n_s][S_m / n_m + S_s / n_s]^{-1}(\overline{X} - \overline{Y})}{T_\bullet^2}. \qquad (39)$$

A 100(1-α)% confidence region for $\mu_m - \mu_s$ is given by

$$C_{\mu_m - \mu_s}^\circ = \left\{ \begin{array}{l} \mu_m - \mu_s : [(\overline{X} - \overline{Y}) - (\mu_m - \mu_s)]' \left(\dfrac{S_m}{n_m} + \dfrac{S_s}{n_s} \right)^{-1} \\[2mm] \times [(\overline{X} - \overline{Y}) - (\mu_m - \mu_s)] \leq \dfrac{p(n_m + n_s - 2)}{n_m + n_s - p - 1} F_{p,\vartheta;1-\alpha} \end{array} \right\}. \qquad (40)$$

Thus, in this case, the decision rule for testing the validity of the model with specified model builder's risk is the following: Accept the validity of the model for the acceptable range of accuracy (5) under the given experimental frame if

$$T_\bullet^2 \leq \frac{p(n_m + n_s - 2)}{n_m + n_s - p - 1} F_{p,\vartheta;1-\alpha} \qquad (41)$$

takes place and

$$|\mu_m - \mu_s| \leq \Delta \text{ for all } (\mu_m - \mu_s) \in C_{\mu_m - \mu_s}^\circ. \qquad (42)$$

The 100(1-α)% simultaneous confidence intervals for the p components of the vector $\mu_m - \mu_s$ are

$$\mu_{mk} - \mu_{sk} : \overline{X}_k - \overline{Y}_k \pm \sqrt{\frac{p(n_m + n_s - 2)}{n_m + n_s - p - 1}} F_{p,\vartheta;1-\alpha} \sqrt{\frac{S_{mk}^2}{n_m} + \frac{S_{sk}^2}{n_s}}, \quad k = 1, ..., p, \qquad (43)$$

where S_{mk}^2 and S_{sk}^2 are the variances for variables X_k and Y_k, respectively.

3.4   Validation of Analytic Model via Testing One Mean Vector

Let us assume that a random sample $\mathbf{Y}_1, \mathbf{Y}_2, \ldots, \mathbf{Y}_{n_s}$ is available from the observed system, where $\mathbf{Y}_i \sim N_p(\boldsymbol{\mu}_s, \mathbf{Q}_s)$, $i \in \{1,2, \ldots, n_s\}$. The problem is to compare the mean vectors $\boldsymbol{\mu}_s$ (from the system) and $\boldsymbol{\mu}_0$ (from the analytic model), i.e., we are interested in the problem of testing

$$H_0: \boldsymbol{\mu}_s = \boldsymbol{\mu}_0 \text{ vs. } H_1: \boldsymbol{\mu}_s \neq \boldsymbol{\mu}_0. \tag{44}$$

Furthermore, let us assume that for the purpose, for which the analytic model is intended, the validity of the model can be determined with respect to its mean response, and the acceptable range of accuracy can be expressed as the difference between the means of the system and the analytic model responses and can be stated as

$$\left| \boldsymbol{\mu}_s - \boldsymbol{\mu}_0 \right| \leq \boldsymbol{\Delta}, \tag{45}$$

where $\boldsymbol{\Delta}$ is a vector of the largest acceptable differences.

If the covariance matrix $\mathbf{Q}_s$ were known, then testing the hypothesis (44) can be based on the chi-squared statistic

$$n_s (\overline{\mathbf{Y}} - \boldsymbol{\mu}_s)' \mathbf{Q}_s^{-1} (\overline{\mathbf{Y}} - \boldsymbol{\mu}_s) \sim \chi_p^2. \tag{46}$$

Typically $\mathbf{Q}_s$ is unknown and so one instead uses

$$T^2 = n_s (\overline{\mathbf{Y}} - \boldsymbol{\mu}_s)' \mathbf{S}_s^{-1} (\overline{\mathbf{Y}} - \boldsymbol{\mu}_s) \tag{47}$$

as the test statistic, which is commonly known as the Hotelling's T^2 statistic with p and $n_s - 1$ degrees of freedom. For T^2 to have an F distribution, the expression for T^2 must be weighted by the factor $(n_s - p)/[p(n_s - 1)]$ so that

$$F = \frac{n_s - p}{p(n_s - 1)} T^2 \sim F_{p, n_s - p}, \tag{48}$$

where $F_{p, n_s - p}$ is the F distribution with degrees of freedom p and $n_s - p$. If

$$F \leq F_{p, n_s - p; 1 - \alpha} \tag{49}$$

or

$$T^2 \leq \frac{p(n_s - 1)}{n_s - p} F_{p, n_s - p; 1 - \alpha}, \tag{50}$$

where

$$\Pr\{F \leq F_{p, n_s - p; 1 - \alpha}\} = 1 - \alpha, \tag{51}$$

the null hypothesis $H_0: \boldsymbol{\mu}_s = \boldsymbol{\mu}_0$ is accepted and rejected otherwise.

A $100(1 - \alpha)\%$ confidence region for $\boldsymbol{\mu}_s$ is given by

$$C_{\boldsymbol{\mu}_s} = \left\{ \boldsymbol{\mu}_s : n_s (\overline{\mathbf{Y}} - \boldsymbol{\mu}_s)' \mathbf{S}_s^{-1} (\overline{\mathbf{Y}} - \boldsymbol{\mu}_s) \leq \frac{p(n_s - 1)}{n_s - p} F_{p, n_s - p; 1 - \alpha} \right\}, \tag{52}$$

where

$$V = \frac{(n_s - p)n_s}{p(n_s - 1)} (\overline{\mathbf{Y}} - \boldsymbol{\mu}_s)' \mathbf{S}_s^{-1} (\overline{\mathbf{Y}} - \boldsymbol{\mu}_s) \sim F_{p, n_s - p} \tag{53}$$

represents a pivotal quantity whose distribution does not depend on unknown parameters.

Thus, the decision rule for testing the validity of the model with specified model user's risk (α) is the following: Accept the validity of the model for the acceptable range of accuracy (45) under the given experimental frame if (50) takes place and

$$|\mu_s - \mu_0| \leq \Delta \text{ for all } \mu_s \in C_{\mu_s}. \tag{54}$$

With

$$\mu'_s = [\mu_{s1}, ..., \mu_{sk}, ..., \mu_{sp}], \tag{55}$$

the 100(1-α)% simultaneous confidence intervals for the p components of the vector μ_s are

$$\mu_{sk} : \overline{Y}_k \pm \sqrt{\frac{p(n_s - 1)}{n_s - p} F_{p, n_s - p, 1-\alpha}} \sqrt{\frac{S_k^2}{n_s}}, \quad k = 1, ..., p, \tag{56}$$

where S_k^2 is the variance for variable Y_k.

4   Application Examples

4.1   Example of Statistical Validation of Simulation Model

Consider airline seat inventory control system [20] and the simulation model, on each of which 4 endpoints (passenger demand) were recorded. Based on the experimental data from the simulation model (n_m=28) and the control system (n_s=28), the summary statistics for the simulation model and the control system follow.

$$\overline{X} = \begin{bmatrix} 29.143 \\ 48.643 \\ 35.571 \\ 86.500 \end{bmatrix}, \quad \overline{Y} = \begin{bmatrix} 28.964 \\ 45.179 \\ 34.679 \\ 81.964 \end{bmatrix} \tag{57}$$

and

$$\overline{X} - \overline{Y} = \begin{bmatrix} 0.179 \\ 3.464 \\ 0.893 \\ 4.536 \end{bmatrix}. \tag{58}$$

The sample covariance matrices defined in (7) and (8) are:

$$S_m = \begin{bmatrix} 22.942 & 30.942 & 4.434 & 21.815 \\ 30.942 & 78.608 & 14.582 & 56.704 \\ 4.434 & 14.582 & 17.513 & 30.519 \\ 21.815 & 56.704 & 30.519 & 91.074 \end{bmatrix} \tag{59}$$

and

$$S_s = \begin{bmatrix} 24.036 & 18.747 & 15.062 & 31.517 \\ 18.747 & 42.374 & 11.726 & 38.451 \\ 15.062 & 11.726 & 20.522 & 31.951 \\ 31.517 & 38.451 & 31.951 & 132.258 \end{bmatrix}, \tag{60}$$

Consider testing

$$H_0: \boldsymbol{\mu}_m - \boldsymbol{\mu}_s = 0 \text{ vs. } H_1: \boldsymbol{\mu}_m - \boldsymbol{\mu}_s \neq 0. \tag{61}$$

Statistical Validation when $\mathbf{Q}_m = \mathbf{Q}_s$. The pooled sample covariance matrix is given by

$$S = \begin{bmatrix} 23.489 & 24.845 & 9.748 & 26.666 \\ 24.845 & 60.491 & 13.154 & 47.577 \\ 9.748 & 13.154 & 19.018 & 31.235 \\ 26.666 & 47.577 & 31.235 & 111.666 \end{bmatrix} \tag{62}$$

and the Hotelling T^2 statistic defined in (15) is computed as

$$T^2 = (\overline{\mathbf{X}} - \overline{\mathbf{Y}})'[(1/n_m + 1/n_s)S]^{-1}(\overline{\mathbf{X}} - \overline{\mathbf{Y}}) = 5.646. \tag{63}$$

Taking $\alpha=0.025$, we have that

$$\frac{p(n_m + n_s - 2)}{n_m + n_s - p - 1} F_{p, n_m + n_s - p - 1; 1 - \alpha} = 12.93. \tag{64}$$

Because $T^2 = 5.646 < 12.93$, we do not reject H_0. Thus, the data do not provide sufficient evidence to indicate that the mean vectors are significantly different.

Testing the Equality of Two Covariance Matrices. It follows from (25) and (28) that

$$M(1 - A) = 19.25377 < \chi^2_{f;1-\alpha} = 20.48, \tag{65}$$

where $M=20.91959$, $A=0.07963$, $f=10$, $\alpha=0.025$. Thus, the data do not provide sufficient evidence to indicate that the covariance matrices are significantly different.

Statistical Validation when $\mathbf{Q}_m \neq \mathbf{Q}_s$. For this example, the values of T^2 and $T_\bullet^2$ must be the same because the sample sizes are equal. Thus,

$$T_\bullet^2 = (\overline{\mathbf{X}} - \overline{\mathbf{Y}})'[\mathbf{S}_m / n_m + \mathbf{S}_s / n_s]^{-1}(\overline{\mathbf{X}} - \overline{\mathbf{Y}}) = 5.646. \tag{66}$$

The approximate degree of freedom

$$v = p(p+1)\left(\frac{\text{tr}(\mathbf{S}_m\mathbf{S}^{-1})^2 + [\text{tr}(\mathbf{S}_m\mathbf{S}^{-1})]^2}{n_m^2(n_m - 1)} \right.$$
$$\left. + \frac{\text{tr}(\mathbf{S}_s\mathbf{S}^{-1})^2 + [\text{tr}(\mathbf{S}_s\mathbf{S}^{-1})]^2}{n_s^2(n_s - 1)} \right)^{-1} = 52.113, \tag{67}$$

and the critical value

$$\frac{pv}{v-p+1}F_{p,v-p+1;1-\alpha} = 12.988.$$

(68)

Because $T_\bullet^2 = 5.646 < 12.988$, we do not reject H_0. Thus, both tests produced similar results and yielded the same conclusions.

It follows from (24) that the $100(1-\alpha)\%$ simultaneous confidence intervals for the p components of the vector $\mu_m - \mu_s$ are

$$\begin{bmatrix} -4.47 \\ -4 \\ -3.3 \\ -5.6 \end{bmatrix} \leq \begin{bmatrix} \mu_{m1} - \mu_{s1} \\ \mu_{m2} - \mu_{s2} \\ \mu_{m3} - \mu_{s3} \\ \mu_{m4} - \mu_{s4} \end{bmatrix} \leq \begin{bmatrix} 4.83 \\ 10.93 \\ 5.1 \\ 14.67 \end{bmatrix}.$$

(69)

Let us assume that $\Delta'=[5, 11, 5.5, 15]$. It follows from (69) that $|\mu_m - \mu_s| < \Delta$. Thus, in this case, the decision rule for testing the validity of the model with specified model builder's risk is the following: Accept the validity of the model for the acceptable range of accuracy

$$|\mu_m - \mu_s| = \begin{vmatrix} \mu_{m1} - \mu_{s1} \\ \mu_{m2} - \mu_{s2} \\ \mu_{m3} - \mu_{s3} \\ \mu_{m4} - \mu_{s4} \end{vmatrix} < \Delta = \begin{bmatrix} 5 \\ 11 \\ 5.5 \\ 15 \end{bmatrix}$$

(70)

under the given experimental frame:

$$T^2 = 5.646 < \frac{p(n_m + n_s - 2)}{n_m + n_s - p - 1}F_{p,n_m+n_s-p-1;1-\alpha} = 12.93$$

(71)

and

$$|\mu_m - \mu_s| < \Delta \text{ for all } (\mu_m - \mu_s) \in C_{\mu_m - \mu_s}$$

$$= \left\{ \begin{aligned} &\mu_m - \mu_s : \frac{n_m n_s}{n_m + n_s}[(\overline{X} - \overline{Y}) - (\mu_m - \mu_s)]'S^{-1}[(\overline{X} - \overline{Y}) - (\mu_m - \mu_s)] \\ &\leq \frac{p(n_m + n_s - 2)}{n_m + n_s - p - 1}F_{p,n_m+n_s-p-1;1-\alpha} \end{aligned} \right\}.$$

(72)

4.2   Example of Statistical Validation of Analytic Model

Performance of mechanical components undergoes a change by uncertainties such as environmental effects, dimensional tolerances, loading conditions, material properties and maintenance processes. Especially when the design criterion is fatigue life, it is significantly affected by system uncertainties. Even with today's modern computing systems, it is infeasible to include all the relevant uncertain variables into the analytical prediction, since many of the potential inputs are not characterized in the design phase. To account for the unknown variables, common practices use so called "safety factors" or statistical minimum properties in conjunction with the analytical prediction when evaluating lifetimes. Due to these conservative estimations, analytical predictions are often in disagreement with field experience, and a gap exists in correlating the field data with the analytical predictions. Thus, there is an increasing need to improve the analytical predictions using field data, which collectively represents the real status of a particular machine.

For example, the expected fatigue life (lifetime) can be determined by the most well known Paris–Erdogan equation [21, 22], which is of the form

$$\frac{da(t)}{dt} = q(a(t))^b \tag{73}$$

in which q and b are constants that are depended on loading conditions, material properties and maintenance processes. The independent variable t can be interpreted as stress cycles, flight hours, or flights depending on the applications. Integrating (73) gives

$$\int_0^{t_f} dt = \int_{a_0}^{a_f} \frac{dv}{qv^b}. \tag{74}$$

Here a_0 is the initial crack length, a_f is the final crack length corresponding to failure, and t_f is the estimated number of flight hours (expected fatigue life) to produce a failure after the initial crack is formed.

Table 1 presents the data of fatigue tests on a particular type of structural components (stringer) of aircraft IL-86.

Table 1. Observed lifetimes of two structural components (stringer) of aircraft IL-86

Components	Observed lifetimes of components (in terms of 10^3 flight hours)								
Component 1	72.8	68.0	59.2	66.7	74.2	70.4	69.6	77.9	63.9	65.1
Component 2	69.9	70.9	68.4	78.2	63.9	64.6	66.5	71.6	77.2	66.8
Y	Y_1	Y_2	Y_3	Y_4	Y_5	Y_6	Y_7	Y_8	Y_9	Y_{10}

It is assumed that $Y_i \sim N_p(\mu_s, Q_s)$, $i=1, \ldots, 10$, where $p=2$, $n_s=10$. Suppose the problem is to test the null hypothesis

$$H_0: \mu_s = \mu_0 \tag{75}$$

against the alternative hypothesis $H_1: \mu_s \neq \mu_0$, where μ_0 (vector of expected lifetimes) obtained from (74) via t_f is given by

$$\mu_0 = \begin{bmatrix} \mu_{01} \\ \mu_{02} \end{bmatrix} = \begin{bmatrix} 69 \\ 69 \end{bmatrix}. \tag{76}$$

The acceptable range of accuracy is expressed as

$$|\boldsymbol{\mu}_s - \boldsymbol{\mu}_0| \le \boldsymbol{\Delta} = \begin{bmatrix} 6 \\ 6 \end{bmatrix}, \tag{77}$$

where $\boldsymbol{\Delta}$ is a vector of the largest acceptable differences.

Since

$$\overline{\mathbf{Y}} = \frac{1}{n_s} \sum_{i=1}^{n_s} \mathbf{Y}_i = \begin{bmatrix} 68.78 \\ 69.80 \end{bmatrix} \tag{78}$$

and

$$\mathbf{S}_s = \frac{1}{n_s - 1} \sum_{i=1}^{n_s} (\mathbf{Y}_i - \overline{\mathbf{Y}})(\mathbf{Y}_i - \overline{\mathbf{Y}})' = \begin{bmatrix} 29.56 & -6.25 \\ -6.25 & 23.72 \end{bmatrix}, \tag{79}$$

it follows from (56) that the $100(1-\alpha)\%$ simultaneous confidence intervals for the components of the vector $\boldsymbol{\mu}_s' = [\mu_{s1}, \mu_{s2}]$ are

$$\mu_{s1} : \overline{Y}_1 \pm \sqrt{\frac{p(n_s - 1)}{n_s - p} F_{p, n_s - p, 1 - \alpha}} \sqrt{\frac{S_1^2}{n_s}} = 68.78 \pm 5.45 \tag{80}$$

and

$$\mu_{s2} : \overline{Y}_k \pm \sqrt{\frac{p(n_s - 1)}{n_s - p} F_{p, n_s - p, 1 - \alpha}} \sqrt{\frac{S_2^2}{n_s}} = 69.80 \pm 4.88, \tag{81}$$

where $\alpha = 0.05$.

It follows from (80) and (81) that

$$\begin{bmatrix} -5.67 \\ -4.08 \end{bmatrix} \le \begin{bmatrix} \mu_{s1} - \mu_{01} \\ \mu_{s2} - \mu_{02} \end{bmatrix} \le \begin{bmatrix} 5.23 \\ 5.68 \end{bmatrix}, \tag{82}$$

i.e., $|\boldsymbol{\mu}_s - \boldsymbol{\mu}_0| < \boldsymbol{\Delta}$. Thus, in this case, the decision rule for testing the validity of the model with specified model builder's risk is the following: Accept the validity of the model for the acceptable range of accuracy

$$|\boldsymbol{\mu}_s - \boldsymbol{\mu}_0| = \begin{vmatrix} \mu_{s1} - \mu_{01} \\ \mu_{s2} - \mu_{02} \end{vmatrix} < \boldsymbol{\Delta} = \begin{bmatrix} 6 \\ 6 \end{bmatrix} \tag{83}$$

under the given experimental frame:

$$T^2 = n_s (\overline{\mathbf{Y}} - \boldsymbol{\mu}_s)' \mathbf{S}_s^{-1} (\overline{\mathbf{Y}} - \boldsymbol{\mu}_s) = 0.27 < \frac{p(n_s - 1)}{n_s - p} F_{p, n_s - p; 1 - \alpha} = 10 \tag{84}$$

and $|\boldsymbol{\mu}_s - \boldsymbol{\mu}_0| < \boldsymbol{\Delta}$ for all

$$\mu_s \in C_{\mu_s} = \left\{ \mu_s : n_s (\overline{\mathbf{Y}} - \mu_s)' \mathbf{S}_s^{-1} (\overline{\mathbf{Y}} - \mu_s) \le \frac{p(n_s - 1)}{n_s - p} F_{p, n_s - p; 1 - \alpha} \right\}. \tag{85}$$

5   Conclusion

Analytical and simulations models have become an important tool for building and testing theories in Cognitive Science during the last years. The area of their applications includes, in particular, business process, resource management, knowledge management systems, transportation systems, service systems, operations research, economics, optimization, operation and production management, supply chain management, work flow management, total quality management, logistics, risk analysis, scheduling, forecasting, etc.

One of the most important steps in the development of a model for observable process in stochastic system is recognition of the model, which is an accurate representation of the process under study. A common form of objective analysis for validating simulation models is statistical hypothesis testing. In this paper, we are concerned about the choice of a test for equality of two multivariate normal mean vectors. The definition of a confidence region is generalized so that problems such as constructing exact confidence regions for the difference in two multivariate normal means can be tackled without the assumption of equal covariance matrices. This allows one to provide satisfactory solutions in a variety of problems, not just the ones reported here.

The authors hope that this work will stimulate further investigation using the approach on specific applications to see whether obtained results with it are feasible for realistic applications.

Acknowledgments. This research was supported in part by Grant No. 06.1936, Grant No. 07.2036, and Grant No. 09.1014 from the Latvian Council of Science and the National Institute of Mathematics and Informatics of Latvia.

References

1. Balci, O., Sargent, R.G.: A Bibliography on the Credibility, Assessment and Validation of Simulation and Mathematical Models. Simuletter 15, 15–27 (1984)
2. Birta, L., Ozmizrak, F.: A Knowledge-Based Approach for the Validation of Simulation Models: The Foundation. ACM Trans. Model. Comput. Simulation 6(1996), 76–98 (1996)
3. Findler, N.V., Mazur, N.M.: A System for Automatic Model Verification and Validation. Transactions of the Society for Computer Simulation 6, 153–172 (1990)
4. Landry, M., Oral, M.: In Search of a Valid View of Model Validation for Operations Research. European Journal of Operational Research 66, 161–167 (1993)
5. Mayer, D.G., Butler, D.: Statistical Validation. Ecol. Model. 68, 21–32 (1993)
6. Nechval, K.N., Nechval, N.A., Vasermanis, E.K.: Technique for Identifying an Observable Process with one of Several Simulation Models. In: Proceedings of the Summer Computer Simulation Conference (SCSC 2003), Montreal, Canada, pp. 70–75 (2003)

7. Vasermanis, E.K., Nechval, K.N., Nechval, N.A.: Statistical Validation of Simulation Models of Observable Systems. Kybernetes (The International Journal of Systems & Cybernetics) 32, 858–869 (2003)
8. Freese, F.: Testing Accuracy. Forest Sci. 6, 139–145 (1960)
9. Ottosson, F., Håkanson, L.: Presentation and Analysis of a Model Simulating the pH Response of Lake Liming. Ecol. Modelling 104, 89–111 (1997)
10. Jans-Hammermeister, D.C., McGill, W.B.: Evaluation of Three Simulation Models Used to Describe Plant Residue Decomposition in Soil. Ecol. Modelling 104, 1–13 (1997)
11. Landsberg, J.J., Waring, R.H., Coops, N.C.: Performance of the Forest Productivity Model 3-PG Applied to a Wide Range of Forest Types. Forest Ecol. Manage. 172, 199–214 (2003)
12. Bartelink, H.H.: Radiation Interception by Forest Trees: A Simulation Study on Effects of Stand Density and Foliage Clustering on Absorption and Transmission. Ecol. Modelling 105, 213–225 (1998)
13. Alewell, C., Manderscheid, B.: Use of Objective Criteria for the Assessment of Biogeochemical Ecosystem Models. Ecol. Modelling 105, 113–124 (1998)
14. Nechval, N.A., Nechval, K.N.: Characterization Theorems for Selecting the Type of Underlying Distribution. In: Abstracts of Communications of the 7th Vilnius Conference on Probability Theory and Mathematical Statistics & the 22nd European Meeting of Statisticians, pp. 352–353. TEV, Vilnius (1998)
15. Nechval, N.A., Nechval, K.N., Vasermanis, E.K.: Technique of Testing for Two-Phase Regressions. In: Proceedings of the Second International Conference on Simulation, Gaming, Training and Business Process Reengineering in Operations, pp. 129–133. RTU, Riga (2000)
16. Nechval, N.A.: A General Method for Constructing Automated Procedures for Testing Quickest Detection of a Change in Quality Control. Computers in Industry 10, 177–183 (1988)
17. Box, G.E.P.: A General Distribution Theory for a Class of Likelihood Criteria. Biometrika 36, 317–346 (1949)
18. Krishnamoorthy, K., Yu, J.: Modified Nel and Van der Merwe Test for the Multivariate Behrens-Fisher Problem. Statistics & Probability Letters 66, 161–169 (2004)
19. Seber, G.A.F.: Multivariate Observations. Wiley, New York (1984)
20. Nechval, N., Purgailis, M., Rozevskis, U., Nechval, K.: Adaptive Stochastic Airline Seat Inventory Control under Parametric Uncertainty. In: Dudin, A., De Turck, K. (eds.) ASMTA 2013. LNCS, vol. 7984, pp. 308–323. Springer, Heidelberg (2013)
21. Nechval, N., Nechval, K., Berzinsh, G., Purgailis, M., Rozevskis, U.: Stochastic Fatigue Models for Efficient Planning Inspections in Service of Aircraft Structures. In: Al-Begain, K., Heindl, A., Telek, M. (eds.) ASMTA 2008. LNCS, vol. 5055, pp. 114–127. Springer, Heidelberg (2008)
22. Nechval, N.A., Nechval, K.N.: Statistical Identification of an Observable Process. Computer Modeling and New Technologies 12, 38–46 (2008)

Detecting Changes in the Scale of Dependent Gaussian Processes: A Large Deviations Approach

Julia Kuhn[1,2], Wendy Ellens[1], and Michel Mandjes[1]

[1] University of Amsterdam, The Netherlands
[2] University of Queensland, Australia

Abstract. This paper devises new hypothesis tests for detecting changes in the *scale* of interdependent and serially correlated data streams, i.e, proportional changes of the mean and (co-)variance. Such procedures are of great importance in various networking contexts, since they enable automatic detection of changes, e.g. in the network load. Assuming the underlying structure is Gaussian, we compute the log-likelihood ratio test statistic, either as a function of the observations themselves or as a function of the innovations (i.e., a sequence of i.i.d. Gaussians, to be extracted from the observations). An alarm is raised if the test statistic exceeds a certain threshold. Based on large deviations techniques, we demonstrate how the threshold is chosen such that the ratio of false alarms is kept at a predefined (low) level. Numerical experiments validate the procedure, and demonstrate the merits of a multidimensional detection approach (over multiple one-dimensional tests). Also a detailed comparison between the observations-based approach and the innovations-based approach is provided.

1 Introduction

Statistical change point detection is an important tool in network control, and has been widely applied in e.g. intrusion detection systems, [17,18,19], and overload detection [13]. In order to enable the network operator to adequately respond to persistent changes in the (inherently random) observations, the main task is to detect persistent changes as quickly as possible while keeping the number of false alarms at a predefined low level (for instance 5%).

Traditionally, in the change point detection literature the main focus has been on detecting a change in the *mean* value corresponding to a sequence of *independent, one-dimensional* observations [4,8,16]. However, in many situations this setting is far from adequate. In the first place, in practice there is typically positive correlation between subsequent data points [20]. Moreover, single data points often consist of multidimensional records, rather than one-dimensional values. In addition, in the context of communication networks, an increase in the number of active users tends to be not reflected by a change in the mean only, but rather as a *change in scale* – a change in both the mean and (proportionally) the corresponding variance. Therefore, to only focus on the detection of mean shifts neglects an additional indicator that a change has taken place [2, Ex. 4.1.9].

B. Sericola, M. Telek, and G. Horváth (Eds.): ASMTA 2014, LNCS 8499, pp. 170–184, 2014.
© Springer International Publishing Switzerland 2014

Motivated by the above considerations, a number of procedures have been proposed that allow for data streams to be either serially correlated [15] or multidimensional [8]. In [19] a detection method for testing serially correlated and multidimensional data streams is presented but the multiple data streams are assumed to be independent. The more general setting of dependent multidimensional data streams is covered in [2], where testing against a change in mean or variance is considered separately. In the current paper, we develop techniques different from those of [2] for the detection of changes in scale in multidimensional and serially correlated sequences, which allow the network operator to limit the false alarm rate to a level of her choice. We focus on *Gaussian* sequences $(X_t)_{t\in\mathbb{Z}}$, where sets of observations have a multivariate Normal distribution. Gaussian time series are popular for modeling network traffic, see e.g. [1] and [12, Part A].

Let us consider the illustrative example of a link of a communication network. If the bandwidth consumed by different users is i.i.d., then the mean and variance of the total bandwidth consumption are both proportional to the number of users. As a consequence, a change in the number of users can be considered as a change of scale, in the sense defined above: the mean and variance exhibit the same relative (i.e., percentage-wise) change. When measuring not only at a single link but at various points in the network, then more information is available, potentially facilitating earlier detection or a lower risk of false alarms. In this case – apart from *serial correlation* (correlation over time) – also *cross-correlation* between data sequences generated by different sensors has to be taken into account, because the same traffic may be captured by several sensors. For large traffic aggregates, the Gaussianity is justified by central-limit type of arguments. Based on the above, we conclude that the setup considered in this paper can be used to detect changes in load, caused by, for instance, a (legal) increase in the number of users, or a DDoS (distributed denial of service) attack.

At the methodological level, the testing procedure we propose is a sequential hypothesis test, in line with the popular CUSUM algorithm [14]. Our procedure monitors a likelihood ratio test statistic, and raises an alarm as soon as it exceeds some predefined threshold. The question arises how this threshold should be chosen so as to ensure that the number of false alarms does not exceed a given (low) level. In case the observations are one-dimensional and independent, the test's false alarm performance can be assessed using a functional central limit theorem to establish the convergence of the test statistic to a Brownian motion [16]. Alternatively, since a false alarm is required to be a rare event, a limit expression for the false alarm probability can be derived using large deviations theory (concerned with the asymptotic behavior of rare event probabilities), see e.g. [10] and [5, Ch. VI.E]. Choosing the smallest threshold that satisfies the predefined level of false alarms ensures that an alarm is raised quickly once a change has occurred.

In [11] we extended the large deviations (LD) approach to detect a change in mean in serially correlated (one-dimensional) autoregressive moving average (ARMA) processes. The main objective of the present paper is to further extend such a LD approach to make it applicable to detect changes in scale in multidimensional correlated data. Furthermore, we compare the method of testing the sequence of *observations* $(X_t)_{t\in\mathbb{N}}$ themselves, with an innovations based approach, where first the observations are transformed into i.i.d. *innovations*, and then change point detection tests are performed on the innovations, see e.g. [2]. Our testing procedure differs from the one considered in [2], which uses the so-called *local approach* [2, 4.2.3] to determine the threshold for testing an i.i.d. sequence, whereas in this paper the false alarm probabilities are evaluated in the large deviations regime to choose the threshold. For the innovations based approach we impose the weak assumption [3, 5.7.1] that the process be linear and invertible, while for the observations based approach we need additional assumptions on the underlying correlation structure. We validate the proposed tests in a series of numerical experiments, which (i) study the trade-off between the detection ratio and the corresponding delay, (ii) assess the gain of multidimensional testing procedures (over multiple one-dimensional tests), and (iii) provide a systematic comparison between (A) the observations-based and (B) the innovations-based method.

This paper is organized as follows. In the next section we explain the change in scale and the set-up of our LD-based hypothesis test in greater detail. In Section 3 we compute the log-likelihood ratio test statistics for the observations and the innovations based detection approach, before we derive the threshold functions in Section 4. The results of the numerical evaluation are presented in Section 5. We conclude in Section 6.

2   Detection Procedure for a Change in Scale

We are concerned with testing a stationary multidimensional Gaussian sequence (X_t) against a change in scale, where after the change both the mean and variance are multiplied by some constant c. Each X_t is a d-dimensional column vector consisting of the measurements of d different 'sensors' at (discrete) time t. In this section we explain the general detection procedure; a more detailed description for the case of a change in scale can be found in the following two sections.

To detect a change in the traffic streams, we monitor windows of size $n \in \mathbb{N}$, i.e., at time t the n most recent observations (in the sequel denoted by $X_1, \ldots, X_n$) are tested in order to decide whether a change has occurred at some point $k \in \{1, \ldots, n\}$. In other words, we consider the hypotheses:

H_0: *No change has occurred within the window.*
H_1: *A change occurred at some point within the window.*

Thus, the alternative hypothesis is essentially the union of hypotheses:

$H_1(k)$: *A change in scale occurred exactly at time k, for a specific $k \in \{1, \ldots, n\}$.*

It will turn out to be convenient to express the *change point* k via the window size n, that is, we write $k = n\beta + 1$, where (throughout the paper) $\beta \in \mathscr{B} = \{0/n, 1/n, \ldots, (n-1)/n\}$.

To set up the testing procedure, we may consider (A) testing the observations directly, or (B) testing the extracted independent sequence of *innovations* – denoted by (ε_t) and defined in Section 3. We list some of the benefits and drawbacks of both approaches in Table 1; the details are explained in the sequel. Our method for testing a window of size n can be summarized as follows.

Table 1. Characteristics of (A) the observations and (B) the innovations based approach

(A)	(B)
Suitable test statistic for changes in mean and variance but also in coefficients	Suitable for detecting changes in mean or variance
Computationally expensive	Recursive computation of LLR and reduced dimensionality
How to define the threshold function in the multidimensional case is unclear, unless there is no shift in mean or data streams are independent	We can compute the threshold function for the change in scale explicitly
The process does not need to be invertible	Requires invertibility
The observations are well-defined test statistics	Since innovations are defined in terms of past observations, initial conditions are required

(i) The log-likelihood ratio (LLR) test statistic $\mathscr{L}_{n\beta}(\cdot)$ for testing H_0 against the simple alternative hypothesis $H_1(n\beta + 1)$ is computed as either (A) $\mathscr{L}_{n\beta}^X(X)$ (when considering the observations) or (B) $\mathscr{L}_{n\beta}^\varepsilon(\varepsilon)$ (when considering the innovations). The two approaches are equivalent under H_0.

(ii) Based on large deviations theory, the threshold $b(\beta)$ is obtained as (A) $b_X(\beta)$ or (B) $b_\varepsilon(\beta)$; it is a function of β such that for any value of β asymptotically (for large n) the probability of raising a false alarm is kept at level α.

(iii) In line with [5, Ch. VI.E, Eqn. (43)] we reject H_0 ("raise an alarm")

(A) as soon as

$$\max_{\beta \in \mathscr{B}} \left(\frac{1}{n} \mathscr{L}_{n\beta}^X(X) - b_X(\beta) \right) := \max_{\beta \in \mathscr{B}} \left(\frac{1}{n} \log \frac{g_{n\beta}(X)}{f_n(X)} - b_X(\beta) \right) > 0, \qquad (1)$$

where f_n and $g_{n\beta}$ are the joint densities of $X_1, \ldots, X_n$ under H_0 and $H_1(n\beta+1)$ respectively.

(B) Accordingly, for the innovations based approach, we raise an alarm when

$$\max_{\beta \in \mathscr{B}} \left(\frac{1}{n} \mathscr{L}_{n\beta}^\varepsilon(\varepsilon) - b_\varepsilon(\beta) \right) > 0. \qquad (2)$$

We explain steps (i) and (ii) in greater detail in Sections 3 and 4.

3   Computation of the Log-Likelihood Ratio Test Statistic

We now formulate the null hypothesis and the alternative hypothesis for the case of a change in scale in terms of an appropriate test statistic, for (A) the observations based approach, and (B) the innovations based approach. Approach (A) can be used to detect a change point in a stationary Gaussian process, for approach (B) we restrict our exposition to linear processes[1], which allows for the rich class of *vector autoregressive moving average* (VARMA) processes. In both cases we may assume, without loss of generality, that the pre-change process has mean vector $\mathbf{0}$ (we may subtract the original mean vector to achieve this).

(A) For the observations based approach, to compute the LLR, we consider the n observations within each window jointly. The joint distribution of $X := (X_1^T, \ldots, X_n^T)^T$ under H_0 is $\mathcal{N}_{nd}(\mathbf{0}, \Sigma)$, a Gaussian distribution of dimension nd. We write the covariance matrix Σ of the joint observations as a block Toeplitz matrix of the individual autocovariance matrices $\Gamma_h = \text{Cov}(X_t, X_{t-h})$.

Now we can formulate H_0 and H_1 more specifically. For all $\beta \in \mathcal{B}$ we want to test

$$H_0 : X \sim \mathcal{N}_{dn}(\mathbf{0}, \Sigma) \quad \text{vs.} \quad H_1(n\beta + 1) : X \sim \mathcal{N}_{dn}(\nu, T),$$

where

$$\nu = \left(\mathbf{0}^T \ldots, \mathbf{0}^T, \bar{\nu}^T, \ldots, \bar{\nu}^T \right)^T, \quad T = \left(\begin{array}{c|c} \Sigma^{(dn\beta)} & 0 \\ \hline 0 & c \cdot \Sigma^{(dn(1-\beta))} \end{array} \right),$$

with $\bar{\nu} = c\mu - \mu$, μ denoting the mean vector before centering, and where m in $\Sigma^{(m)}$ denotes the dimension of the matrix. For method (A), we assume that the sequence before $n\beta + 1$ is independent of the sequence afterward. This assumption enables computations, and is reasonable if a change has taken place, and the cause of the change is 'external' (as in the examples mentioned in the introduction).

The LLR for testing $X \sim \mathcal{N}_{nd}(\mathbf{0}, \Sigma)$ against the simple alternative hypothesis $X \sim \mathcal{N}_{nd}(\nu, T)$ can be computed as

$$\mathcal{L}_n^X(X) = \frac{1}{2} \log|\Sigma| - \frac{1}{2} \log|T| + \frac{1}{2} X^T \Sigma^{-1} X - \frac{1}{2}(X - \nu)^T T^{-1}(X - \nu).$$

Filling in ν, Σ, T, the LLR for testing against a change in scale at a specific point $n\beta + 1$ becomes

$$\mathcal{L}_{n\beta}^X(X) = -\frac{1}{2} dn(1 - \beta) \log c + \frac{1}{2} \check{X}^T \left(\Sigma^{(dn(1-\beta))} \right)^{-1} \check{X}$$
$$- \frac{1}{2c} \left(\check{X} - \nu^{(dn(1-\beta))} \right)^T \left(\Sigma^{(dn(1-\beta))} \right)^{-1} \left(\check{X} - \nu^{(dn(1-\beta))} \right), \qquad (3)$$

where $\check{X} := (X_{n\beta+1}^T, \ldots, X_n^T)^T$.

(B) For the innovations based approach we need to impose further assumptions (see also Table 1). We focus on *linear* processes, i.e., we assume that X_t can be modeled as

$$X_t = \sum_{j=0}^{\infty} \Psi_j Z_{t-j} =: \Psi(L) Z_t, \qquad (4)$$

---

[1] Generalization may be possible using Wold's decomposition theorem.

(L denoting the lag operator: $LZ_t := Z_{t-1}$), with uncorrelated error terms $Z_t \sim N_d(\mathbf{0}, \Omega)$, and where the Ψ_j form an absolutely summable sequence of coefficient matrices [3].

We further need to assume that the process be *invertible*, i.e., that i.i.d. the sequence of *innovations*

$$\varepsilon_t := X_t - \mathbb{E}(X_t \mid X_{t-1}, \dots, X_1) \tag{5}$$

can be extracted as a well-defined function of present and past observations (lie in their closed linear span). If X_t is given by a VARMA(p,q) process

$$X_t = \sum_{i=1}^{p} A_i X_{t-i} + \sum_{j=1}^{q} B_j Z_{t-j} + Z_t,$$

then a well-known sufficient condition for invertibility is that $|B(z)|$ has no roots on the unit circle, where $B(z) = I + \sum_{j=1}^{q} B_j z^j$ denotes the MA-polynomial [3].

Given such an invertibility assumption holds, a proportional change in the co-variance matrix of the observations (i.e. covariances are inflated by c) can be detected as a proportional change in the covariance matrix of the innovations, as it is known [3, Eqn. (11.1.13)] that under H_0 the autocovariances of X_t are given by $\Gamma_h = \sum_j \Psi_j \Omega \Psi_{j-h}^{\mathrm{T}}$. It has been shown in [2] that (for VARMA processes) the sequence of innovations is a sufficient statistic for detecting a change in the mean value.

Then, defining $\theta = \Psi(L)^{-1}\bar{v}$, the above hypotheses can equivalently be formulated as

$$H_0: \varepsilon_t \sim N_d(\mathbf{0}, \Omega), \, t = 1, \dots, n \quad \text{vs.} \quad H_1(n\beta + 1): \begin{cases} \varepsilon_t \sim N_d(\mathbf{0}, \Omega), & t \le n\beta, \\ \varepsilon_t \sim N_d(\theta, c\Omega), & t > n\beta. \end{cases}$$

Since the innovations are independent, the LLR $\mathscr{L}_{n\beta}^{\varepsilon}(\varepsilon)$ for testing H_0 against $H_1(n\beta + 1)$ can be expressed as the sum of the LLRs at time $t > n\beta$ (since the LLR is zero for $t \le n\beta$). Therefore, $\mathscr{L}_{n\beta}^{\varepsilon}$ becomes

$$\mathscr{L}_{n\beta}^{\varepsilon}(\varepsilon) = \sum_{t=n\beta+1}^{n} \frac{1}{2} \log \frac{1}{c^d} + \frac{1}{2} \varepsilon_t^{\mathrm{T}} \Omega^{-1} \varepsilon_t - \frac{1}{2c}(\varepsilon_t - \theta)^{\mathrm{T}} \Omega^{-1} (\varepsilon_t - \theta). \tag{6}$$

Note that in this case we can compute the LLR for each new window recursively (for details, see the literature on CUSUM, e.g., [6]). On the other hand, in practice the true innovations after the change points can only be estimated as the recursion (5) requires initial conditions. The effect is minor if the order of the process is small (see Section 5).

The LLR test statistics obtained for approach (A) and (B) are compared with the associated threshold functions as derived in the next section.

4   Derivation of the Threshold Function

In this section we show how to obtain the threshold function as $b_X(\beta)$ for the observations based or $b_\varepsilon(\beta)$ for the innovations based approach. We first outline the main idea behind the derivation of the threshold function for both approaches (therefore, the subscripts X and ε are omitted).

Let $\mathbb{P}_0, \mathbb{E}_0$ denote probability and expectation under H_0. When testing H_0 against $H_1(n\beta+1)$ for any fixed $\beta \in \mathcal{B}$, the probability of a type I error is given by $\mathbb{P}_0(\mathscr{L}_{n\beta}(\cdot)/n > b(\beta))$. Since we wish this probability to be *small*, it certainly holds that $b(\beta) > \mathbb{E}_0\mathscr{L}_{n\beta}(\cdot)/n$, so that we are indeed concerned with a rare event. LD theory suggests that for fixed β the false alarm probability can be approximated by

$$\mathbb{P}_0\left(\frac{1}{n}\mathscr{L}_{n\beta}(\cdot) > b(\beta)\right) \approx \exp\left(-n\mathscr{I}(b(\beta))\right),$$

where $\mathscr{I}$ denotes a function specified below. Recall that we wish the false alarm probability on the left hand side to be kept at a small level α. This suggests to pick the threshold function b such that it satisfies

$$\alpha = \exp\left(-n\mathscr{I}(b(\beta))\right) \tag{7}$$

for all $\beta \in \mathcal{B}$. This choice entails that raising a false alarm is essentially equally likely irrespective of the supposed location of the change point within the window.

Now let us make the above more rigorous. The *limiting logarithmic moment generating function* $\Lambda(\lambda)$ associated with the distribution of the LLR is defined as

$$\Lambda(\lambda) := \lim_{n\to\infty} \frac{1}{n} \log M_n(\lambda) := \lim_{n\to\infty} \frac{1}{n} \log \mathbb{E}_0\left(e^{\lambda\mathscr{L}_{n\beta}(\cdot)}\right); \tag{8}$$

we assume for now that this function exists and is finite for every $\lambda \in \mathbb{R}$. Define $\mathscr{I}$ as the Fenchel-Legendre transform of $\Lambda(\lambda)$, that is,

$$\mathscr{I}(b(\beta)) = \sup_{\lambda\in\mathbb{R}} (\lambda b(\beta) - \Lambda(\lambda)). \tag{9}$$

Provided that Λ exists and is finite, by the Gärtner-Ellis theorem [5,9], it holds that

$$\lim_{n\to\infty} \frac{1}{n} \log \mathbb{P}_0(\mathscr{L}_{n\beta}(\cdot) > nb(\beta)) = -\mathscr{I}(b(\beta)).$$

In accordance with the idea expressed in (7), we choose the threshold function $b(\beta)$ such that it satisfies

$$-\mathscr{I}(b(\beta)) = \lim_{n\to\infty} \frac{1}{n} \log \mathbb{P}_0\left(\frac{1}{n}\mathscr{L}_{n\beta}(\cdot) - b(\beta) > 0\right) = -\gamma \tag{10}$$

for some positive $\gamma = -1/n \log \alpha$, across all $\beta \in \mathcal{B}$. Asymptotically, as $n \to \infty$, the probability of raising a false alarm within the window is then kept at level α.

To be able to obtain $b(\beta)$ from (10), we need to compute the limiting log-moment generating function $\Lambda(\lambda)$ in more explicit terms (this way we also check that it indeed exists and is finite for all λ).

(A) In Section 3 of [11] we outlined how to compute the moment generating function $M_n(\lambda)$ for testing $X \sim \mathcal{N}_{nd}(0, \Sigma)$ against $X \sim \mathcal{N}_{nd}(\nu, T)$ (for arbitrary ν, Σ, T):

$$M_n(\lambda) = \left(\frac{|\Sigma|}{|T|}\right)^{\lambda/2} \frac{1}{|\lambda T^{-1}\Sigma + (1-\lambda)I_{dn}|^{1/2}}$$
$$\times \exp\left(-\frac{\lambda}{2}\nu^T T^{-1}\nu + \frac{\lambda^2}{2}\nu^T T^{-1}\left(\lambda T^{-1} + (1-\lambda)\Sigma^{-1}\right)^{-1} T^{-1}\nu\right).$$

Filling in the specific ν, Σ, T for testing against a change in scale, this expression reduces to

$$M_{n\beta}(\lambda) = c^{-\lambda dn(1-\beta)/2}\left(\frac{\lambda}{c} + 1 - \lambda\right)^{-dn(1-\beta)/2} \times \exp\left(\bar{\nu}^T s_{n\beta}\bar{\nu}\frac{\lambda^2 - \lambda}{2(\lambda + c - \lambda c)}\right),$$

where $s_{n\beta}$ denotes the sum of all d dimensional covariance matrices within the lower right $dn(1-\beta) \times dn(1-\beta)$ dimensional block matrix in Σ^{-1}.
Using the expression we obtained for $M_n(\lambda)$, the limiting log-moment generating function as defined in (8) becomes

$$\Lambda(\lambda) = -\frac{1}{2}\lambda d(1-\beta)\log(c) - \frac{1}{2}d(1-\beta)\log\left(\frac{\lambda}{c} + 1 - \lambda\right) + \lim_{n\to\infty}\frac{1}{n}\bar{\nu}^T s_{n\beta}\bar{\nu}\frac{\lambda^2 - \lambda}{2(\lambda + c - \lambda c)}.$$

We can evaluate the limit in the specific cases (i) X_t can be modeled as d independent ARMA processes

$$X_{it} = Z_{it} + \sum_{j=1}^{p} a_{ij}X_{i,t-j} + \sum_{j=1}^{q} b_{ij}Z_{i,t-j},$$

(i.e., the d monitored traffic streams are independent), or (ii) there is no shift in mean, i.e. $\bar{\nu} = 0$. The latter may happen, for example, if the number of users stays constant while the variance of their load changes (e.g. due to application changes).

(i) In the first case, the autocovariance matrices Γ_h are diagonal, and thus the expression $\bar{\nu}^T s_{n\beta}\bar{\nu}$ reduces to $\sum_{i=1}^{d} \bar{\nu}_i^2 t_{i,n\beta}$, where $\bar{\nu}_i$ is the size of the mean shift of X_{it}, and $t_{i,n\beta}$ denotes the sum of the entries of the lower right $n(1-\beta) \times n(1-\beta)$-dimensional block matrix of Σ_i^{-1}, the inverse covariance matrix of X_{it}. From [11, Lemma 1] we have

$$\lim_{n\to\infty}\frac{t_{i,n\beta}}{n(1-\beta)} = \left(\frac{1 - \sum_{j=1}^{p} a_{ij}}{\sigma_i\left(1 + \sum_{j=1}^{q} b_{ij}\right)}\right)^2 =: \tau_i,$$

and hence, the limiting log-moment generating function exists and is finite. The threshold $b_X(\beta)$ can then be evaluated by putting the resulting rate function

$$\sup_{\lambda}\left\{\lambda b_X(\beta) + \frac{1}{2}(1-\beta)\left[\lambda d\log c + d\log\left(\frac{\lambda}{c} + 1 - \lambda\right) - \frac{\lambda^2 - \lambda}{\lambda + c - \lambda c}\sum_{i=1}^{d} \bar{\nu}_i^2\tau_i\right]\right\}$$

equal to γ. Defining $\eta = -d(1-c)^2/2\sum_{i=1}^d \bar{v}_i^2 \tau_i$, we compute the optimizing λ to be

$$\frac{c}{1-c}\left[\left(\eta + \sqrt{\eta^2 + c - d + 1 + \frac{4c\eta}{1-c}\left(\frac{b(\beta)}{1-\beta} + \frac{1}{2}\log c\right)}\right)^{-1} - 1\right]. \quad (11)$$

The threshold function $b_X(\beta)$ can be evaluated using standard numerical procedures.

(ii) If there is no shift in mean, then $M_n(\lambda)$ does not depend on $s_{n\beta}$. Hence the limiting log-moment generating function always exists, and $b_X(\beta)$ follows from

$$\gamma = \mathscr{I}(b_X(\beta)) = \sup_\lambda \left(\lambda b_X(\beta) + \frac{1}{2}d(1-\beta)\left[\lambda \log c + \log\left(\frac{\lambda}{c} + 1 - \lambda\right)\right]\right).$$

The optimizing λ is

$$-\left(\frac{d(1-\beta)}{2b_X(\beta) + d(1-\beta)\log c} + \frac{c}{1-c}\right).$$

(B) When using the innovations based approach, we may make use of the fact that innovations are independent, in which case the LLR can be written as a sum of the form $\sum_{t=n\beta+1}^n s_t$ as given in (6). It follows that $\Lambda(\lambda)$ exists as a finite number:

$$\Lambda(\lambda) = \lim_{n\to\infty} \frac{1}{n} \log\left[\mathbb{E}_0 \exp(\lambda s_1)\right]^{n(1-\beta)} = (1-\beta)\log \mathbb{E}_0 \exp(\lambda s_1).$$

The threshold can be found from putting

$$\sup_\lambda \left[\lambda b_\varepsilon(\beta) + \frac{1}{2}(1-\beta)\left(\lambda d \log c + d\log\left(\frac{\lambda}{c} + 1 - \lambda\right) - \frac{\lambda^2 - \lambda}{\lambda + c - \lambda c}\theta^T \Omega^{-1}\theta\right)\right] \quad (12)$$

equal to γ.

The optimizing λ is similar to (11) (replace η by $-d(1-c)^2/2\theta^T \Omega^{-1}\theta$).

As expected both approaches yield the same threshold function in case there is no shift in mean. We now know how to compute the LLR and the threshold function either using the observations or the innovations based approach. In the next section we evaluate the performance of the resulting detection procedures (1) and (2) respectively.

5 Numerical Evaluation

In this section we summarize the results of our numerical experimentation, carried out with MATLAB. We investigate the performance of detection methods (A) and (B) with respect to the false alarm rate and the detection delay, when testing vector autoregressive (VAR) processes against a change in scale.

We begin in Section 5.1 with an illustrative example which outlines how the testing methods (A) and (B) could be applied in practice. Then, in Section 5.2, we explain how the performance measures, false alarm rate and detection delay, are evaluated. Finally, in Section 5.3, we demonstrate the potential gain from using multidimensional detection procedures by comparing the multidimensional procedure to the corresponding one-dimensional procedure that tests each data stream individually.

5.1   On-line Detection

Let us first explain how to apply the detection methods set up in this paper for on-line detection of changes in scale in multidimensional Gaussian processes. We assume that one new observation arrives at a time, and the n most recent observations are being tested against a change with scaling factor c. As an illustrative example, we run the following procedure.

- We simulate a VAR(1) process of length N according to

$$X_t = AX_{t-1} + Z_t, \tag{13}$$

  where Z_t is Gaussian white noise with $Z_t \sim N(0, \Omega)$ for $t = 1, \dots, k-1$, $0 < k < N$, and $Z_t \sim N(\theta, c\Omega)$ afterward.
- We consider windows of size $n < k$, adding one new observation at a time while deleting the oldest.
- In order to test whether a change in scale with scaling factor c has occurred in a particular window, we determine whether (A) criterion (1) holds true if the LLR is computed as a function of observations, or (B) criterion (2) holds true if the LLR is expressed as a function of innovations. In the latter case, the innovations are extracted as $X_t - AX_{t-1}$ for all t, and thus, the assumed independence between pre- and post-change observations is neglected. We do so to account for the fact that in practice the true value of ε_k is not known as it depends on unknown initial values.
- We repeat the above steps 15,000 times, and divide the total number of alarms raised for each window by 15,000 so as to obtain the alarm ratio for each window.

Two examples are presented in Fig. 1. It can be seen that the false alarm rate (the ratio of alarms before the change point as indicated by the vertical line) is indeed kept at a low level, whereas the alarm rate increases gradually to 1 after the change has occurred. It is not surprising that the detection ratio depends on the position of the change point within the window – the more observations have been affected by the change, the easier the change can be detected.

The figure shows that method (B) results in a slightly higher detection rate than method (A). This may be due to the fact that in the test set-up for approach (A) we neglected the dependence between $X_1, \dots, X_{k-1}$ and $X_k, \dots, X_N$ under H_0.

As expected, we also see that if $\bar{v} \neq 0$, i.e., if there is a change in the mean value also, then both false alarm rate and detection rate improve; the shift in mean is an additional indicator that a change has occurred (for a formal proof of this intuitive result, see [2, Ex. 4.1.9]). In the following, we focus on the worst-case setting $\bar{v} = 0$ when evaluating the performance measures, false alarm ratio and the detection delay, in the next section.

5.2   Performance Measures

To evaluate the *false alarm rate*, we perform the above experiment; however, instead of shifting windows along a series of length $N > n$, we now consider a single window of

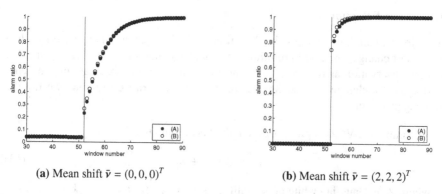

(a) Mean shift $\bar{\nu} = (0, 0, 0)^T$ **(b)** Mean shift $\bar{\nu} = (2, 2, 2)^T$

Fig. 1. Alarm ratios obtained when testing a three-dimensional AR(1) sequence of observations, simulated according to (13) with diagonal coefficient matrix A with diagonal entries 0.5 and diagonal input variance matrix Ω with diagonal entries 1, against a change in scale with $c = 2$, $\alpha = 0.01$, window size $n = 50$. The first window containing the change is indicated by a vertical line.

observations that all correspond to H_0. Then every alarm that is raised in 15,000 runs is a false alarm, and hence, the number of change points detected on average gives an estimate for the false alarm rate. The significance level is set to $\alpha \in \{0.01, 0.05\}$, and we pick $c = 2$ (as no change is simulated, the choice of c has little impact on the test results).

In order to evaluate the *detection delay*, we simulate a VAR(1) sequence where the first 49 observations correspond to H_0 while all later observations have been affected by the change. We test windows of size 50, at each point in time adding one new observation and dropping the oldest (thus, in window i only i out of 50 observations have been affected by the change). The procedure is stopped as soon as the change has been recognized, i.e., when the first alarm was raised. We then take the number of the first window for which this happened, averaged over 30,000 runs (to obtain an estimate for the *average run length* under H_1, i.e. the number of decisions that have to be taken before the change is detected), and subtract one to obtain the detection delay.

The results of these experiments, where data streams are tested *jointly*, are presented in Table 2 for a number of two-dimensional examples (next to the results from testing the streams separately as explained in Section 5.3). It can be seen that – as expected – the outcome of the experiments is similar for methods (A) and (B), and the false alarm rate is generally close to the significance level α as desired. Table 2 also shows that the detection delay is small, and provides quantitative insight into the the trade-off between the false alarm rate and the detection delay: It suffices if 22% of the observations have been affected by the change when $\alpha = 0.01$ while less than 12% need to be affected when $\alpha = 0.05$.

These and similar examples suggest that the test performance is affected neither by the sign (positive or negative) nor by the magnitude of the correlation induced by Ω

Table 2. False alarm rates and detection delays obtained from testing two-dimensional VAR(1) sequences, using (A) the observations-based approach and (B) the innovations-based approach, with $c = 2$, window size $n = 50$, mean zero. Streams are tested jointly with significance level α, and separately (ignoring interdependence) with significance level $\alpha/2$. In the latter case an alarm is raised as soon as a change point is found in any of the d streams. The standard error is given in parentheses.

Example	α	Testing	False alarm rate (A)	(B)	Delay (A)	(B)
$A = \begin{pmatrix} 0.5 & 0 \\ 0 & 0.5 \end{pmatrix}$, $\Omega = \begin{pmatrix} 1.0 & 0 \\ 0 & 1.0 \end{pmatrix}$	0.01	separately	0.007 (0.0006)	0.007 (0.0006)	14.278 (0.075)	14.139 (0.075)
		jointly	0.008 (0.0007)	0.007 (0.0007)	10.510 (0.058)	10.289 (0.058)
	0.05	separately	0.031 (0.0014)	0.032 (0.0015)	7.998 (0.050)	7.818 (0.050)
		jointly	0.038 (0.0016)	0.038 (0.0016)	5.992 (0.040)	5.802 (0.040)
$A = \begin{pmatrix} 0.5 & 0.4 \\ 0.4 & 0.5 \end{pmatrix}$, $\Omega = \begin{pmatrix} 1.0 & 0 \\ 0 & 1.0 \end{pmatrix}$	0.01	separately	0.397 (0.0040)	0.374 (0.0040)	3.264 (0.036)	3.438 (0.037)
		jointly	0.008 (0.0007)	0.007 (0.0007)	7.384 (0.055)	6.970 (0.054)
	0.05	separately	0.552 (0.0041)	0.529 (0.0041)	1.527 (0.022)	1.625 (0.023)
		jointly	0.038 (0.0016)	0.038 (0.0016)	4.105 (0.037)	3.768 (0.036)
$A = \begin{pmatrix} 0.5 & 0 \\ 0 & 0.5 \end{pmatrix}$, $\Omega = \begin{pmatrix} 1.0 & 0.5 \\ 0.5 & 1.0 \end{pmatrix}$	0.01	separately	0.006 (0.0006)	0.006 (0.0006)	15.502 (0.082)	15.340 (0.082)
		jointly	0.008 (0.0007)	0.007 (0.0007)	10.509 (0.058)	10.289 (0.058)
	0.05	separately	0.031 (0.0014)	0.031 (0.0014)	8.782 (0.055)	8.634 (0.055)
		jointly	0.038 (0.0016)	0.038 (0.0016)	5.992 (0.040)	5.802 (0.040)
$A = \begin{pmatrix} 0.5 & 0.4 \\ 0.4 & 0.5 \end{pmatrix}$, $\Omega = \begin{pmatrix} 1.0 & 0.5 \\ 0.5 & 1.0 \end{pmatrix}$	0.01	separately	0.515 (0.0041)	0.485 (0.0041)	2.674 (0.035)	2.919 (0.037)
		jointly	0.008 (0.0007)	0.007 (0.0007)	7.458 (0.055)	7.023 (0.055)
	0.05	separately	0.640 (0.0039)	0.610 (0.0040)	1.295 (0.023)	1.428 (0.022)
		jointly	0.038 (0.0016)	0.038 (0.0016)	4.146 (0.037)	3.796 (0.036)

because the change size is relative to the size of the covariances if $\bar{v} = 0$. (The effect of the shift size $\bar{v}$ has been investigated in [11] for the case of a change in mean only.) A higher correlation via A on the other hand seems to have a positive effect on the delay – the effect of a change is enhanced due to the cross correlation.

5.3 A Case for Multidimensional Testing Procedures

In this section we demonstrate the merits of multidimensional detection procedures. In general, the signature of a change in scale is stronger when it affects $d > 1$ data streams simultaneously. In fact, in case the d tested data streams are independent, and the detection probability for each of them is p, then the detection probability when testing the d streams simultaneously is $1 - (1 - p)^d$. For example, if the detection probability for one data stream is 0.8, then the detection probability for testing three i.i.d. data streams simultaneously is 0.992. As a consequence, the multidimensional procedure outperforms a procedure that tests one of the individual data streams.

The more interesting question is whether the multidimensional procedure (testing data streams jointly) performs better than a one-dimensional approach where each of the d data streams is tested *separately* but an alarm is raised as soon as a change has been detected in *any* of the streams. In the latter case the significance level is corrected using the (conservative) Bonferroni method [7], that is, it is put to α/d for each one-dimensional testing procedure.

The main conclusion we draw from the results presented in Table 2 is that indeed the multidimensional detection procedure outperforms the method of separate testing of data streams in terms of false alarm rate and detection delay, even if the sequences are independent. However, it should be noted that this benefit comes at the cost of a longer computation time.

Furthermore, it can be seen that testing the data streams separately results in a considerably larger false alarm rate as soon as the data streams are mutually dependent via the coefficient matrix A; due to the increased correlation, the process X_t makes larger jumps, but the separate testing does not account for this. It is surprising that the performance in terms of detection delay is good when streams are tested separately, but this may be explained by the high false alarm rate.

Cross-correlations in the covariance matrix of the innovations process on the other hand have a negative impact on the detection delay when testing the streams separately, whereas the false alarm rate remains low. This is because the fluctuations of the process X_t are of smaller magnitude if the error terms Z_{it} are cross-correlated. (In the example given in the table, Z_t is generated as $Z_t = \Omega^{1/2} Y_t$, where the two components of Y_t are independent standard Normals. Therefore, $Z_{1t} = Y_{1t}$ and $Z_{2t} = 0.5Y_{1t} + 0.866Y_{2t}$. This way it can be seen that jumps of Z_t are more moderate than when there is no cross-correlation in Ω.)

6 Conclusion

In this paper we explained how to set up a testing procedure for detecting a change in scale within multidimensional serially correlated Gaussian processes, and found appropriate threshold functions. In the networking context, this type of change may occur for instance as a change in scale in correlated traffic streams due to an increase in the number of users, or due to an attack on the network.

We applied the testing procedure to (A) the sequence of observations and (B) the sequence of innovations. We listed benefits and drawbacks of each approach, and saw that both performed well in numerical experiments. We also demonstrated the supremacy of multidimensional detection procedures – compared to one-dimensional testing methods – for detecting changes that affect multiple data streams simultaneously, even if the data streams are independent.

A number of interesting questions arise. For example, can we quantify the advantage of approach (B) over (A) in terms of running time? Can we compute the threshold function in more general cases? How can we generalize the LD testing procedure, for example, to detect changes in processes that are not purely indeterministic, or to detect different types of changes, such as changes in correlation structure? We hope to address these questions in future research.

Acknowledgements. Julia Kuhn is supported by Australian Research Council (ARC) grant DP130100156. The authors thank Yoni Nazarathy for many helpful comments.

References

1. Addie, R., Mannersalo, P., Norros, I.: Most probable paths and performance formulae for buffers with Gaussian input traffic. European Transactions on Telecommunications 13, 183–196 (2002)
2. Basseville, M., Nikiforov, I.: Detection of Abrupt Changes: Theory and Application, vol. 104. Prentice Hall, Englewood Cliffs, NJ, USA (1993)
3. Brockwell, P., Davis, R.: Time Series: Theory and Methods. Springer, Berlin (1987)
4. Brodsky, B.E., Darkhovsky, B.S.: Nonparametric Methods in Change-Point Problems. Kluwer Academic Publishers, The Netherlands (1993)
5. Bucklew, J.: Large Deviation Techniques in Decision, Simulation, and Estimation. Wiley Series in Probability and Mathematical Statistics. Wiley, New York (1990)
6. Callegari, C., Coluccia, A., D'Alconzo, A., Ellens, W., Giordano, S., Mandjes, M., Pagano, M., Pepe, T., Ricciato, F., Żuraniewski, P.: A methodological overview on anomaly detection. In: Biersack, E., Callegari, C., Matijasevic, M. (eds.) Data Traffic Monitoring and Analysis, pp. 148–183. Springer, Berlin (2013)
7. Casella, G., Berger, R.L.: Statistical Inference, vol. 70. Duxbury Press, Belmont (1990)
8. Chen, J., Gupta, A.: Parametric Statistical Change Point Analysis: With Applications to Genetics, Medicine, and Finance. Springer, Berlin (2012)
9. Dembo, A., Zeitouni, O.: Large Deviations Techniques and Applications, 2nd edn. Springer, New York (1998)
10. Deshayes, J., Picard, D.: Off-line statistical analysis of change-point models using non parametric and likelihood methods. In: Basseville, M., Benveniste, A. (eds.) Detection of Abrupt Changes in Signals and Dynamical Systems. LNCS, vol. 77, pp. 103–168. Springer, Heidelberg (1986)
11. Ellens, W., Kuhn, J., Mandjes, M., Żuraniewski, P.: Changepoint detection for dependent Gaussian sequences. arXiv:1307.0938 (2013) (submitted)
12. Mandjes, M.: Large Deviations for Gaussian Queues. John Wiley & Sons, Chichester (2007)
13. Mandjes, M., Żuraniewski, P.: M/G/∞ transience, and its applications to overload detection. Performance Evaluation, 507–527 (2011)

14. Page, E.: Continuous inspection scheme. Biometrika 41, 100–115 (1954)
15. Robbins, M., Gallagher, C., Lund, R., Aue, A.: Mean shift testing in correlated data. Journal of Time Series Analysis 32, 498–511 (2011)
16. Siegmund, D.: Sequential Analysis. Springer, New York (1985)
17. Sperotto, A., Mandjes, M., Sadre, R., de Boer, P.T., Pras, A.: Autonomic parameter tuning of anomaly-based IDSs: An SSH case study. IEEE Transactions on Network and Service Management 9, 128–141 (2012)
18. Tartakovsky, A.G., Rozovskii, B.L., Blazek, R.B., Kim, H.: A novel approach to detection of intrusions in computer networks via adaptive sequential and batch-sequential change-point detection methods. IEEE Transactions on Signal Processing 54, 3372–3382 (2006)
19. Tartakovsky, A.G., Veeravalli, V.: Change-point detection in multichannel and distributed systems. In: Mukhopadhyay, N., Datta, S., Chattopadhyay, S. (eds.) Applied Sequential Methodologies: Real-World Examples with Data Analysis, pp. 339–370. Marcel Dekker, NY (2004)
20. Wilson, M. (2006), A historical view of network traffic models. Unpublished survey paper, http://www.arl.wustl.edu/~mlw2/classpubs/traffic_models/

Impatient Customers in Power-Saving Data Centers

Tuan Phung-Duc

Department of Mathematical and Computing Sciences
Tokyo Institute of Technology
Ookayama, Tokyo 152-8552, Japan
tuan@is.titech.ac.jp

Abstract. In data centers, there are a huge number of servers which consume a large amount of energy. Reducing a few percent of the power consumption leads to saving a large amount of money and also saving our environment. In the current technology, an idle server still consumes about 60% of its peak and thus a simple way to save energy is shutdown of idle servers. However, when the workload increases, we have to turn on the OFF servers. A server needs some setup time to be active during which the server consumes energy but cannot process any job. Furthermore, a waiting job may abandon without service after a long waiting time which may be incurred by setup times. In this paper, we consider the power saving and the performance trade-off in data centers through a multiserver queueing model with setup time and impatient customers. We formulate the system by a level-dependent QBD process obtaining the stationary distribution and some performance measures. Our numerical results provide various insights into the performance of the system.

Keywords: multiserver queues, cloud computing, data centers, power-saving, setup time, abandonment, ON-OFF policy.

1 Introduction

1.1 Motivation

Cloud computing is a new paradigm where companies make money by providing computing services through the Internet [12]. Customers use software and hardware from a provider through the Internet so they do not have to maintain the resources by themselves. The core part of cloud computing is data center where a huge number of servers are available. The key issue for the management of data centers is to minimize the power consumption while keeping an acceptable service level for customers. It is reported that under the current technology an idle server still consumes about 60% of its peak processing jobs [2]. A simple way to save energy is to turn off immediately idle servers. This method is referred to as ON-OFF policy in the literature [6]. However, if the workload increases, OFF servers should be turned on so as to serve awaiting customers. Furthermore,

B. Sericola, M. Telek, and G. Horváth (Eds.): ASMTA 2014, LNCS 8499, pp. 185–199, 2014.

servers need some setup time to be active during which they consume power but cannot process jobs. Therefore, customers may have to wait a longer time in comparison with the case where all the servers are always ON. Furthermore, customers tend to be impatient if the waiting time is too long. If the waiting time is greater than the patience time of a customer, the customer abandons the system. From a management point of view, the abandonment of customers implies the loss of profit for the provider. Thus, setup time may incur both extra waiting time and abandonment. It is important to understand the situations under which the ON-OFF policy outperforms the ON-IDLE policy where a server is always ON.

Furthermore, the switching rate, i.e., the mean number of ON-OFF switches per unit time is also an important performance measure. This is because a server instantaneously consumes a large amount of energy when it is switched from OFF to ON. This incurs an instantaneous increase in power consumption which has a negative effect on the whole system [6]. Therefore, the relations between, power consumption, abandonment rate and switching rate on various parameters should be carefully evaluated. This motivates us to consider multiserver queue with setup cost and impatient customers.

1.2 Related Work

Although queues with setup time have been extensively investigated in the literature, most of papers deal with single server case without impatient customers [3,4,5,17]. These papers deal with single server queues with general service time distribution. Artalejo et al. [1] present a thorough analysis for multiserver queues with setup time in which the authors consider the case where at most one server can be in setup mode at a time. This mechanism is referred to as staggered setup policy in the literature [6]. It should be noted that the model in [1] is formulated by a level-independent quasi-birth-and-death process (QBD) for which an iterative solution is available [9]. Using some special structure, Artalejo et al. [1] show an analytical solution where the stationary distribution is recursively obtained without any approximation. Artalejo et al. [1] also point out that the rate matrix of the underlying QBD is explicitly obtained. Recently, motivated by applications in data centers, multiserver queues with setup time have been extensively investigated in the literature. In particular, Gandhi et al. [6,7,8] analyze multiserver queues with setup time. They derive explicit solution for the staggered setup policy and some closed form approximations for the ON-OFF policy where any number of servers can be in the setup mode.

Mitrani [12] considers models for server farms with setup cost. The author deals with the case where a group of reserve servers are shutdown concurrently if the workload is lower than some lower threshold and are powered up concurrently when the workload exceeds some upper threshold. Due to this concurrent shutdown and setup, the underlying Markov chains in [12] have a homogeneous birth-and-death structure which allows closed form solutions. The author investigates the optimal lower and upper thresholds for the system. Mitrani [10] extends his analysis to the case where each customer has an exponentially distributed random timer exceeding which the customer abandons the system. Mitrani [11] also considers a more complicated control with multiple lower and upper thresholds. Slegers et al. [16] formulate power-saving server farms using a Markov decision process (MDP) and propose some heuristic policies.

It should be noted that the threshold control policy in [10,11,12] operates in a centralized manner while the ON-OFF policy works on a decentralized manner. Thus the implementation of the latter is easier than that of the former.

1.3   Contribution

As is pointed out in Gandhi et al. [6], from an analytical point of view the most challenging model is the ON-OFF policy where the number of servers in setup mode is not limited. Gandhi et al. [6] present some approximation models for the ON-OFF policy which work well in practical parameter settings. Because setup time not only incurs extra waiting time but also causes abandonments, the impatient behavior of customers should be carefully taken into account. The main aim of this paper is a numerical investigation of the multiserver queueing model with setup time under the ON-OFF policy and impatient customers. The abandonment of customers leads to the inhomogeneity in the underlying Markov chain which allows analytical solutions for only some special cases. More specifically, the underlying Markov chain is a level-dependent quasi-birth-and-death process (LDQBD) whose stationary distribution can be numerically obtained by an efficient algorithm by Phung-Duc et al. [13]. In this paper, we adopt the algorithm in [13] to investigate the stationary distribution of the model. A comparison with some exact formulae reveals that numerical results are highly accurate. Furthermore, for the single server case, we obtain exact solutions for the partial generating functions of the joint stationary distribution of the state of the server and the number of waiting customers.

The rest of this paper is organized as follows. Section 2 describes the model in details. Section 3 presents a level-dependent QBD formulation for the model and its analysis. Section 4 is devoted to an analytical solution for the single server case. Section 5 presents some numerical results and Section 6 concludes the paper and gives some remarks.

2   Model

We consider $M/M/c$ queueing systems with setup time and abandonment. Customers arrive at the system according to a Poisson process with rate λ. In this system, a server is shutdown immediately if it has no job to do. An arriving job seeing an OFF server turns on the server. However, a server needs some setup time to be active so as to serve a waiting customer. We assume that the setup time follows an exponential distribution with mean $1/\nu$. Let j denotes the number of customers in the system and i denotes the number of active servers. The number of servers in setup process is given by $\min(j - i, c - i)$. Under these assumptions, the number of active servers does not exceed the number of customers in the system. It should be noted that in this model a server is in either

busy, off or setup. Furthermore, we assume that each waiting job has an exponentially distributed timer with mean $1/\gamma$. If the waiting time is larger than the
timer, awaiting job leaves the system without receiving a service. We assume
that the service time of jobs follows an exponential distribution with mean $1/\mu$.

3   Level-Dependent QBD Process

In this section, we present the level-dependent quasi-birth-and-death process for
the model.

Let $C(t)$ and $N(t)$ denote the number of busy servers and the number of jobs
in the system (including those in service), respectively. Under the assumptions
made in Section 2, it is easy to confirm that $\{X(t) = (N(t), C(t)); t \geq 0\}$ forms
a Markov chain in the state space

$$S = \{(i, j); j = 0, 1, \ldots, \min(i, c), i \in \mathbb{Z}_+\},$$

where $\mathbb{Z}_+$ denotes the set of non-negative integer. The infinitesimal generator of
$\{X(t); t \geq 0\}$ is given by

$$Q = \begin{pmatrix} Q_1^{(0)} & Q_0^{(0)} & O & O & \cdots \\ Q_2^{(1)} & Q_1^{(1)} & Q_0^{(1)} & O & \cdots \\ O & Q_2^{(2)} & Q_1^{(2)} & Q_0^{(2)} & \cdots \\ O & O & Q_2^{(3)} & Q_1^{(3)} & \cdots \\ \vdots & \vdots & \vdots & \vdots & \ddots \end{pmatrix},$$

where O denotes a zero matrix with an appropriate dimension. The block matrices $Q_2^{(i)}$, $Q_1^{(i)}$ and $Q_0^{(i)}$ $(i \geq c)$ are explicitly given as follows.

$$Q_0^{(i)} = \lambda I,$$

$$Q_1^{(i)} = \begin{pmatrix} -q_0^{(i)} & c\nu & 0 & \cdots & \cdots & 0 \\ 0 & -q_1^{(i)} & (c-1)\nu & \ddots & & \vdots \\ 0 & 0 & -q_2^{(i)} & \ddots & \ddots & \vdots \\ \vdots & \ddots & \ddots & \ddots & \ddots & 0 \\ \vdots & & \ddots & \ddots & -q_{c-1}^{(i)} & \nu \\ 0 & \cdots & & 0 & 0 & -q_c^{(i)} \end{pmatrix},$$

$$Q_2^{(i)} = diag(i\gamma, (i-1)\gamma + \mu, \ldots, (i-c)\gamma + c\mu),$$

where $q_j^{(i)} = \lambda + (c-j)\nu + (i-j)\gamma + j\mu$ and I is the identity matrix of an
appropriate size.

For $i = 0, 1, \ldots, c - 1$, $Q_2^{(i)}$, $Q_1^{(i)}$ and $Q_0^{(i)}$ are $(i + 1) \times (i + 2)$, $(i + 1) \times (i + 1)$ and $(i + 1) \times i$ matrices whose contents are given as follows.

$$Q_0^{(i)} = \begin{pmatrix} \lambda & 0 & \cdots & 0 & 0 \\ 0 & \lambda & \ddots & \vdots & \vdots \\ \vdots & \ddots & \ddots & 0 & 0 \\ 0 & \cdots & 0 & \lambda & 0 \end{pmatrix},$$

$$Q_1^{(i)} = \begin{pmatrix} -q_0^{(i)} & i\nu & 0 & \cdots & \cdots & 0 \\ 0 & -q_1^{(i)} & (i-1)\nu & \ddots & & \vdots \\ 0 & 0 & -q_2^{(i)} & \ddots & \ddots & \vdots \\ \vdots & \ddots & \ddots & \ddots & \ddots & 0 \\ \vdots & & & \ddots & -q_{i-1}^{(i)} & \nu \\ 0 & \cdots & & \cdots & 0 & -q_i^{(i)} \end{pmatrix},$$

$$Q_2^{(i)} = \begin{pmatrix} i\gamma & 0 & \cdots & \cdots & & 0 \\ 0 & (i-1)\gamma + \mu & \ddots & \ddots & & \vdots \\ 0 & 0 & & \ddots & & \vdots \\ \vdots & \ddots & & \ddots & \ddots & 0 \\ \vdots & & & \ddots & \gamma + (i-1)\mu \\ 0 & \cdots & & \cdots & 0 & i\mu \end{pmatrix},$$

where $q_j^{(i)} = (i - j)\nu + (i - j)\gamma + j\mu$ $(j = 0, 1, \ldots, i)$.

In what follows we assume that $\{X(t); t \geq 0\}$ is positive recurrent and thus the stationary distribution uniquely exists. Indeed, the Markov chain is always ergodic if $\gamma > 0$ while if $\gamma = 0$, the ergodic condition is simply $\lambda < c\mu$ due to the fact that eventually all the servers are active if the queue length is long enough.

We define the stationary probabilities and vectors as follows.

$$\pi_{i,j} = \lim_{t \to \infty} \Pr(N(t) = i, C(t) = j), \qquad (i, j) \in \mathcal{S},$$
$$\boldsymbol{\pi}_i = (\pi_{i,0}, \pi_{i,1}, \ldots, \pi_{i,\min(i,c)}), \qquad i \in \mathbb{Z}_+,$$
$$\boldsymbol{\pi} = (\boldsymbol{\pi}_0, \boldsymbol{\pi}_1, \ldots).$$

The stationary distribution $\boldsymbol{\pi}$ is the unique solution of

$$\boldsymbol{\pi} Q = \mathbf{0}, \qquad \boldsymbol{\pi} e = 1,$$

where $\mathbf{0}$ and e represent a row vector of zeros and a column vector of ones with an appropriate size. According to the matrix analytic methods [15], we have

$$\boldsymbol{\pi}_i = \boldsymbol{\pi}_{i-1} R^{(i)}, \qquad i \in \mathbb{N},$$

where $\mathbb{N} = \{1, 2, \ldots\}$ and π_0 is the solution of the boundary equation

$$\pi_0(Q_1^{(0)} + R^{(1)}Q_2^{(1)}) = 0, \qquad \pi_0(e + R^{(1)}e + R^{(1)}R^{(2)}e + \cdots) = 1.$$

Here $\{R^{(i)}; i \in \mathbb{N}\}$ is the minimal nonnegative solution of the following equation

$$Q_0^{(i-1)} + R^{(i)}Q_1^{(i)} + R^{(i)}R^{(i+1)}Q_2^{(i+1)} = O. \qquad (1)$$

Equation (1) is equivalent to

$$R^{(i)}\left(Q_1^{(i)} + R^{(i+1)}Q_2^{(i+1)}\right) = -Q_0^{(i-1)}, \qquad (2)$$

or

$$R^{(i)} = -Q_0^{(i-1)}\left(Q_1^{(i)} + R^{(i+1)}Q_2^{(i+1)}\right)^{-1}. \qquad (3)$$

Letting $Q^{(i)} = Q_1^{(i)} + R^{(i+1)}Q_2^{(i+1)}$, we observe that $Q^{(i)}$ is invertible because it represents the defective infinitesimal generator of a Markov chain. We define the function $R_i(\cdot)$ as follows.

$$R_i(X) = -Q_0^{(i-1)}\left(Q_1^{(i)} + XQ_2^{(i+1)}\right)^{-1}.$$

Phung-Duc et al. (2010) propose an algorithm for $R^{(i)}$ based on Lemma 1 below.

Lemma 1 (Proposition 2.4 in [13]). it We define $\{R_k^{(i)}; k \in \mathbb{Z}_+\}$ as follows.

$$R_0^{(i)} = O, \qquad R_k^{(i)} = R_i\left(R_{k-1}^{(i+1)}\right), \qquad k \in \mathbb{N},$$

or equivalently

$$R_k^{(i)} = R_i \circ R_{i+1} \circ \cdots \circ R_{i+k-1}(O), \qquad k \in \mathbb{N},$$

where $f \circ g(\cdot) = f(g(\cdot))$. We then have $\lim_{k\to\infty} R_k^{(i)} = R^{(i)}$.

Corollary 2. $\{R^{(n)}; n \in \mathbb{N}\}$ are upper diagonal matrices.

Proof. We observe that $Q_0^{(i)}, Q_1^{(i)}$ ($i \in \mathbb{Z}_+$) and $Q_2^{(i)}$ ($i \in \mathbb{N}$) are upper triangular matrices. Thus, it is easy to check that $R_1^{(n)}$ is an upper diagonal matrix too. Consequently, we confirm from the definition that $R_k^{(n)}$ is an upper diagonal matrix $\forall n, k \in \mathbb{N}$. Therefore, it follows from Lemma 1 that $R^{(n)}$ is also an upper diagonal matrix.

Remark 1. Based on Lemma 1, approximations to the rate matrices and the stationary distribution can be calculated by the algorithms in [13]. The main step is the backward calculation in (3) and Lemma 1 whose complexity might be improved using the sparsity presented in Corollary 2. We refer to Appendix A for details. A similar technique is also used in [14].

4   Single Server Case

In this section, we derive the analytical solution for the single server case. Let $S(t)$ denote the number of active servers, i.e., $S(t) = 0$ if the server is not active and $S(t) = 1$ if the server is active. Let $Q(t)$ denote the number of waiting customers. Let $p_{i,j}$ denote the stationary probability that $S(t) = i$ and $Q(t) = j$. We write down the balance equations as follows.

$$\lambda p_{0,0} = \gamma p_{0,1} + \mu p_{1,0}, \tag{4}$$
$$(\lambda + i\gamma + \nu)p_{0,i} = \lambda p_{0,i-1} + (i+1)\gamma p_{0,i+1}, \qquad i = 1, 2, \ldots, \tag{5}$$
$$(\lambda + \mu)p_{1,0} = \nu p_{0,1} + (\gamma + \mu)p_{1,1}, \tag{6}$$
$$(\lambda + \nu + i\gamma)p_{1,i} = \lambda p_{1,i-1} + \nu p_{0,i+1} + [(i+1)\gamma + \mu]p_{1,i+1}, \quad i = 1, 2, \ldots. \tag{7}$$

We define the partial generating functions as follows.

$$p_0(z) = \sum_{i=0}^{\infty} p_{0,i} z^i, \qquad p_1(z) = \sum_{i=0}^{\infty} p_{1,i} z^i.$$

Transforming the balance equations (4) and (5), we obtain

$$\lambda p_0(z) + \nu(p_0(z) - p_{0,0}) + \gamma z p_0'(z) = \lambda z p_0(z) + \gamma p_0'(z) + \mu p_{1,0},$$

yielding

$$p_0'(z) = \left[\frac{\lambda}{\gamma} - \frac{\nu}{\gamma(z-1)} \right] p_0(z) + \frac{\nu p_{0,0} + \mu p_{1,0}}{\gamma(z-1)}. \tag{8}$$

Similarly, we obtain the following differential equation for $p_1(z)$ by transforming equations (6) and (7).

$$(\lambda + \mu)p_1(z) + \gamma z p_1'(z) = \lambda z p_1(z) + \frac{\nu}{z}(p_0(z) - p_{0,0}) + \frac{\mu}{z}(p_1(z) - p_{1,0}) + \gamma p_1'(z),$$

or equivalently

$$p_1'(z) = \left(\frac{\lambda}{\gamma} - \frac{\mu}{\gamma z} \right) p_1(z) + \frac{\nu p_0(z) - (\nu p_{0,0} + \mu p_{1,0})}{\gamma z(z-1)}. \tag{9}$$

First, we solve equation (8). The general solution of (8) is given as follows.

$$p_0(z) = B(z) \exp\left(\frac{\lambda}{\gamma} z \right) (1 - z)^{-\frac{\nu}{\gamma}}, \tag{10}$$

where

$$B(z) = C - \int_0^z \exp\left(-\frac{\lambda}{\gamma} u \right) (1 - u)^{\frac{\nu}{\gamma}} \frac{\nu p_{0,0} + \mu p_{1,0}}{\gamma(1 - u)} du.$$

Because $p_0(z)$ is analytic at $z = 1$, we have $B(1) = 0$, i.e.,

$$C = \int_0^1 \exp\left(-\frac{\lambda}{\gamma} u \right) (1 - u)^{\frac{\nu}{\gamma}} \frac{\nu p_{0,0} + \mu p_{1,0}}{\gamma(1 - u)} du.$$

and thus

$$B(z) = \int_z^1 \exp\left(-\frac{\lambda}{\gamma}u\right)(1-u)^{\frac{\nu}{\gamma}-1}\frac{\nu p_{0,0} + \mu p_{1,0}}{\gamma}du. \qquad (11)$$

Therefore, it follows from (10) and (11) that

$$p_0(z) = (\nu p_{0,0} + \mu p_{1,0})\widehat{p}_0(z),$$

where $\widehat{p}_0(z)$ is a known function. Next, we find explicit expression for $p_1(z)$. The general solution for (9) is given by

$$p_1(z) = \exp\left(\frac{\lambda}{\gamma}z\right)z^{-\frac{\mu}{\gamma}}\int_0^z \exp\left(-\frac{\lambda}{\gamma}u\right)u^{\frac{\mu}{\gamma}-1}\frac{\nu p_{0,0} + \mu p_{1,0} - \nu p_0(u)}{\gamma(1-u)}du. \qquad (12)$$

Substituting $p_0(z)$ into (12) and arranging the result we eventually obtain

$$p_1(z) = (\nu p_{0,0} + \mu p_{1,0})\widehat{p}_1(z),$$

where

$$\widehat{p}_1(z) = \exp\left(\frac{\lambda}{\gamma}z\right)z^{-\frac{\mu}{\gamma}}\int_0^z \exp\left(-\frac{\lambda}{\gamma}u\right)u^{\frac{\mu}{\gamma}-1}\frac{1 - \nu\widehat{p}_0(u)}{\gamma(1-u)}du.$$

It follows from the normalization condition

$$p_0(1) + p_1(1) = 1,$$

that

$$\nu p_{0,0} + \mu p_{1,0} = \frac{1}{\widehat{p}_0(1) + \widehat{p}_1(1)}.$$

Therefore, we have

$$p_0(z) = \frac{\widehat{p}_0(z)}{\widehat{p}_0(1) + \widehat{p}_1(1)}, \qquad p_1(z) = \frac{\widehat{p}_1(z)}{\widehat{p}_0(1) + \widehat{p}_1(1)}.$$

As a byproduct, we can obtain

$$p_{0,0} = \frac{\widehat{p}_0(0)}{\widehat{p}_0(1) + \widehat{p}_1(1)}, \qquad p_{1,0} = \frac{\widehat{p}_1(0)}{\widehat{p}_0(1) + \widehat{p}_1(1)}.$$

We derive simple formulae for the mean number of waiting customers in the system. To this end, we calculate $p_0'(1)$ and $p_1'(1)$. Taking the limit of equation (8) as $z \to 1$, we obtain

$$p_0'(1) = \frac{\lambda}{\nu + \gamma}p_0(1).$$

On the other hand, taking the limit of equation (9) as $z \to 1$, we obtain

$$p_1'(1) = \frac{(\lambda - \mu)p_1(1) + \nu p_0'(1)}{\gamma}.$$

The mean number of waiting customers is given by $p_0'(1) + p_1'(1)$.

Remark 2. Although we obtain closed form expressions for the partial generating functions, it is not convenient to compute the stationary distribution from these expressions. The level-dependent QBD formulation in the Section 3 is more efficient and can apply for general case with any c.

5   Performance Measures and Numerical Results

5.1   Performance Measures

First of all, let $E[C]$ denote the mean number of busy servers (serving a job). We then have

$$E[C] = \sum_{i=0}^{\infty} \sum_{j=0}^{\min(c,i)} \pi_{i,j} j.$$

Let $E[S]$ denote the mean number of setting up servers, i.e.,

$$E[S] = \sum_{i=0}^{\infty} \sum_{j=0}^{\min(c,i)} \pi_{i,j} (\min(i,c) - j).$$

Let $E[Q]$ denote the mean number of waiting customers. We then have

$$E[Q] = \sum_{i=0}^{\infty} \sum_{j=0}^{\min(c,i)} \pi_{i,j} (i - j).$$

We define the virtual waiting time as the time from the arrival epoch until the time the customer either abandons or is served by the server. The mean virtual waiting time is given by $E[Q]/\lambda$ due to Little's law. The mean number of OFF servers by $c - E[C] - E[S]$.

Let $E[S_w]$ denote the mean number of switches per a time unit. It should be noted that the number of switches from OFF to ON is equal to that from ON to OFF. Thus, we have

$$E[S_w] = \sum_{i=1}^{c} \pi_{i,i} \times i\mu.$$

We also have $E[S_w] = E[S]\nu$ according to the definition.

5.2   Numerical Result

In this section, we present numerical results to show the effect of parameters on performance measures. We fix $\mu = 1$ and $c = 10$ in all of our numerical experiments. The stationary distribution is obtained by Algorithm 3 in [13] with the truncation point $N = 100$, $\epsilon = 10^{-10}$ and $k_n = 2^n - 1$.

5.3   Non-abandonment Case

We observe from Figure 1 that the number of busy servers $E[C]$ is insensitive to ν and is equal to $c\rho$. This fact agrees with Little's law implying that the truncation point $N = 100$ is large enough and that our algorithm yields accurate result.

Figure 2 presents the mean virtual waiting time against the traffic intensity. We observe a very interesting phenomenon that the mean virtual waiting time is not monotonic increase with the traffic intensity. In particular, the mean virtual waiting time decreases with ρ when ρ is small and increases with ρ when ρ is relatively large. The reasons are as follows. The waiting time is influenced by the number of servers that are in setup process. When the traffic is light, increasing ρ leads to the increase in the number of setting up servers. As a result, the mean virtual waiting time decreases. However, when the traffic intensity is large enough, this influence is small, i.e., all the servers are likely ON. Thus, the mean virtual waiting time depends on only the amount of offered traffic. As a result, the mean virtual waiting time increases with the traffic intensity.

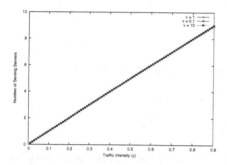

Fig. 1. Mean # of serving customers   **Fig. 2.** Mean virtual waiting time

5.4 Total Cost

As we have observed in previous sections, the ON-OFF policy may reduce the power consumption. At the same time it also incurs some extra impatience of customers leading to the loss of profit. Thus, in order to have a fair comparison between the ON-OFF and the ON-IDLE policies we need a cost function taking into account both the power consumption and the abandonment rate.

In the ON-OFF model, we assume that the power consumption per a time unit for an ON server and a setting up server is $C_a = 1$. In the ON-IDLE model, we assume that the power consumption per a time unit is C_a for an ON server while it is $C_i = 0.6 \times C_a$ for an idle server. In addition, the cost per an abandoned customer for both policies is $C_r = 3$. Under these assumptions, the cost for ON-OFF model and ON-IDLE model are given in (13) and (14) as follows.

$$C_{ON-OFF} = C_a(\mathrm{E}[C] + \mathrm{E}[S]) + C_r\mathrm{E}[Q]\gamma, \qquad (13)$$
$$C_{ON-IDLE} = C_a\mathrm{E}[C] + C_i(c - \mathrm{E}[C]) + C_r\mathrm{E}[Q]\gamma. \qquad (14)$$

It should be noted that equations (13) and (14) are calculated based on the model in this paper and the corresponding M/M/c model with impatient customers and without setup time, respectively.

We observe from Figures 3 and 4 that there exists some $\nu_{\gamma,\rho}$ such that ON-IDLE outperforms the ON-OFF policy when $\nu < \nu_{\gamma,\rho}$ while the latter is superior to the former for $\nu > \nu_{\gamma,\rho}$. This suggests that when the setup time is short enough the ON-OFF policy outperforms the ON-IDLE one while it is better to keep the servers always ON if the setup time is relatively long.

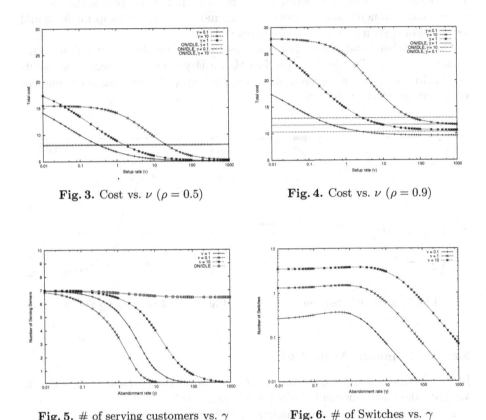

Fig. 3. Cost vs. ν ($\rho = 0.5$)   **Fig. 4.** Cost vs. ν ($\rho = 0.9$)

Fig. 5. # of serving customers vs. γ **Fig. 6.** # of Switches vs. γ

5.5 Performance Against γ

We investigate the performance measures against the abandonment rate (γ). We also fix the traffic intensity $\rho = \lambda/(c\mu) = 0.7$, i.e., $\lambda = 7$.

We observe from Figure 5 that the mean number of busy servers decreases with γ as is expected. This is because a large impatient rate γ implies the loss of traffic and thus the mean number of busy servers decreases.

Figure 6 represents the switching rate against the abandonment rate. We also observe that the switching rate increases and then decreases with γ when γ is relatively small and relatively large, respectively. Figure 7 shows that the number of OFF servers increases with γ as expected. An interesting point is that all the curves cross at $\gamma = 1.0$.

Figure 8 presents the number of starting up servers against γ. We observe an interesting phenomenon where the mean number of servers in setup mode increases with a relatively small γ and while decreasing with a relatively large γ. The reason is given as follows. The number of servers in setup model is influenced by two factors. First, the number of servers in setup mode decreases with the abandonment rate when γ is large enough. This is because the increase in the abandonment rate γ means that the more servers in setup mode should be shutdown. As a result, $E[S]$ decreases with γ.

On the other hand, increasing γ means that the number of abandoned customers increases. This leads to the fact that a large amount of servers are shutdown. However, they are eventually setup, leading to the increase in the number of servers in setup mode $E[S]$.

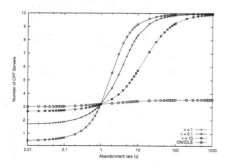

Fig. 7. # of OFF Servers vs. γ

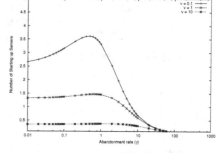

Fig. 8. # of Starting up Servers vs. γ

5.6   Performance Against ν

In this section, we investigate the influence of ν on the performance measures. We keep the traffic intensity $\lambda/(c\mu) = 0.7$, i.e., $\lambda = 7$.

Figure 9 presents the mean number of busy servers against ν for $\gamma = 0.1, 1, 10$ and $\gamma = 0$. We observe that the number of busy servers increases with ν for the cases $\gamma = 0.1, 1$ and 10 while keeping constant for the case $\gamma = 0$. It should be noted that $\nu = 0$ corresponds to the case where a customer never abandons the system. In this case, the number of busy servers is equal to the offered load $\lambda/\mu = 7$. We also observe that when $\nu \to \infty$, the mean number of busy servers tends to some fixed value. The reason is that when $\nu \to \infty$ our system tends to that without setup time.

Figure 10 shows the mean number of waiting customers against ν. We observe from all four curves that the mean number of waiting customers decreases with ν as expected. This is because a fast setup time results in increasing the number of busy servers. As a result, the mean number of waiting customers decreases. We also observe that the mean number of waiting customers decreases with the abandonment rate γ as is expected. We further observe that the mean number of waiting customers tends to some fixed value as $\nu \to \infty$.

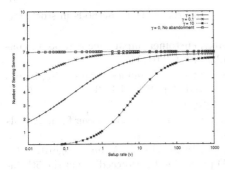

Fig. 9. Mean # of serving customers

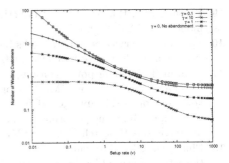

Fig. 10. Mean # of waiting customers

6 Conclusion

In this paper, we have analyzed the multiserver queue with setup time and impatient customers for data centers. We have formulated the model using a level-dependent QBD process whose rate matrices have an upper diagonal structure. This structure might be used to develop a special algorithm to compute the rate matrices and then the stationary distribution. We have analyzed the influence of various parameters on the power consumption, the queue length and the mean number of switches. Our numerical results have shown that a decision whether to choose the ON-OFF policy or the ON-IDLE policy should be carefully considered based on not only the power consumption but also the abandonment rate and the switching rate. For future work, we may use Lemma 1 to improve reduce the complexity of the algorithms so as to deal with large-scale systems.

References

1. Artalejo, J.R., Economou, A., Lopez-Herrero, M.J.: Analysis of a multiserver queue with setup times. Queueing Systems 51, 53–76 (2005)
2. Barroso, L.A., Holzle, U.: The case for energy-proportional computing. Computer 40, 33–37 (2007)
3. Bischof, W.: Analysis of M/G/1-Queues with Setup Times and Vacations under Six Different Service Disciplines. Queueing Systems: Theory and Applications 39, 265–301 (2001)
4. Choudhury, G.: On a batch arrival Poisson queue with a random setup time and vacation period. Computers and Operations Research 25, 1013–1026 (1998)
5. Choudhury, G.: An MX/G/1 queueing system with a setup period and a vacation period. Queueing Systems: Theory and Applications 36, 23–38 (2000)
6. Gandhi, A., Harchol-Balter, M., Adan, I.: Server farms with setup costs. Performance Evaluation 67, 1123–1138 (2010)
7. Gandhi, A., Gupta, V., Harchol-Balter, M., Kozuch, M.A.: Optimality analysis of energy-performance trade-off for server farm management. Performance Evaluation 67, 1155–1171 (2010)
8. Gandhi, A., Harchol-Balter, M.: M/G/k with staggered setup. Operations Research Letters 41, 317–320 (2013)

9. Latouche, G.: Ramaswami, Introduction to matrix analytic methods in stochastic modeling. SIAM (1999)
10. Mitrani, I.: Service center trade-offs between customer impatience and power consumption. Performance Evaluation 68, 1222–1231 (2011)
11. Mitrani, I.: Trading power consumption against performance by reserving blocks of servers. In: Tribastone, M., Gilmore, S. (eds.) EPEW/UKPEW 2012. LNCS, vol. 7587, pp. 1–15. Springer, Heidelberg (2013)
12. Mitrani, I.: Managing performance and power consumption in a server farm. Annals of Operations Research 202, 121–134 (2013)
13. Phung-Duc, T., Masuyama, H., Kasahara, S., Takahashi, Y.: A simple algorithm for the rate matrices of level-dependent QBD processes. In: Proceedings of the 5th International Conference on Queueing Theory and Network Applications, pp. 46–52. ACM (2010)
14. Phung-Duc, T., Masuyama, H., Kasahara, S., Takahashi, Y.: A matrix continued fraction approach to multiserver retrial queues. Annals of Operations Research 202, 161–183 (2013)
15. Ramaswami, V., Taylor, P.G.: Some properties of the rate operations in level dependent quasi-birth-and-death processes with countable number of phases. Stochastic Models 12, 143–164 (1996)
16. Slegers, J., Thomas, N., Mitrani, I.: Dynamic server allocation for power and performance. In: Kounev, S., Gorton, I., Sachs, K. (eds.) SIPEW 2008. LNCS, vol. 5119, pp. 247–261. Springer, Heidelberg (2008)
17. Takagi, H.: Priority queues with setup times. Operations Research 38, 667–677 (1990)

A   Efficient Method for Feedback Computation

The main step for the calculation of the rate matrices and the stationary distribution is to solve the equation

$$X \left(Q_1^{(i)} + R^{(i+1)} Q_2^{(i+1)} \right) = -Q_0^{(i-1)}, \qquad i \in \mathbb{N}. \tag{15}$$

provided that $R^{(i+1)}$ is given. With the observation that X and $R^{(i+1)}$ are upper triangular matrices, (15) can be solved effectively by a simple recursive scheme.

First, we consider the case where $i \geq c + 1$. In this case, the size of involved matrices is $(c + 1) \times (c + 1)$. For simplicity, we consider a problem solving the following equation

$$XA = -\lambda I,$$

where both X and A are upper triangular matrices of size $(c + 1) \times (c + 1)$. Let

$$x_i = (0, 0, \ldots, 0, x_{i,i}, x_{i,i+1}, \ldots, x_{i,c})$$

denote the i-th row of X. The above equation is equivalent to

$$x_i A = (0, 0, \ldots, 0, -\lambda, 0, \ldots, 0), \qquad i = 0, 1, \ldots, c,$$

where $-\lambda$ is the $(i+1)$-th entry in the vector of the left hand side. The solution of this equation is given by the following recursive formulae.

$$x_{i,i} = -\frac{\lambda}{a_{i,i}}, \qquad x_{i,j} = -\frac{\sum_{k=i}^{j-1} x_{i,k} a_{k,j}}{a_{j,j}}, \qquad j = i+1, i+2, \ldots, c.$$

Second, we consider the problem obtaining $R^{(i)}$ provided that $R^{(i+1)}$ is given for the case $i = 1, 2, \ldots, c$. In this case, we need to solve the following equation

$$XA = -Q_0^{(i-1)},$$

where A is an $(i+1) \times (i+1)$ matrix and X is an $i \times (i+1)$ matrix. Let $\boldsymbol{x}_j = (0, 0, \ldots, x_{j,j}, x_{j,j+1}, \ldots, x_{j,i})$ $(j = 0, 1, \ldots, i-1)$ denote the j-th row vector of X. The above equation is equivalent to

$$\boldsymbol{x}_j A = (0, 0, \ldots, -\lambda, 0, \ldots, 0), \qquad j = 0, 1, \ldots, i-1,$$

where the $-\lambda$ is the $(j+1)$-th entry of the vector in the right hand side. The solution of this equation is given by

$$x_{j,j} = -\frac{\lambda}{a_{j,j}}, \qquad x_{j,l} = -\frac{\sum_{k=j}^{l-1} x_{j,k} a_{k,l}}{a_{l,l}}, \qquad l = j+1, j+2, \ldots, i.$$

We also can easily find X using the fundamental transform (Gauss-Seidel). Indeed, we consider the following matrix.

$$\left(Q_1^{(i)} + R^{(i+1)} Q_2^{(i+1)}; -Q_0^{(i-1)} \right).$$

We use fundamental transforms to obtain the following form,

$$(I; X),$$

where I is the identity matrix. We then have $X = -Q_0^{(i-1)} (Q_1^{(i)} + R^{(i+1)} Q_2^{(i+1)})^{-1}$.

Remark 3. The computational complexity for (15) is $O(c^2)$ by the method presented in this section while that by a conventional method is $O(c^3)$.

Convexity Results for Queueing System with Variable Service Rate*

Xuelu Zhang[1,2], Jinting Wang[1,**], and Qing Ma[3]

[1] Department of Mathematics, Beijing Jiaotong University, 100044, Beijing, China
{11118382,jtwang}@bjtu.edu.cn
[2] College of Science, Qilu University of Technology, 250353, Jinan, China
[3] Department of basic courses, Shandong Women's University, 250002, Jinan, China

Abstract. In this paper we consider a single server queueing system with variable service rate. If the number of customers in system is less than a threshold, the service rate is set in a low value and it also can be switched to a high value once the number reaches to the threshold. We study five performance measures: the probability that the system is empty, the expected number in system and in queue, as well as the expected sojourn time in system and waiting time in queue. And we primarily show that these performance measures have the monotonicity or convexity with respect to the traffic intensity. These results are useful to the optimization problem in queueing system.

Keywords: Queueing system, Variable service rate, Monotonicity, Convexity.

1 Introduction

Convexity results have very extensive applications in the queueing system because these results can guarantee the unique global optimal solution which is very important for optimization models. Some performance measures are used to obtain the optimal arrival rate, service rate, the number of server and the traffic intensity. The performance measures mainly include the probability that the system is empty, the expected number in system and in queue, as well as the expected sojourn time in system and waiting time in queue.

Rolfe [1] established that the expected sojourn time is decreasing and convex in C for a $M/D/C$ queue and Dyer and Proll [2] proved Rolfe's conjecture that the result is also correct for a $M/M/C$ queue. Grassmann [3] proved that the expected number in system and in queue are convex with respect to the traffic intensity in a $M/M/C$ queue and Lee and Cohen got the same results in Ref [4] independently. In Ref [5] Harel and Zipkin gave the results of the convexity: the

---

* This work was supported by the National Natural Science Foundation of China (Nos. 11171019, 71390334), and the Program for New Century Excellent Talents in University (NCET-11-0568).
** Corresponding author. Fax: +86-10-51840433.

B. Sericola, M. Telek, and G. Horváth (Eds.): ASMTA 2014, LNCS 8499, pp. 200–207, 2014.

reciprocal of the average sojourn time and the standard deviation in a $M/M/C$ queue and generalized the similar results in a $M/G/C$ queue. Finally, they got the mean and stand deviation of the sojourn time are jointly strictly convex in arrival and service rate. Moreover Harel and Zipkin [6] established the convexity of the expected number in system and in queue, the expected sojourn time in system and waiting time in queue with respect to the S in a $M/M/S$ queue .

Recently, Dimitrakopoulos and Burnetas [7] presented an $M/M/1$ queue based on a T-threshold strategy, where the service rate switches between a low and a high value according to the number of customers in system T. Except for some equilibrium strategies, they also derived some convexity results of the waiting time and the social welfare function with respect to the arrival rate in the case of $T = 1$. In this paper we consider the $M/M/1$ queueing system with variable service rate and concentrate on properties of five main performance measures. For any $T \geq 1$, we establish the monotonicity or convexity of the performance measures with respect to the traffic intensity.

The organization of this paper is as follows. The model under consideration is described in Section 2 and we give some notations and preliminary results in Section 3. Then according to the value of traffic intensity, we study the properties of these performance measures in Section 4 and Section 5 respectively.

2   Model Description

In this model we consider an $M/M/1$ queueing system with T-threshold service policy in which all the customers arrive according to a Poisson process with rate λ. If the customers find the server is free upon arrival, they immediately occupy the server and leave the system after the service completion. Otherwise, if the server is busy upon arrival, the customers have to enter the queue to wait for service and would not leave the system without service. When the number of customers in system is less than a natural number $T(\geq 1)$, the service rate is set in a low value with rate μ_0 and it is switched to a high value with rate $\mu(> \mu_0)$ if the number is equal to or greater than the threshold T. If $\mu_0 = \mu$, the model reduce to the normal $M/M/1$ queue. Here we neglect the switching period. We assume all the customers are informed to the service policy and without loss of generality, the inter-arrival time and the service time are mutually independent. The system consists of a set of states $\{0, 1, \cdots, T - 1, T, T + 1, \cdots\}$ and the transition rate diagram is shown in Fig.1.

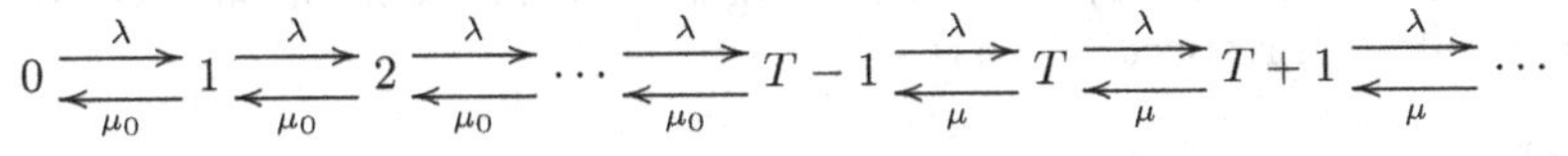

Fig. 1. Transition rate diagram of the original model

3   Notations and Preliminary Results

We study this system under the T-threshold strategy. First, we give some performance measures that have been proven in [8]. The steady-state probabilities are

$$p_n = \begin{cases} \rho_0^n p_0, & \text{if } 1 \le n < T, \\ \rho_0^{T-1} \rho^{n-T+1} p_0, & \text{if } n \ge T, \end{cases}$$

and the probability that the system is empty

$$p_0 = \begin{cases} \left(T + \frac{\rho}{1-\rho}\right)^{-1}, & \text{if } \rho_0 = 1, \rho < 1, \\ \left(\frac{1-\rho_0^T}{1-\rho_0} + \frac{\rho\rho_0^{T-1}}{1-\rho}\right)^{-1}, & \text{if } \rho_0 \ne 1, \rho < 1, \end{cases} \tag{1}$$

where $\rho_0 = \lambda/\mu_0$, $\rho = \lambda/\mu < 1$ and $\rho_0 > \rho$. The expected number in system is then

$$L = \begin{cases} p_0\left(\frac{T(T-1)}{2} + \frac{\rho[T(1-\rho)+\rho]}{(1-\rho)^2}\right), & \text{if } \rho_0 = 1, \rho < 1, \\ p_0\left(\frac{\rho_0[1+(T-1)\rho_0^T - T\rho_0^{T-1}]}{(1-\rho_0)^2} + \frac{\rho\rho_0^{T-1}[T-(T-1)\rho]}{(1-\rho)^2}\right), & \text{if } \rho_0 \ne 1, \rho < 1, \end{cases} \tag{2}$$

and the expected number in queue is

$$L_q = L - (1 - p_0). \tag{3}$$

Finally, according to Little's formulae, we find the expected sojourn time in system W and the expected waiting time in queue W_q as

$$W = L/\lambda \quad \text{and} \quad W_q = L_q/\lambda. \tag{4}$$

In the following two sections, we establish the properties of the five performance measures p_0, L, L_q, W, W_q for two cases: $\rho_0 = 1$ and $\rho_0 \ne 1$.

4   The Properties for Case $\rho_0 = 1$

Theorem 1. The probability $p_0(\rho)$ is strictly monotone decreasing and concave in ρ, $0 < \rho < 1, \forall\, T \ge 1$.

Proof. Using equation (1), we can easily get the conclusion through the derivative of $p_0(\rho)$ with respect to ρ as

$$\frac{dp_0(\rho)}{d\rho} = -\frac{1}{[T(1-\rho)+\rho]^2} < 0,$$

$$\frac{d^2 p_0(\rho)}{d\rho^2} = -\frac{2(T-1)}{[T(1-\rho)+\rho]^3} \le 0. \qquad \square$$

Theorem 2. The expected number in system $L(\rho)$ is strictly monotone increasing and strictly convex in ρ, $0 < \rho < 1$, $\forall\, T \geq 1$.

Proof. Using equations 1-2, we have

$$L(\rho) = \frac{T(T-1)}{2}p_0 + \frac{\rho}{1-\rho}.$$

So we can get the conclusions through the derivative of $L(\rho)$ with respect to ρ as

$$\frac{dL(\rho)}{d\rho} = \frac{T(T-1)}{2}\frac{dp_0(\rho)}{d\rho} + \frac{d}{d\rho}\left(\frac{\rho}{1-\rho}\right)$$

$$= \frac{2[\rho^2 + 2T\rho(1-\rho)] + T(T+1)(1-\rho)^2}{2(1-\rho)^2[T(1-\rho)+\rho]^2} > 0,$$

$$\frac{d^2L(\rho)}{d\rho^2} = \frac{T(T-1)}{2}\frac{d^2p_0(\rho)}{d\rho^2} + \frac{d^2}{d\rho^2}\left(\frac{\rho}{1-\rho}\right)$$

$$= \frac{2[3T^2\rho(1-\rho)^2 + 3T(1-\rho)\rho^2 + \rho^3] + T(T^2 + 2T - 1)(1-\rho)^3}{(1-\rho)^3[T(1-\rho)+\rho]^3} > 0.$$

$\square$

Similar to the Theorem 2, using equations (3)-(4) it is evident to get the following corollaries about $L_q(\rho)$, $W(\rho)$ and $W_q(\rho)$.

Corollary 3. The expected number in queue $L_q(\rho)$ is strictly monotone increasing and strictly convex in ρ, $0 < \rho < 1$, $\forall\, T \geq 1$.

Corollary 4. $W(\rho)$ and $W_q(\rho)$ are all strictly monotone increasing and strictly convex in ρ, $0 < \rho < 1$, $\forall\, T \geq 1$ when λ is held constant.

5   The Properties for Case $\rho_0 \neq 1$

For convenience, we introduce some notations as follows and it is not difficult to determine the sign of them:

$$C_1 = \frac{1 - \rho_0^T}{1 - \rho_0} > 0, \quad C_2 = \rho_0^{T-1} > 0, \quad C_2 - C_1 = -\frac{1 - \rho_0^{T-1}}{1 - \rho_0} \leq 0, \qquad (5)$$

$$C_1(1-\rho) + C_2\rho > 0, \quad C_3 = \frac{\rho_0[1 + (T-1)\rho_0^T - T\rho_0^{T-1}]}{(1-\rho_0)^2} \geq 0. \qquad (6)$$

Theorem 5. The probability $p_0(\rho)$ is strictly monotone decreasing and concave in ρ, $0 < \rho < 1$, $\forall\, T \geq 1$.

Proof. Using equations (1), (5) and

$$p_0(\rho) = \frac{1 - \rho}{C_1(1-\rho) + C_2\rho},$$

we can easily get the conclusion through the derivative of $p_0(\rho)$ with respect to ρ as

$$\frac{dp_0(\rho)}{d\rho} = -\frac{C_2}{[C_1(1-\rho) + C_2\rho]^2} < 0,$$

$$\frac{d^2 p_0(\rho)}{d\rho^2} = \frac{2C_2(C_2 - C_1)}{[C_1(1-\rho) + C_2\rho]^3} \leq 0.$$

□

In order to give the monotonicity of $L(\rho)$, the following Lemma is necessary.

Lemma 6. For $\forall\, T \geq 1, \rho_0 \neq 1$, the following three conclusions hold true:
(1) $A(\rho_0, T) = \rho_0^{T-1} - (T-1)\rho_0 + (T-2) \geq 0$, more specifically, $A(\rho_0, 1) = A(\rho_0, 2) = 0$ and $A(\rho_0, T) > 0$ for $T \geq 3$;
(2) $B(\rho_0, T) = 2[-\rho_0^T + T\rho_0 - (T-1)] \leq 0$, more specifically, $B(\rho_0, 1) = 0$ and $B(\rho_0, T) < 0$ for $T \geq 2$;
(3) $C(\rho_0, T) = \rho_0^{T+1} - (T+1)\rho_0 + T > 0$.

Proof. (1) It is obvious that, for any $\rho_0 \neq 1$, $A(\rho_0, 1) = A(\rho_0, 2) = 0$ and then for $T \geq 3$,

$$A(\rho_0, T+1) - A(\rho_0, T) = (1-\rho_0)(1 - \rho_0^{T-1}) = (1-\rho_0)^2(1 + \rho_0 + \cdots + \rho_0^{T-2}) > 0,$$

so $A(\rho_0, T)$ is strictly monotone increasing in T and $A(\rho_0, T) \geq A(\rho_0, 3) = (\rho_0 - 1)^2 > 0$.
(2)(3) Be the same as (1).

□

Theorem 7. The expected number in system $L(\rho)$ is strictly monotone increasing in ρ, $0 < \rho < 1, \forall\, T \geq 1$.

Proof. Using equations (1)-(2), (5)-(6) and

$$L(\rho) = \frac{1-\rho}{C_1(1-\rho) + C_2\rho} \cdot \left(C_3 + \frac{C_2\rho[T - (T-1)\rho]}{(1-\rho)^2}\right),$$

the derivative of $L(\rho)$ with respect to ρ is

$$\frac{dL(\rho)}{d\rho} = -\frac{C_2}{[C_1(1-\rho) + C_2\rho]^2}\left(C_3 + \frac{C_2\rho[T-(T-1)\rho]}{(1-\rho)^2}\right) + \frac{C_2}{C_1(1-\rho) + C_2\rho} \cdot \frac{T + (2-T)\rho}{(1-\rho)^2}.$$

So we just need to determine the sign of $\frac{dL(\rho)}{d\rho}$. Multiplying $\frac{dL(\rho)}{d\rho}$ by $\frac{[C_1(1-\rho)+C_2\rho]^2(1-\rho)^2}{C_2}$ and using equations (5)-(6), we have

$$L_1(\rho) = \frac{[C_1(1-\rho) + C_2\rho]^2(1-\rho)^2}{C_2} \cdot \frac{dL(\rho)}{d\rho}$$
$$= [C_1(1-\rho) + C_2\rho][T + (2-T)\rho] - C_3(1-\rho)^2 - C_2\rho[T - (T-1)\rho]$$
$$= (C_1 T + C_2 - 2C_1 - C_3)\rho^2 + (2C_1 - 2C_1 T + 2C_3)\rho + C_1 T - C_3$$

$$= \frac{\rho_0^{T-1} - (T-1)\rho_0 + (T-2)}{(1-\rho_0)^2}\rho^2 + \frac{2[-\rho_0^T + T\rho_0 - (T-1)]}{(1-\rho_0)^2}\rho$$

$$+ \frac{\rho_0^{T+1} - (T+1)\rho_0 + T}{(1-\rho_0)^2}$$

$$= \frac{1}{(1-\rho_0)^2}[A(\rho_0, T)\rho^2 + B(\rho_0, T)\rho + C(\rho_0, T)] = \frac{1}{(1-\rho_0)^2}L_2(\rho).$$

Next, we analyze the sign of function $L_2(\rho) = A(\rho_0, T)\rho^2 + B(\rho_0, T)\rho + C(\rho_0, T)$ based on T.

Case 1: $T = 1$. In this case, according to the Lemma 6, $A(\rho_0, T) = B(\rho_0, T) = 0, C(\rho_0, T) = (1 - \rho_0)^2$, so $L_2(\rho) = (1 - \rho_0)^2 > 0$, $L_1(\rho) = 1 > 0$ and then $L(\rho)$ is strictly monotone increasing in $\rho \in (0, 1)$.

Case 2: $T = 2$. In this case, according to the Lemma 6, $A(\rho_0, T) = 0$, $B(\rho_0, T) = -2(\rho_0 - 1)^2, C(\rho_0, T) = \rho_0^3 - 3\rho_0 + 2$, so $L_2(\rho) = [-2(1 - \rho_0)^2\rho + (\rho_0^3 - 3\rho_0 + 2)] > 0$, $L_1(\rho) = \frac{1}{(1-\rho_0)^2}L_2(\rho) > 0$. The first inequality follows from the terms $\rho \in (0, 1)$ and $\rho_0^3 - 3\rho_0 + 2 > 2(1 - \rho_0)^2$ for all $\rho_0 \neq 1$. Therefore $L(\rho)$ is strictly monotone increasing in $\rho \in (0, 1)$.

Case 3: $T \geq 3$. In this case, according to the Lemma 6, $A(\rho_0, T) > 0$, $B(\rho_0, T) < 0$ and $C(\rho_0, T) > 0$. So $L_2(\rho)$ is a normal quadratic function.

$$\Delta = [B(\rho_0, T)]^2 - 4A(\rho_0, T)C(\rho_0, T)$$
$$= 4[(T-1)\rho_0^{T+2} - (3T-2)\rho_0^{T+1} + (3T-1)\rho_0^T - T\rho_0^{T-1} + (\rho_0 - 1)^2]$$
$$= 4[T\rho_0^{T-1}(\rho_0^3 - 3\rho_0^2 + 3\rho_0 - 1) + (\rho_0 - 1)^2 - \rho_0^T(\rho_0 - 1)^2]$$
$$= 4[T\rho_0^{T-1}(\rho_0 - 1)^3 + (\rho_0 - 1)^2(1 - \rho_0^T)]$$
$$= 4(1 - \rho_0)^2[1 - \rho_0^T - (1 - \rho_0)T\rho_0^{T-1}]$$
$$= 4(1 - \rho_0)^3(1 + \rho_0 + \cdots + \rho_0^{T-1} - T\rho_0^{T-1}) > 0.$$

The above inequality follows from the term $(1-\rho_0)^3(1+\rho_0+\cdots+\rho_0^{T-1} - T\rho_0^{T-1})$ is positive for $\forall\rho_0 \neq 1$. Hence the equation $L_2(\rho) = 0$ has two distinct roots $\hat{\rho}$, $\bar{\rho}$ and by means of Lemma 6 it is easy to show that $\min\{\hat{\rho}, \bar{\rho}\} > 1$. Therefore for $\forall\rho \in (0, 1)$, $L_2(\rho) > 0$, $L_1(\rho) > 0$ and then $L(\rho)$ is strictly monotone increasing in $\rho \in (0, 1)$.

Generally speaking, $L(\rho)$ is strictly monotone increasing in ρ, $0 < \rho < 1, \forall T \geq 1$. □

Corollary 8. $W(\rho)$ is strictly monotone increasing in ρ, $0 < \rho < 1, \forall T \geq 1$ when λ is held constant.

Finally we state the convexity of $L(\rho)$ with the help of Theorem 7.

Theorem 9. The expected number in system $L(\rho)$ is strictly convex in ρ, $0 < \rho < 1, \forall T \geq 1$.

Proof. It is necessary to show that $\frac{d^2 L(\rho)}{d\rho^2} > 0$. According to Theorem 7 and Lemma 6, we have

$$
\begin{aligned}
\frac{d^2 L(\rho)}{d\rho^2} &= \frac{2C_2(C_2 - C_1)}{[C_1(1-\rho) + C_2\rho]^3}\left(C_3 + \frac{C_2\rho[T - (T-1)\rho]}{(1-\rho)^2}\right) \\
&\quad - \frac{C_2^2}{[C_1(1-\rho) + C_2\rho]^2} \cdot \frac{T + (2-T)\rho}{(1-\rho)^3} - \frac{C_2(C_2 - C_1)}{[C_1(1-\rho) + C_2\rho]^2} \cdot \frac{T + (2-T)\rho}{(1-\rho)^2} \\
&\quad + \frac{C_2}{C_1(1-\rho) + C_2\rho} \cdot \frac{(2+T) + (2-T)\rho}{(1-\rho)^3} \\
&= \frac{C_2}{[C_1(1-\rho) + C_2\rho]^3(1-\rho)^3}\left\{2(C_2 - C_1)(C_1 T + C_2 - 2C_1 - C_3)\rho^3\right. \\
&\quad + 3(C_2 - C_1)(2C_1 - 2C_1 T + 2C_3)\rho^2 + 6(C_2 - C_1)(C_1 T - C_3)\rho \\
&\quad \left. + 2[(C_2 - C_1)(C_3 - C_1 T) + C_1^2]\right\} \\
&= \frac{C_2}{[C_1(1-\rho) + C_2\rho]^3(1-\rho)^3(1-\rho_0)^2}\left\{2(C_2 - C_1)A(\rho_0, T)\rho^3\right. \\
&\quad + 3(C_2 - C_1)B(\rho_0, T)\rho^2 + 6(C_2 - C_1)C(\rho_0, T)\rho \\
&\quad \left. + 2[-(C_2 - C_1)C(\rho_0, T) + (1 - \rho_0^T)^2]\right\} \\
&= \frac{C_2}{[C_1(1-\rho) + C_2\rho]^3(1-\rho)^3(1-\rho_0)^2}L_3(\rho).
\end{aligned}
$$

So we just need to determine $L_3(\rho) > 0$ by virtue of equations (5)-(6).

Case 1: $T = 1$. In this case, according to the equations (5)-(6), $C_2 - C_1 = 0$, $L_3(\rho) = 2(1 - \rho_0)^2 > 0$ and then $\frac{d^2 L(\rho)}{d\rho^2} = \frac{2}{(1-\rho)^3} > 0$, so $L(\rho)$ is strictly convex in $\rho \in (0, 1)$.

Case 2: $T \geq 2$. In this case, according to the equation (5), $C_2 - C_1 < 0$ and considering the results $L_2(\rho) > 0$ in Theorem 7, we have

$$
\frac{dL_3(\rho)}{d\rho} = 6(C_2 - C_1)[A(\rho_0, T)\rho^2 + B(\rho_0, T)\rho + C(\rho_0, T)] = 6(C_2 - C_1)L_2(\rho) < 0.
$$

So $L_3(\rho)$ is strictly monotone decreasing in $\rho \in (0, 1)$, furthermore, according to equations (5)-(6) and Lemma 6, we get

$$
L_3(0) = 2[-(C_2 - C_1)C(\rho_0, T) + (1 - \rho_0^T)^2] > 0, \quad L_3(1) = 2C_2^2(1 - \rho_0)^2 > 0,
$$

and

$$
L_3(0) - L_3(1) = 2(C_1 - C_2)[C(\rho_0, T) + (1 - \rho_0^2)(C_1 + C_2)] > 0.
$$

Hence, $L_3(\rho) > 0$ and $\frac{d^2 L(\rho)}{d\rho^2} > 0$ for all $\rho \in (0, 1)$, so $L(\rho)$ is strictly convex in $\rho \in (0, 1)$.

Generally speaking, $L(\rho)$ is strictly convex in ρ, $0 < \rho < 1, \forall T \geq 1$. $\square$

Corollary 10. $W(\rho)$ is strictly convex in ρ, $0 < \rho < 1, \forall T \geq 1$ when λ is held constant.

References

1. Rolfe, A.J.: A Note on Marginal Allocation in Multiple-Server Service Systems. Management Science 16, 656–658 (1971)
2. Dyer, M.E., Proll, L.G.: On the Validity of Marginal Analysis for Allocating Servers in M/M/C Queues. Management Science 23, 1019–1022 (1977)
3. Grassmann, W.: The Convexity of the Mean Queue Size of the M/M/C Queue with respect to the Traffic Intensity. Journal of Applied Probability 20, 916–919 (1983)
4. Lee, H.L., Cohen, M.A.: A Note on the Convexity of Performance Measures of M/M/C Queueing Systems. Journal of Applied Probability 20, 920–923 (1983)
5. Harel, A., Zipkin, P.H.: Strong Convexity Results for Queueing System. Operations Research 35, 405–418 (1987)
6. Harel, A., Zipkin, P.H.: Convexity Results for the Erlang Delay and Loss Formulae When the Server Unilization Is Held Constant. Operations Research 59, 1420–1426 (2011)
7. Dimitrakopoulos, Y., Burnetas, A.: Customer Equilibrium and Optimal Strategies in an M/M/1 Queue with Dynamic Service Control. arXiv:1112.1372v1
8. Gross, D., Shortle, J.F., Thompson, J.M., Harris, C.M.: Fundamental of Queueing Theory. A John wiley & Sons, Inc., Publication, Hoboken (2008)

On the Performance of Secondary Users in a Cognitive Radio Network

Osama I. Salameh[1], Koen De Turck[1], Herwig Bruneel[1],
Chris Blondia[2], and Sabine Wittevrongel[1]

[1] Department of Telecommunications and Information Processing (TELIN),
Ghent University,
Sint-Pietersnieuwstraat 41, B-9000 Gent, Belgium
{osalameh,kdeturck,hb,sw}@telin.ugent.be
[2] Department of Mathematics and Computer Science,
University of Antwerp,
Middelheimlaan 1, B-2020 Antwerpen, Belgium
chris.blondia@ua.ac.be

Abstract. Cognitive radio networks employ opportunistic scheduling of secondary (unlicensed) users for the efficient use of the scarce radio spectrum resources. The main idea is that secondary users (SUs) transmit data opportunistically by utilizing idle licensed frequency bands. The transmission of a SU may get interrupted several times due to the arrival of primary (licensed) users; the SU then needs to sense the spectrum to determine another available channel for retransmission. In this paper, we investigate the performance of SUs in a cognitive radio network. To this end, we develop a three-dimensional continuous-time Markov chain (CTMC) model of the system. We present an efficient method to compute steady-state probabilities by exploiting the specific Quasi-Birth-Death structure of the CTMC. Based on this, several SU performance measures are evaluated such as the mean delay of a SU, the SU interruption probability, the probability of a SU getting discarded from the system after an interruption and the SU blocking probability upon arrival. Numerical examples illustrate the influence of system parameters such as the sensing rate on the SU performance.

Keywords: Cognitive radio, Opportunistic scheduling, Markov chain, Performance evaluation.

1 Introduction

In wireless communication, the radio spectrum is a limited resource by its nature. Traditionally, spectrum regulators use a fixed spectrum allocation policy to assign each spectrum band to dedicated (licensed) users. In recent years, the demand for wireless communication is constantly increasing due to the increasing number of wireless devices and services. This leads to the fact that the spectrum is fully allocated in many countries [1] on one hand. On the other hand, it has been shown through several spectrum measurement studies [2–4] that the

B. Sericola, M. Telek, and G. Horváth (Eds.): ASMTA 2014, LNCS 8499, pp. 208–222, 2014.
© Springer International Publishing Switzerland 2014

spectrum is heavily underutilized, which in turn leads to wasted bandwidth of wireless channels.

Cognitive radio networks (CRNs) [5] are a radical solution to spectrum scarcity. The aim of CRNs is to use the free spectrum gaps without causing any harm to licensed transmissions. For that purpose cognitive radios should be able to adapt their transmission parameters to the changing spectrum opportunities.

In CRNs, two kinds of spectrum handoff for secondary users are differentiated: proactive and reactive spectrum handoff [6]. In proactive spectrum handoff, a secondary user (SU) evacuates the current channel upon the arrival of a primary user (PU) and the interrupted SU switches to a new channel based on a pre-determined channel hopping sequence. This sequence is obtained through the analysis of the traffic statistics and the interrupted SU does not perform any spectrum sensing. In reactive spectrum handoff, an interrupted SU is required to sense the spectrum to determine an idle channel to retransmit.

The performance of CRNs has been extensively studied in the last years. A Markovian multiserver model with a random preemptive discipline is presented in [7]. The case with r kinds of user classes is considered where within each class, customers are served according to their arrival order. The moments of the sojourn time distribution for lower priority customers are derived. An assumption of this model is that higher priority customers can interrupt lower priority customers only when all servers are busy. This means that higher priority customers are aware of the presence of lower priority customers, which does not correctly depict an important aspect of the CRN paradigm where PUs are completely unaware of SU actions. In [8], continuous-time Markov chains (CTMCs) with and without queueing are proposed to analyse the performance of SUs in a CRN. The case where multiple SUs can simultaneously share a spectrum band is considered. A limitation of this work is that it assumes that an interrupted SU should wait on the same channel to complete unfinished service when the channel becomes available again. A loss model with finite population for spectrum access is presented in [9]. In this work, the delay of SUs is investigated based on a CTMC. A CTMC model to assess the maximum throughput of SUs in a heterogeneous CRN is developed in [10]. The behavior of a CRN system with both PUs and SUs is modeled using a two dimensional Markov chain in [11–14]. Throughput and forced termination of a SU are derived in [11] and blocking probabilities for PUs and SUs are calculated in [12–14]. In [13] spectrum sensing errors are considered. None of the papers discussed above however take the sensing time into consideration. M/G/1 queueing models are proposed in [15, 16], where the authors investigate the case where each SU can transmit on all channels simultaneously. These models also do not take the effect of the handoff processing time into consideration.

Other studies use the ON/OFF random process to describe the behavior of PUs on each channel, where the OFF period represents a spectrum opportunity to SUs [17, 18]. In [17] the spectrum utilization and blocking time are derived but the effect of the sensing time is not addressed. The influence of the sensing time on the data delivery time is examined in [18]. An assumption of this work is that at least one channel is always available for SUs, and the case where the system is blocked is not investigated.

Some recent analytical models that include sensing time are proposed in [6,19], but with different sensing time definitions. The authors in [19] refer to in-Band sensing where the sensing delay is the time from a collision moment between a PU and SU until the moment that the collision is detected by the SU. In [6], sensing time is defined as the time from the moment a SU is interrupted by a PU until the moment an idle channel is found. Both these definitions do not consider the need of a SU to sense upon arrival.

The goal of this paper is to analyze the performance of SUs in a CRN with reactive spectrum handoff. The sensing time is defined as the time used by a SU to scan the spectrum and detect the first idle channel. Also, each SU performs spectrum sensing every time it attempts to access a channel including the case when a SU has just arrived. Upon the arrival of a PU, the SU instantly evacuates the occupied channel, i.e. no collision between a SU and a PU can happen as in [20]. The contributions of this paper are summarized as follows. A CTMC of a CRN with reactive-decision spectrum handoff is developed. This model describes the interactions of PUs and SUs where PUs exercise preemptive priority over SUs. Based on this model, a wide range of performance measures are evaluated including the mean SU delay, the SU interruption probability, the SU discard probability and the SU blocking probability.

The structure of this paper is as follows. In Section 2 we describe the CTMC model of the system. In Section 3 we focus on the calculation of several SU performance measures. Next, the computation methodology is explained in Section 4. We provide some numerical examples in Section 5. Finally, we draw conclusions in Section 6.

2   Model Description

We consider a spectrum divided into N frequency bands forming N identical channels, i.e. channels with the same radio characteristics. Each channel can be accessed by PUs and SUs. PUs and SUs arrive according to a Poisson process with rates λ_1 and λ_2 respectively. Upon arrival, a PU is assigned to an idle channel (i.e., a channel not occupied by another PU) randomly, i.e. no default channel is allocated. The PU transmission time is exponentially distributed with rate μ_1. If all channels are occupied by PUs, a new arriving PU is blocked. Arriving SUs enter into a sensing state. We assume that a SU senses all channels one by one until it detects the first idle channel. This sensing procedure seems reasonable and has been used in [21, 22]. Also each SU chooses its own sensing order randomly. In this setting, as the sensing time can be different for each SU, we assume it is exponentially distibuted with rate σ. In case all channels are busy, a SU stays in the sensing state until at least one channel is detected idle (i.e., not occupied by any PU or other SU). After sensing, a SU enters into a transmission state. The SU transmission time is exponentially distributed with rate μ_2. A SU transmission can be interrupted by the arrival of a PU on the same channel, and in this case the SU is transferred into the sensing state again for retransmission. Thus, and different from [6], the assumptions that a default channel is assigned to SUs upon arrival and the interrupted SU has to stay on its channel if all the other channels are busy, are relaxed. We assume no collisions between SUs trying to transmit on the same channel can happen as in [6,8] and the sensing room is limited to N^1 SUs. Arriving and interrupted SUs who find a full sensing room are blocked and discarded from the system respectively.

In order to analyze this system, we create a CTMC where a state x is given as $x = (x_1, x_2, x_3)$. Here, x_1 and x_2 are the numbers of PUs and transmitting SUs and x_3 is the number of sensing SUs. The state space S contains all states such that

$$x_1 + x_2 \leq N, x_3 \leq N^1. \tag{1}$$

The transition rates $q_{x,y}$ from one state x into another state y $(x \neq y)$ are given as

$$q_{x,y} = \begin{cases} \lambda_1 \dfrac{N - x_1 - x_2}{N - x_1} & \text{if } y = (x_1 + 1, x_2, x_3), \\ & x_1 < N, \\ \lambda_1 \dfrac{x_2}{N - x_1} & \text{if } y = (x_1 + 1, x_2 - 1, x_3 + 1), \\ & x_3 < N^1, x_1 < N, \\ \lambda_1 \dfrac{x_2}{N - x_1} & \text{if } y = (x_1 + 1, x_2 - 1, x_3), \\ & x_3 = N^1, x_1 < N, \\ \lambda_2 & \text{if } y = (x_1, x_2, x_3 + 1), \\ & x_3 < N^1, \\ x_1 \mu_1 & \text{if } y = (x_1 - 1, x_2, x_3), \\ x_2 \mu_2 & \text{if } y = (x_1, x_2 - 1, x_3), \\ x_3 \sigma & \text{if } y = (x_1, x_2 + 1, x_3 - 1), \\ & (x_1 + x_2) < N, \\ 0, & \text{otherwise.} \end{cases} \tag{2}$$

To explain the above equation, we differentiate 8 transition cases. In case 1, an arriving PU does not interrupt a transmitting SU. The rate $\lambda_1(x_2)/(N - x_1)$ is the fraction of λ_1 where a transmitting SU is interrupted and transferred into the sensing state (case 2) or lost due to lack of sensing room (case 3). An arriving SU starts to sense if there is a place in the sensing room (case 4). In cases 5 and 6, a PU and a SU finish transmission respectively. In case 7, a SU leaves the sensing state into the transmitting state with rate $x_3 \sigma$ if there is at least one idle channel.

We show next that the above CTMC has a Quasi-Birth-Death (QBD) structure. This allows us to use a specific methodology to efficiently compute steady-state probabilities as described later in the computation methodology section. In order to display the infinitesimal generator Q of the chain in a QBD format, we define the QBD level as the number of sensing SUs x_3, whereas the

QBD phase corresponds to the pair (x_1, x_2) of transmitting PUs and transmitting SUs. The possible phases are ordered lexicographically for (x_1, x_2), i.e. in the order $(0,0),(0,1),\cdots,(0,N),\ (1,0),(1,1),\ \cdots,(1,N-1),\cdots,\ (i,0),\ (i,1),\ \cdots,$ $(i,N-i),\cdots,\ (N,0)$. In view of the transition rates of equation (2), the generator matrix Q has the following block structure:

$$Q = \begin{bmatrix} Q_0 & \Lambda & & & & \\ \Sigma & Q_1 & \Lambda & & & \\ & 2\Sigma & Q_2 & \Lambda & & \\ & & & \ddots & & \\ & & & (N^1-1)\Sigma & Q_{N^1-1} & \Lambda \\ & & & & N^1\Sigma & Q^*_{N^1} \end{bmatrix}, \tag{3}$$

where the submatrices $\Lambda, j\Sigma, Q_j$ and $Q^*_{N^1}$ are derived as follows. We let the subscripts i, m, j denote the number of PUs x_1, transmitting SUs x_2 and sensing SUs x_3. In the sequel, we also use the concept of horizontal ($\sim$) and vertical ($|$) matrix concatenation. Submatrix Λ then corresponds to transitions where the level is increased by 1; these are due either to the interruption of a SU by the arrival of a PU (case 2) or the arrival of a new SU (case 4). Matrix Λ is therefore given as

$$\Lambda = \begin{bmatrix} \Lambda_{2,0} & I_0 & & & \\ & \Lambda_{2,1} & I_1 & & \\ & & \ddots & & \\ & & & \Lambda_{2,N-1} & I_{N-1} \\ & & & & \Lambda_{2,N} \end{bmatrix},$$

with as component matrices the diagonal matrix $\Lambda_{2,i} = \lambda_2 I$, where I is the identity matrix, and I_i defined as

$$I_i = \begin{bmatrix} 0 & & & & \\ \lambda^*_{i,1} & & & & \\ & \lambda^*_{i,2} & & & \\ & & \ddots & & \\ & & & \lambda^*_{i,N-i} \end{bmatrix},$$

with elements $\lambda^*_{i,m} = \dfrac{\lambda_1 m}{N-i}$. Submatrix $j\Sigma$ corresponds to transitions from level j to level $j-1$ due to a SU moving from the sensing to the transmitting state after finding an idle channel (case 7). Matrix Σ is therefore structured as follows:

$$\Sigma = \begin{bmatrix} \Sigma_0 & & & \\ & \Sigma_1 & & \\ & & \ddots & \\ & & & \Sigma_{N-1} \\ & & & & 0 \end{bmatrix},$$

where $\Sigma_i = v_1 \sim \sigma * I | v2$, with v_1 a column vector of zeros concatenated to the left side of $\sigma * I$ and v_2 a row vector of zeros concatenated vertically below $v_1 \sim \sigma * I$. The non-diagonal elements in Q_j correspond to transitions where the level j remains unchanged. For $j < N^1$, such transitions are due to the arrival of a PU without an interruption of a SU (case 1), the end of a transmission of a PU (case 5) or the end of a transmission of a SU (case 6). Matrix Q_j, for $j < N^1$, and its components are hence given by

$$
Q_j =
\begin{bmatrix}
\overline{Q_{0,j}} & \Lambda_{1,0} & & & & \\
M_{1,1} & \overline{Q_{1,j}} & \Lambda_{1,1} & & & \\
& 2M_{1,2} & \overline{Q_{2,j}} & & \Lambda_{1,2} & \\
& & & \ddots & & \\
& & & (N-1)M_{1,N-1} & \overline{Q_{N-1,j}} & \Lambda_{1,N-1} \\
& & & & NM_{1,N} & \overline{Q_{N,j}}
\end{bmatrix},
$$

$$
\Lambda_{1,i} =
\begin{bmatrix}
\lambda_{i,0}^{**} & & & & \\
& \lambda_{i,1}^{**} & & & \\
& & \ddots & & \\
& & & \lambda_{i,(N-i-1)}^{**} \\
& & & & 0
\end{bmatrix},
$$

$$
\overline{Q_{i,j}} =
\begin{bmatrix}
s_{i,0,j} & & & \\
\mu_2 & s_{i,1,j} & & \\
& 2\mu_2 & s_{i,2,j} & \\
& & \ddots & \\
& & & (N-i)\mu_2 \; s_{i,N-i,j}
\end{bmatrix},
$$

$M_{1,i} = \mu_1 I \sim v_3$, where a column vector v_3 of zeros is concatenated to the right side of $\mu_1 I$, and $\lambda_{i,m}^{**} = \lambda_1(N - i - m)/(N - i)$. The diagonal elements $s_{i,m,j}$ of the matrix $\overline{Q_{i,j}}$ are such that the row sums in the generator matrix Q equal

Table 1. Diagonal elements of $\overline{Q_{i,j}}$

Elements $s_{i,m,j}$	Condition
$-(\lambda_1 + \lambda_2 + m\mu_2 + i\mu_1 + j\sigma)$	$i + m < N$
$-(\lambda_1 + \lambda_2 + m\mu_2 + i\mu_1)$	$i + m = N,\ i \neq N$
$-(\lambda_2 + i\mu_1)$	$i = N$

Table 2. Component matrices description

Matrix	Type	Size
Σ_i	square	$N - i + 1, N - i + 1$
$\Lambda_{2,i}$	square	$N - i + 1, N - i + 1$
I_i	non square	$N - i + 1, N - i$
$\overline{Q_{i,j}}$	square	$N - i + 1, N - i + 1$
$\Lambda_{1,i}$	non square	$N - i + 1, N - i$
$M_{1,i}$	non square	$N - i + 1, N - i + 2$

zero. Their values are given in Table 1. A description of the various component matrices is given in Table 2. Finally, the last diagonal block of Q is defined as $Q^*_{N^1} = Q_{N^1} + \Lambda$. This is due to case 3, where SUs can still be interrupted while the sensing room is full.

3   SU Performance Measures

From this CTMC model, we can compute various performance measures, which we detail below. All these measures are based on the steady-state vector π, which we solve from the set of equations $\pi Q = 0$. We describe in Section 4 how this can be done in an efficient manner.

First, the expected number of transmitting SUs $E[\text{UTransmit}_{\text{su}}]$ and the expected number of sensing SUs $E[\text{USense}_{\text{su}}]$ are respectively given by

$$E[\text{UTransmit}_{\text{su}}] = \sum_{x_3=0}^{N^1} \sum_{x \in S_{x_3}} x_2 \pi_x \,, \tag{4}$$

$$E[\text{USense}_{\text{su}}] = \sum_{x_3=1}^{N^1} \sum_{x \in S_{x_3}} x_3 \pi_x \,, \tag{5}$$

where S_{x_3} is the set of all states within level x_3 and π_x is the probability that the system is in state x. The blocking probability γ of SUs, i.e. the probability that an arriving SU finds a full sensing room, is given as

$$\gamma = \sum_{x \in S_{N^1}} \pi_x.$$

With these results and based on Little's law, the expected delay $E[D_{su}]$ of a SU is then calculated as

$$E[D_{su}] = (E[\text{USense}_{\text{su}}] + E[\text{UTransmit}_{\text{su}}])/(\lambda_2 - \lambda_2 \gamma) \,. \tag{6}$$

Secondly, the SU interruption probability α, i.e. the probability that a transmitting SU is interrupted upon arrival of a PU, is given as follows:

$$\alpha = \sum_{x_3=0}^{N^1} \sum_{x \in S^*_{x_3}} \pi_x \frac{\lambda_1 x_2/(N - x_1)}{D_1(x)} \,, \tag{7}$$

where $\lambda_1 x_2/(N - x_1)$ is the transition rate from state x where a SU gets interrupted due to a PU arrival, $D_1(x)$ denotes the total transition rate from state x, which is given by

$$D_1(x) = \begin{cases} x_1\mu_1 + x_2\mu_2 + \lambda_1 + \lambda_2 & \text{if } x_3 < N^1, \\ & x_1 + x_2 = N, \\ x_1\mu_1 + x_2\mu_2 + \lambda_1 + \lambda_2 + x_3\sigma & \text{if } x_3 < N^1, \\ & x_1 + x_2 < N, \\ x_1\mu_1 + x_2\mu_2 + \lambda_1 & \text{if } x_3 = N^1, \\ & x_1 + x_2 = N, \\ x_1\mu_1 + x_2\mu_2 + \lambda_1 + x_3\sigma & \text{if } x_3 = N^1, \\ & x_1 + x_2 < N, \end{cases}$$

and $S_{x_3}^* = S_{x_3} \setminus \{(N, 0, x_3)\}$ denotes the set of all states within level x_3 except the state where $x_1 = N$ (since in this state all N channels are occupied by PUs, there are no transmitting SU to interrupt and a new PU is always blocked).

Similarly, the SU discard probability β, i.e. the probability that a transmitting SU is interrupted and discarded upon arrival of a PU because of a full sensing room, is computed as

$$\beta = \sum_{x \in S_{N1}^*} \pi_x \frac{\lambda_1 x_2/(N - x_1)}{D_2(x)}, \tag{8}$$

where

$$D_2(x) = \begin{cases} x_1\mu_1 + x_2\mu_2 + \lambda_1 & \text{if } x_1 + x_2 = N, \\ x_1\mu_1 + x_2\mu_2 + \lambda_1 + x_3\sigma & \text{if } x_1 + x_2 < N. \end{cases}$$

4   Computation Methodology

Here we show how to solve the equation $\pi Q = 0$ for a CTMC with QBD structure. We solve this equation for the CTMC of Section 2.

To obtain the stationary probability vector π, we employ the Gaussian elimination technique and the concept of censored Markov chains applied to a block structured tridiagonal QBD using a similar approach as in [23]. To illustrate the major steps of the technique, let a CTMC $X(t)$ have the following generator matrix:

$$Q = \begin{bmatrix} L_0 & F_0 & & & & \\ B_1 & L_1 & F_1 & & & \\ & B_2 & L_2 & F_2 & & \\ & & \ddots & \ddots & \ddots & \\ & & & B_{n-1} & L_{n-1} & F_{n-1} \\ & & & & B_n & L_n \end{bmatrix}. \tag{9}$$

Let $L_1^u = L_1 - B_1 L_0^{-1} F_0$ be the Schur complement of L_0 in Q. Let $Q^{i \to n}$ denote the generator matrix of the Markov chain censored to the levels i to n, and let $\pi_{i \to n}$ be the corresponding (partial) distribution. We apply Schur complementation level by level: first eliminating level 0, then level 1, etc. For example, eliminating the level 0 results in

$$Q^{1\to n} = \begin{bmatrix} L_1^u & F_1 & & & \\ B_2 & L_2 & F_2 & & \\ & \ddots & \ddots & \ddots & \\ & & B_{n-1} & L_{n-1} & F_{n-1} \\ & & & B_n & L_n \end{bmatrix}. \tag{10}$$

We keep folding up the state space in this manner:

$$Q^{i\to n} = \begin{bmatrix} L_i^u & F_i & & & \\ B_{i+1} & L_{i+1} & F_{i+1} & & \\ & \ddots & \ddots & \ddots & \\ & & B_{n-1} & L_{n-1} & F_{n-1} \\ & & & B_n & L_n \end{bmatrix}, \tag{11}$$

where L_i^u is recursively given as $L_i^u = L_i - B_i L_{i-1}^u{}^{-1} F_{i-1}$. We end up with $Q^{n\to n} = L_n^u$. Finding the stationary solution then proceeds in a backwards fashion. We have that $\pi_n L_n^u = 0$, and recursively from the first block row of equation $\pi_{i\to n} Q^{i\to n} = 0$,

$$\pi_i L_i^u + \pi_{i+1} B_{i+1} = 0. \tag{12}$$

Finally, we normalize the obtained stationary probability vectors π_i corresponding to level i using $\sum_{i=0}^{n} \pi_i \mathbf{1}$, where $\mathbf{1}$ is the column vector of ones. This leads to the steady-state vector π:

$$\pi = \frac{1}{\sum_{i=0}^{n} \pi_i \mathbf{1}} \begin{bmatrix} \pi_0 & \pi_1 & \dots & \pi_n \end{bmatrix}. \tag{13}$$

5   Numerical Examples

To investigate the SU performance, we consider a cognitive radio system with $N = 20$ channels. The average transmission time for both PUs and SUs equals $1/\mu_1 = 1/\mu_2 = 10$ ms. The offered PU load ρ_{pu} and the offered SU load ρ_{su} are defined as $\lambda_1/(N\mu_1)$ and $\lambda_2/(N\mu_2)$ respectively. We consider the case where $\rho_{pu} = 0.3$ as CRNs are expected to operate under light PU load. Of particular interest is the effect of the sensing rate σ of SUs and the maximum number of sensing SUs N^1 on the performance measures derived above.

Fig. 1 shows the SU discard probability β as a function of N^1. As can be seen the curves have two parts. In the first part, for increasing N^1 the SU discard probability will increase as well. This is because an increase of N^1 for low values of N^1 will increase the number of SUs that can access the system, and hence (for a given σ) also the number of transmitting SUs, so more SUs can get interrupted by a PU, while the probability that an interrupted SU finds a full sensing room

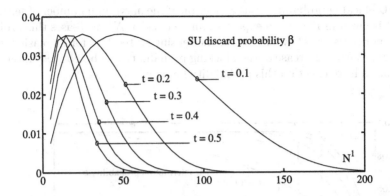

Fig. 1. SU discard probability β versus maximum number of sensing SUs N^1 for various sensing rates normalized to the average SU transmission rate $t = \sigma/\mu_2 = 0.1$, 0.2, 0.3, 0.4, 0.5, $\rho_{pu} = 0.3$, $\rho_{su} = 0.5$, $N = 20$ and $1/\mu_1 = 1/\mu_2 = 10$ ms

still remains high. In the second part, an increase in N^1 decreases the probability β because the load offered from both PUs and SUs is fixed and for high N^1 it is more likely that an interrupted SU will be able to sense again until eventually no interrupted SUs are discarded. We observe that for higher values of σ, this behavior is repeated over a smaller range of N^1 and clearly the maximum discard probability is not affected. Indeed, for higher σ, sensing SUs leave the sensing room and enter the transmitting state at a higher rate, and therefore already for lower values of N^1 there will likely remain space in the sensing room for an interrupted SU, so the SU doesn't need to be discarded.

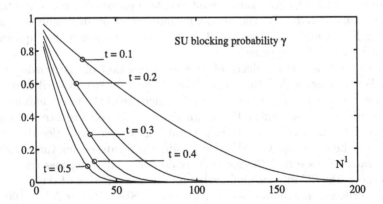

Fig. 2. SU blocking probability γ versus maximum number of sensing SUs N^1 for various sensing rates normalized to the SU average transmission rate $t = \sigma/\mu_2 = 0.1$, 0.2, 0.3, 0.4, 0.5, $\rho_{pu} = 0.3$, $\rho_{su} = 0.5$, $N = 20$ and $1/\mu_1 = 1/\mu_2 = 10$ ms

The SU blocking probability γ as a function of the maximum number of sensing users N^1 is given in Fig. 2. As expected, an increase of N^1 decreases this probability as more SUs are allowed to enter the sensing state. Also for increasing σ, a further and sharper decrease of the blocking probability can be noticed. Interestingly, σ has a huge effect on this probability for some values of N^1.

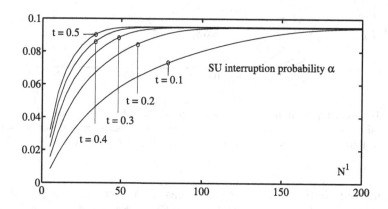

Fig. 3. SU interruption probability α versus maximum number of sensing SUs N^1 for various sensing rates normalized to the average SU transmission rate $t = \sigma/\mu_2 = 0.1$, 0.2, 0.3, 0.4, 0.5, $\rho_{pu} = 0.3$, $\rho_{su} = 0.5$, $N = 20$ and $1/\mu_1 = 1/\mu_2 = 10\ ms$

In Fig. 3, the SU interruption probability α is plotted as a function of N^1. Increasing the maximum number of sensing users N^1, we observe an increase of the SU interruption probability. For increasing σ, an even sharper increase of this probability can be seen. These observations are intuitively clear as explained before. Also we notice that the interruption probability converges to the same value for different σ as the maximum number of sensing users increases. Comparing Fig. 1 and Fig. 3 we see that there is convergence of α upwards a value of N^1 for which β approaches zero.

Fig. 4 shows the average delay of SUs as a function of the sensing room size N^1. For increasing N^1, the mean delay linearly increases until it reaches some convergence point. Afterwards, a further increase of the maximum number of sensing users will not affect the mean delay. This convergence is fully in accordance with the observation in Fig. 1 and Fig. 3. Also we notice the strong effect of σ on the mean delay of SUs. Decreasing the mean sensing time $1/\sigma$ over the given range, we see that the mean delay doubles and sometimes triples.

Fig. 5 shows the SU interruption probability α as a function of the SU arrival rate λ_2 for different maximum numbers of sensing SUs N^1. For $N^1 = 100$ and increasing λ_2, we can see that the SU interruption probability increases steadily until the sensing room is almost full of sensing users; afterwards, it increases more slowly. The same behavior can be noticed for other values of N^1. The sharp increase in the SU interruption probability when the cumulative offered

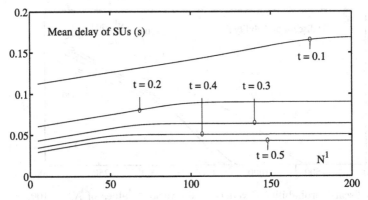

Fig. 4. Average delay of SUs $E[D_{su}]$ versus maximum number of sensing SUs N^1 for various sensing rates normalized to the average SU transmission rates $t = \sigma/\mu_2 = 0.1$, 0.2, 0.3, 0.4, 0.5, $\rho_{pu} = 0.3$, $\rho_{su} = 0.5$, $N = 20$ and $1/\mu_1 = 1/\mu_2 = 10\ ms$

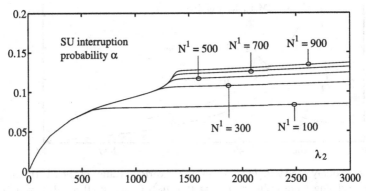

Fig. 5. SU interruption probability α versus λ_2 for various values of $N^1 = 100, 300,$ 500, 700, 900, $\rho_{pu} = 0.3$, $N = 20$, $1/\sigma = 100\ ms$ and $1/\mu_1 = 1/\mu_2 = 10\ ms$

load from both SUs and PUs approaches one (in the range for λ_2 between 1200 and 1400) is explained by the sudden increase of the number of sensing users in this region.

Fig. 6 displays the SU discard probability β as a function of the SU arrival rate λ_2 for different maximum numbers of sensing SUs N^1. As expected, for small values of λ_2 and increasing N^1, the discard probability decreases since there are more possibilities for an interrupted SU to sense again. But surprisingly, when λ_2 increases further, we observe that the discard probability is higher for larger sensing rooms, which is not intuitively expected. Based on Fig. 5 and Fig. 6, we observe that for the highest value of $N^1 = 900$, the maximum values of the interruption and the discard probabilities are below 15% and 10% respectively.

Fig. 7 shows the average delay of a SU as a function of the SU arrival rate λ_2 for different values of the transmission time $1/\mu_2 = 100, 50\ ms$ when the sum of

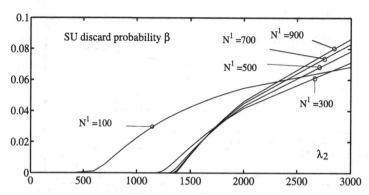

Fig. 6. SU discard probability β versus λ_2 for various values of $N^1 = 100, 300, 500,$ 700, 900, $\rho_{pu} = 0.3$, $N = 20$, $1/\sigma = 100$ ms and $1/\mu_1 = 1/\mu_2 = 10$ ms

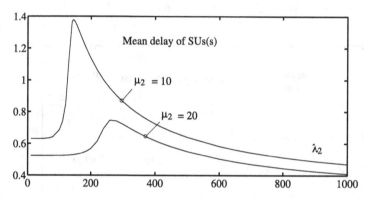

Fig. 7. Average delay of SUs $E[D_{su}]$ versus λ_2 for various values of the SU transmission time and SU sensing time $(1/\mu_2 = 1/\sigma = 100$ $ms)$, $(1/\mu_2 = 50$ $ms, 1/\sigma = 150$ $ms)$, $N^1 = 200$, $\rho_{pu} = 0.3$, $N = 20$ and $1/\mu_1 = 10$ ms

both transmission and sensing times is 200 ms. This plot can be divided into two parts. In the first part, for an increasing SU arrival rate λ_2 the average SU delay increases as the number of SUs (sensing and transmitting) increases. In this part, no SUs are discarded from the system as the sensing room is not full yet. In the second part, the sensing room is almost full and an increase of λ_2 increases the probability that a SU is discarded from the system before successful completion and consequently decreases the time the SU spends in the system.

6   Conclusion

In this paper, we investigated the impact of different parameters on SU performance measures in a CRN using a finite Quasi-Birth-Death CTMC. It has been shown that an increase of the sensing rate σ increases the SU interruption probability, but considerably reduces the mean SU delay. Further research would be needed to investigate the effect of sensing errors on these SU performance measures.

Acknowledgement. The second author is a Postdoctoral Fellow with the Research Foundation, Flanders (FWO-Vlaanderen), Belgium. This research has been funded by the Interuniversity Attraction Poles Programme initiated by the Belgian Science Policy Office.

References

1. Liang, Y.-C., Chen, K.-C., Li, G.Y., Mähönen, P.: Cognitive Radio Networking and Communications: An Overview. IEEE Transactions on Vehicular Technology 60(7), 3386–3407 (2011)
2. Datla, D., Wyglinski, A.M., Minden, G.J.: A Spectrum Surveying Framework for Dynamic Spectrum Access Networks. IEEE Transactions on Vehicular Technology 58(8), 4158–4168 (2009)
3. Spectrum Policy Task Force Report, Federal Commun. Comm., Washington, DC, ET Docket No. 02-135 (November 2002)
4. Islam, M.H., Koh, C.L., Oh, S.W., Qing, X., Lai, Y.Y., Wang, C., Liang, Y.-C., Toh, B.E., Chin, F., Tan, G.L., Toh, W.: Spectrum Survey in Singapore: Occupancy Measurements and Analyses. In: 3rd International Conference on Cognitive Radio Oriented Wireless Networks and Communications, CrownCom 2008, Singapore, pp. 1–7 (May 2008)
5. Mitola, J.: Cognitive radio: An integrated agent architecture for software defined radio, PhD Dissertation. KTH, Stockholm, Sweden (December 2000)
6. Wang, C.-W., Wang, L.-C., Adachi, F.: Analysis of reactive spectrum handoff in cognitive radio networks. IEEE Journal on Selected Areas in Communications 30(10), 2016–2028 (2012)
7. Huang, Y., Wang, J.: A multi-class preemptive priority cognitive radio system with random interruption discipline. In: Proceedings of the 5th International Conference on Queueing Theory and Network Applications, pp. 162–168 (2010)
8. Wang, B., Ji, Z., Liu, K.J.R., Clancy, T.C.: Primary-Prioritized Markov Approach for Dynamic Spectrum Allocation. IEEE Transactions on Wireless Commununications 8(4), 1854–1865 (2009)
9. Wong, E., Foh, C.: Analysis of Cognitive Radio spectrum access with finite user population. IEEE Communications Letters 13(5), 294–296 (2009)
10. Pla, V., De Vuyst, S., De Turck, K., Bernal-Mor, E., Martinez-Bauset, J., Wittevrongel, S.: Saturation Throughput in a Heterogeneous Multi-channel Cognitive Radio Network. In: Proceedings of the IEEE International Conference on Communications, ICC 2011, Kyoto, pp. 1–5 (June 2011)
11. Pacheco-Paramo, D., Pla, V., Martinez-Bauset, J.: Optimal Admission Control in Cognitive Radio Networks. In: 4th International Conference on Cognitive Radio Oriented Wireless Networks and Communications, CrownCom 2009, Hannover (June 2009)
12. Tang, S., Mark, B.L.: Performance Analysis of a Wireless Network with Opportunistic Spectrum Sharing. In: IEEE Global Communications Conference, GLOBECOM 2007 (November 2007)

13. Tang, S., Mark, B.L.: Modeling and Analysis of Opportunistic Spectrum Sharing with Unreliable Spectrum Sensing. IEEE Transactions on Wireless Communications 8, 1934–1943 (2009)
14. Tang, S., Mark, B.L.: An Analytical Performance Model of Opportunistic Spectrum Access in a Military Environment. In: IEEE Wireless Communications and Networks Conference, WCNC 2008 (April 2008)
15. Sai Shankar, N.: Squeezing the Most Out of Cognitive Radio: A joint MAC/PHY Perspective. In: IEEE International Conference on Acoustics, Speech and Signal Processing, vol. 4 (April 2007)
16. Sai Shankar, N., Chou, C.-T., Challapali, K., Mangold, S.: Spectrum Agile Radio: Capacity and QoS Implications of Dynamic Spectrum Assignment. In: IEEE Global Communications Conference, GLOBECOM 2005 (November 2005)
17. Chou, C.-T., Sai Shankar, N., Kim, H., Shin, K.G.: What and how much to gain by spectrum agility? IEEE Journal on Selected Areas in Communications 25(3), 576–588 (2007)
18. Wang, L.-C., Chen, A.: On the Performance of Spectrum Handoff for Link Maintenance in Cognitive Radio. In: IEEE International Symposium on Wireless Pervasive Computing, ISWPC (May 2008)
19. Song, Y., Xie, J.: Performance analysis of spectrum handoff for cognitive radio ad hoc networks without common control channel under homogeneous primary traffic. In: Proceedings IEEE INFOCOM 2011, pp. 3011–3019 (April 2011)
20. Laourine, A., Chen, S., Tong, L.: Queuing analysis in multichannel cognitive spectrum access: A large deviation approach. In: Proceedings IEEE INFOCOM 2010, pp. 1–9 (March 2010)
21. Li, B., Yang, P., Li, X.-Y., Wang, J., Wu, Q.: Finding Optimal Action Point for Multi-Stage Spectrum Access in Cognitive Radio Networks. In: IEEE International Conference on Communications, ICC 2011, pp. 1–5 (June 2011)
22. Yuan, G., Grammenos, R.C., Yang, Y., Wang, W.: Performance Analysis of Selective Opportunistic Spectrum Access With Traffic Prediction. IEEE Transactions on Vehicular Technology 59(4), 1949–1959 (2010)
23. Tang, S., Mark, B.L.: Modeling an Opportunistic Spectrum Sharing System with a Correlated Arrival Process. In: IEEE Wireless Communications and Networking Conference, WCNC 2008, pp. 3297–3302 (April 2008)

Statistical Analysis of the Workload of a Video Hosting Server

Christoph Möbius and Waltenegus Dargie

Technical University of Dresden, 01062 Dresden, Germany
{christoph.moebius,waltenegus.dargie}@tu-dresden.de

Abstract. The amount of data hosted by Internet servers and data centers is increasing at a remarkable pace requiring more capable and more efficient servers. However, physical efficiency does not necessarily correlate with computational efficiency. In fact, independent studies reveal that Internet servers are mostly over provisioned and still additional servers are deployed each year. Understanding the characteristics of the workload of servers is an essential step to efficiently manage them. For example, from the workload statistics, it is possible to predict idle or underutilized states and to consolidate workload, so that the idle or underutilized servers can be switched off. In this paper, we systematically analyze the characteristics of video servers – since they are responsible for producing the largest Internet traffic – and provide an insight into the relationship between the statistics pertaining to workload, the size of videos, and service time. We shall show that from the distribution of the video sizes on host servers, it is possible to estimate the distribution of the workload size produced by clients and the distribution of the time required to process individual requests.

Keywords: Service time, workload characterization, workload generation, workload size, workload statistics, video server, video size.

1 Introduction

The amount of data hosted, processed, and communicated by Internet-based servers and data centers is increasing at a remarkable pace. According to a recent report by Cisco Global Cloud Index[1], the global data center IP traffic will be 554 exabyte per month by 2016. In comparison, this has been 146 exabyte per month in 2011. Likewise, the global cloud IP traffic will reach 355 exabyte per month by 2016 (from 57 exabyte per month in 2011). The corresponding amount of workload per installed cloud server will increase by more than twofold by 2016 compared to the workload per installed server in 2011.

The research community as well as the IT industry approaches this phenomenon in a number of ways. Two of them, and perhaps the most ubiquitous ones, involve (1) the replacement of existing servers with more capable servers

---

[1] Cisco Global Cloud Index: Forecast and Methodology, 2011–2016, Cisco Inc. (http://www.cisco.com)

B. Sericola, M. Telek, and G. Horváth (Eds.): ASMTA 2014, LNCS 8499, pp. 223–237, 2014.

and (2) the deployment of additional servers. The estimated worldwide server deployment in 2010 was 40 million units [10], but additional servers have been steadily deployed since then. The latest statement from the International Data Corporation (IDC)[2] reveals that 1.9 million server units have been shipped in the first quarter of 2013 alone. Unfortunately, these approaches alone do not ensure a sustainable computing due to the fact that a rapid growth in the number and capacity of installed servers results in an equally rapid growth in power consumption [1,15,19]. Moreover, physical efficiency does not necessarily correlate with computational efficiency. In fact, independent studies reveal that Internet servers are mostly over provisioned yet additional servers are deployed each year. For example:

- At Twitter CPU utilization is less than 20% per server even though resource reservation reaches up to 80%. Similarly it uses between 40 and 50% of the available memory but memory reservation is approximately 80% [9].
- In a Google cluster CPU utilization is between 25 and 35% per server but reservation is approximately 75%. Likewise, memory utilization is approximately 40% but memory reservation is approximately 60% [25].
- In Amazon's EC2 cloud environment, the CPU utilization per server is between 3 and 7% [17].

The third approach presently adopted by the industry combines server virtualization with cloud computing, so that Internet services encapsulated in virtualized machines can share hardware resources, but each virtual machine has its own dedicated execution space. Moreover, the virtual machines can be freely migrated from one physical machine (server) to another at runtime. This feature has two advantages: Firstly, virtual machines are not bound to any specific server; their owners can change host servers whenever they wish to. Secondly, infrastructure owners can freely decide where and for how long individual virtual machines should execute, so that they can efficiently utilize hardware resources – this aspect is known in the literature as service or workload consolidation [4] as well as server consolidation [2].

Whether in a virtualized environment or otherwise, understanding the characteristics of the workload of servers is useful for efficiently managing hardware resources [8]. Firstly, the workload of underutilized or overloaded servers can be timely migrated to servers which can be loaded optimally. Secondly, from the statistics of resource utilization, it is possible to determine whether and for how long idle servers can be switched off to save power [26,27]. Thirdly, services that consume complementary resources can be scheduled on the same machine whereas services known for competing for similar resources can be scheduled to execute on separate servers [29].

The workload of an Internet server is primarily generated by users issuing requests. Hence, it consists of two independent quantities which cannot be known in advance except in a probabilistic sense. The first quantity refers to the arrival

---

[2] http://www.idc.com/getdoc.jsp?containerId=prUS24136113 (last visited August 20, 2013).

pattern of the requests (request arrival rate) while the second refers to the size of each request or the computational complexity each request induces on the server. Most existing probabilistic models for managing the resources and predicting the performance of Internet services rely on these two quantities.

In this paper, we shall experimentally demonstrate that for online video hosting services, such as Metacafe and YouTube, the statistics of the workload as well as the time needed to serve individual requests can be sufficiently determined from the statistics of the videos they host. The justification for our assertion is that for a large number of videos, there is a strong correlation between the preference of the users who generate videos and the users who view these videos. This knowledge is useful because service providers can estimate (1) the amount of resources they should make available to accommodate user requests and (2) the quality of service they can achieve for a given resource configuration. In other words, service providers need only examine how the statistics of the videos they host change over time to balance the supply of resources (for example, the leasing of network bandwidth or storage) with the demand for resources and to make a desirable trade-off between performance and resource consumption (including power). Since the required statistics is always available to them on the server side, they can make decisions without the influence of external entities.

We summarize our contribution as follows:

1. We develop a realistic and comprehensive workload generation model for a video hosting server. Unlike previous models which address partial aspects of video hosting servers, our model attempts to provide a more complete and a more unified aspect.
2. We set up a realistic server environment to test the workload generator; and,
3. We experimentally examine the relationship between the statistics of the request size received by the server, the video size hosted by the server, and service time needed to process user requests.

The rest of this paper is structured as follows: In section 2, we analyze related work. In section 3, we describe our experimental setting and how we generate workload. In section 4, we analyze our measurement data and discuss our observation. Finally, in section 5, we provide summary and conclusion.

2   Related Work

The term *workload* is understood in the literature in one of the following two ways: In the first, it refers to the magnitude of client requests processed by an Internet server[3] [6,11,20,21,33]; and in the second, it refers to the magnitude of utilization of hardware resources (such as CPU and memory) [4,14,23,24,31]. The main difference lies in the quantity that is available for modeling and analyzing the characteristic of a service. In this paper, we adopt the first association.

---

[3] As long as the context is clear, we use the terms *service* and *server* interchangeably. We use the term *physical machine* when we wish to put the emphasis on the hardware server.

Regardless of the way a workload is understood, obtaining sufficient statistics to accurately model and analyze Internet services is a difficult task because of privacy concerns and business secrecy. In the past, researchers have tried to piece together several parameters that can characterize the workload of web servers, particularly, video hosting applications. Some have made use of publicly available data, such as traces of CPU utilization of real-world web servers, so as to model and reason about similar web servers running on different platforms [16], [31]. Others have employed either web crawling to obtain metadata of files hosted by Internet services or filtered and analyzed Internet packets destined to or arriving from hosting sites at particular gateways. Evidently, all of these approaches can only provide partial views of the real workloads.

Barford et al. [3] identify seven statistical properties that characterize (conventional) HTTP traffic. These are the probability distribution of file sizes at the server side, the file popularity, temporal locality, the request sizes, active OFF times, inactive OFF times, and the number of embedded references. They assert that file sizes follow heavy-tailed probability density functions.

Tang et al. [28] analyze the traffic of a media streaming server. Their model builds on the idea of Barford et al. but relaxes the assumption that file popularity is statistically stationary. Instead they define a life-span distribution to account for a file popularity that changes over time. Moreover, they determine two types of life-spans: a regular life-span following a log-normal distribution and a news-like life-span following a Pareto distribution. The parameters for both types of life-span distributions are normally distributed. Their approach is the only approach known to us which examines prefix durations (i.e., aborted sessions).

Gill et al. [12] investigate the traffic of YouTube at a campus network. The central finding of their work is understanding the relationship between file types and traffic size: Whereas only 3% of all requests were for video files, 98.6% of the traffic was caused by video files. The majority of requests, i.e. 86%, were for images and text files which account for less than 1% of all traffic. The remaining 11% of requests were for applications and script data which account for 0.5% of all traffic. In addition, the study reveals that video file sizes are not considerably variable and therefore, cannot be modeled as long-tailed random variables. This is most likely due to the 10-minutes duration restriction for videos existing at the time which has been increased to 15 minutes as of July 2010. Today it is possible to upload videos larger than 20 GB as a result of which the probability density function of video size can be expected to be heavy-tailed.

Similarly, Cheng et al. [7], Cha et al. [5] and Mitra et al. [18] analyze the traffic of several video hosting applications. One of the observations common to all is that the popularity of a video does not follow a purely Zipf function. Instead, it exhibits a cutoff at the lower end. In other words, less popular videos still receive more views than assumed by a purely Zipf function. Like Tang et al. these researchers emphasize the need to capture a change in file popularity (life span). The analysis of Cheng et al. reveals that the life span of a video follows a Pareto density function. The most important parameter for the life-span is the growth trend factor, p. A value of $p > 1$ indicates a rise in popularity while

$0 < p < 1$ indicates a decline in popularity. According to Cheng et al. in 70% of all the videos they considered, $p < 1$. Based on this observation, Cheng et al. propose a model to predict the amount of additional views a video receives in future, which takes video age, current popularity, and p as its input parameters.

Cha et al. [5] analyze the popularity patterns of videos in YouTube, Daum, and Lovefilm. They find no correlation between video length and video popularity. The popularity follows a power law with an exponential cutoff, an observation confirmed by Mitra et al. [18]. Gummadi et al. [13] attribute this cutoff to a post-filtering process by recommendation systems. According to Cha et al. 99% of all videos of the video hosting applications are shorter than 10 minutes (which, again, is most probably influenced by the then existing 10-minute upload limit).

Finally, the investigation of Cha et al. reveals that the share of workload generated by users' activities (ratings and comments) is almost negligible: For YouTube only 0.22% of all views result in a rating and only 0.16% of all views result in a comment. This observation agrees with an earlier observation [12]. Similar observations are made by Mitra et al. – the workload due to ratings, comments, and uploads is typically several orders of magnitudes less than the workload generated by video views.

In the following sections, we build on the ideas and concepts discussed in this section to generate realistic workloads for a video hosting server.

3   Workload Generation

Whether running on a privately owned server or on a leased public cloud platform, understanding the workload of a server is vital for planning and managing resources (for example, resource-efficient schedulers employ workload statistics to determine where a given request should be processed). We assert that some of the statistics of a workload can be established from the statistics of the files the server hosts. We shall analytically as well as experimentally illustrate the correctness of this assertion by taking a video hosting server as an example. We assume that the videos hosted by the server are generated and viewed by users who are independent of the service management.

Our server consists of four quad core Intel Xeon E5-4603 processors, 16 GB memory, 10 Gbps Intel NIC, and an XFS-formatted 6 TB hard disk RAID with a theoretical sustained data rate of $\sim$465 MB/s (the data rate reduces to $\sim$300 MB/s during heavy contention). The server hosts 5000 videos of different sizes and user requests are served by an Apache2 web server.

The first step towards examining our assertion is to generate a realistic workload and to feed this workload to the video hosting server. We combine together the different models we reviewed in section 2 to generate the workload. As can be recalled, the models are developed by independent researchers who had access to actual Internet workloads (they employ traces or web crawlers). These models refer to: (1) The distribution of file sizes on the host server, (2) the file popularity at the start of the experiment, (3) the popularity growth factor, (4) the age of the files, and (5) the distribution of weekly views. We explain these features in more detail in the subsequent subsections and report how they relate to the workload of our server.

3.1 Video Size

As we already mentioned earlier, determining the distribution of the sizes of files in existing video hosting platforms is difficult for lack of access to the actual servers and because file sizes are not parts of publicly available meta data. In addition, most sites do not allow web crawling. However, early analysis of Internet traffic shows that the density of data size exhibits a heavy-tailed density function [22,32,3]. Studies contending this assertion (for example, [12,7]) often refer to restrictions made by service providers on the size of videos that can be uploaded on their servers. For example, YouTube currently limits the uploaded video duration to 15 minutes for most users, but for users with *good conduct record* this limit is pushed to 12 hours[4]. Likewise, Vimeo currently allows uploads of up to 5 GB for standard users and up to 20 GB per upload for Pro users[5]. Even so, it is reasonable to assume a heavy-tailed density for video file size.

We therefore fix the maximum permissible file size to 25 GB[6] and take a previously published value (median = 8.215 MB) to determine the minimum median value for the video size [12]. Even though results published by Barford et al. [3] suggest a Pareto distribution for the density of the traffic size, the rpareto function from GNU R's VGAM package we employ for our analysis produces hardly controllable variates. We therefore decide to replace it by the Weibull density which is implemented by the rweibull function in GNU R's stats package. We choose the parameter values $k = 0.3$ (shape) and $\lambda = 30$ (scale) to produce variates comparable to the above mentioned medians and maximum values. We then generate 5000 variates with $M_s = 8.514$, $\mu_s = 255.8$, and $max_s = 24680$, where M_s refers to the median video size; μ_s the mean video size, and max_s the maximum video size.

We then convert these figures to bytes and add an offset term to avoid a 0 byte video size. Gill et al. [12] found minimum payload sizes ranging from 452 to 95760 bytes for four different Youtube traces. We pick a random clip of 1s duration from Youtube and determine the minimum video size accordingly; it is 11500 bytes.

Based on these specifications we fragment a large video file of approximately 25 GB into 5000 randomly generated video clips. The sizes of these clips follow a Weibull distribution.

3.2 Video Popularity

Popularity refers to the number of times a particular video has been viewed in the past. Researchers who study the statistics of video popularity assert that

---

[4] https://support.google.com/youtube/answer/71673?hl=en
[5] https://vimeo.com/pro
[6] Vimeo has reduced file upload limit from 25 GB to 20 GB only recently.

it follows a power law[7] [12], but because most existing video servers employ recommendation systems the distribution function experiences an exponential cutoff at the lower end of the density function [5,7,18,28]. For our case, we do not employ a recommendation system and, therefore, the video popularity is assumed to obey a power law. As a basis for establishing the parameter values of the power-law variates, we use actual values from Dailymotion as presented in [18] with $\alpha = -1.72$, and the $max_v = 2,895,396$.

3.3   View Gain

The video popularity serves as a basis for estimating the additional number of views a video gains in future. Cheng et al. [7] derive a quantitative expression for the view gain after x additional weeks as follows:

$$v(x) = v_0 \cdot \frac{(x + a)^p}{a^p} \tag{1}$$

where v_0 refers to the present popularity of the video; a refers to the age of the video in weeks at the beginning of the observation period and p refers to popularity growth factor. Cheng et al. provide the plot of the CDF of p but left out its mathematical expression. We perform a graphical analysis and estimate the CDF with a Weibull distribution with $W(2, 0.9)$. With this knowledge we calculate the additional view gains for each video our server hosts. We will use the terms *view gain* and *popularity gain* interchangeably.

3.4   Video Age

To determine a video's age, we use a parameter called video upload trend, α. The upload trend of a video hosting server refers to the number of videos it hosts each week: $n = w^\alpha$, where n refers to the number of videos currently hosted by the server and w refers to the number of weeks the server has been active. The upload trend of YouTube in 2008 has been estimated to be 2.61 [7]. Since our video server hosts 5000 videos at the time of our experiment, for $\alpha = 2.61$, the oldest video should be 16 weeks old whereas the newest video is 1 week old. Hence, "16 weeks ago" one video was uploaded and the first file size variate is associated with an age of 1. By applying the upload trend like this we calculate $\lceil 2^{2.61} \rceil = 7$ uploads for the second week and the next 7 request size variates are associated with an age of 2. This procedure is repeated until the age of all videos are determined. After every request size variates is associated with an age, we calculate the video gain for each video using Equation 1.

---

[7] The probability density function of a random variable x obeying a power law is expressed as: $f(x) = Cx^{-\alpha}$ with $x > x_{min}$ and.

3.5 Request Distribution

The view gain expresses the number of additional views a video receives on a weekly basis. This term has to be broken down into days and the time of a day to estimate the workload size per unit time. For our experiment, we consider it sufficient to generate requests for a time span of one day. Therefore, we evenly distribute the view gain (i.e., the number of requests) to the seven days of the week, but a further even distribution of the daily requests to the 24 hourly slots is not plausible, because repeated observations indicate that the daily load of a multimedia server exhibits rather a wave-like distribution [12,34]. However, results in [28,30] show that stationary can be assumed if the day is split into time slots of one hour or less. We thus employ a y-shifted cosine function to determine the portion of workload for each time slot.

Mathematically, this portion is determined as:

$$A_s = \int_l^u cos(x) + 1.1 \; dx \tag{2}$$

where $u = 2\pi\frac{s}{24}$, and $l = 2\pi\frac{s-1}{24}$ with $s = \{1, \ldots, 24\}$. Then, the amount of requests a video v receives in the time slot s is expressed as: $r_s^v = \frac{A_s}{A} \times g_d$, where $A = \int_0^{2\pi} cos(x) + 1.1 \; dx$ and g_d is the view gain for day d.

3.6 Test Cases

It turned out that generating requests for all the files the video server hosts overwhelms the physical machine due to contention at the disk drive. This problem can be addressed in two different ways: (1) By reducing the number of available videos on the server, since the user request rate depends on this quantity (the larger the number of videos the server hosts, the larger the number of users it attracts); or (2) by scaling down the initial popularity of the video files since the view gain per week (and thus, per time slot) directly depends on this quantity. Similarly, the view gain for each video can be scaled down. For our experiment we adopt the first approach.

With our server configuration, a workload generated for 300 videos slightly overloads the sever (in terms of system load average[8]). When the number of videos is reduced to 200 a low to medium load is generated on the server. When the number of videos is reduced to 100 then the server has sufficient resources to accommodate user requests (low system load average). We consider these three scenarios to generate different request distributions and to analyze the relationship between the statistics of the workload generated by users and the statistics of the videos hosted by the server.

---

[8] The load average refers to the average length of the CPU run queue. If this length is greater than the number of logical cores, we consider the server to be overloaded.

4   Observation and Analysis

In this section, we analyze the relationship between the statistics of (1) the service time for individual requests and the size of individual requests (workload size), and (2) the workload size and the video size. In all our investigation, we shall focus on the probability distribution function (CDF) because this function sufficiently expresses a random variable.

As mentioned in section 1, the workload of a video server is the convolution of two random variables, namely, the request arrival rate and the size of individual requests. The distribution of request sizes is in turn influenced by (1) the distribution of sizes of available videos on the server and (2) the popularity (view) gain of each video in each time slot. We consider both parameters and generate six different types of workloads: We vary the number of available videos to 100, 200, and 300 and we consider four different types of workloads which are produced under the assumption that the hosted videos have initial video popularity (v_0 in Equation 1) distributions obeying power law (for all the video sets), normal, uniform, and gamma distributions (for the 200 video set). Fig. 1 shows the rank-frequency plots of the different view gains we consider to produce the workloads.

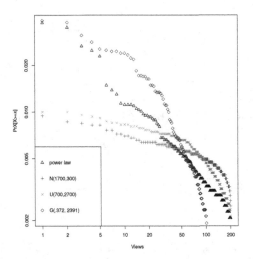

Fig. 1. The rank-frequency plot of the initial video popularity

Varying the amount of available videos changes the parameters of the video size distribution while varying the popularity distribution changes the popularity gain and thus, the distribution of the request size. We perform the experiments in the time slots between 16:00 and 19:00 o'clock. However, in this paper, we shall limit ourselves to the analysis of the experiment results of the time slot between 18:00 and 19:00 o'clock (this is the time slot with the highest request rate). The experiment data for the other time slots do not lead to different results.

4.1 Service Time vs Request Size

One of the most critical parameters to evaluate the performance of a server is the service time as seen by clients. We define this time as the time span beginning from sending a request up to the time the requested video is downloaded completely. Technically, it is the time span beginning from starting to establish an HTTP session (wait time) until the HTTP session termination (download time).

The wait time (the time needed to establish a session) does not much depend on the request size; instead, it depends on the request arrival rate. Even so, the wait time is very small compared to the download time and can be neglected – the mean wait time that can be experienced under a heavy load is below 0.08 second whereas the mean service time under the same condition is 13.18 seconds. Therefore, the service time can be approximated by the download time, which depends on the size of the video being downloaded. We propose to estimate the service time in terms of the request size (video size) using a linear model.

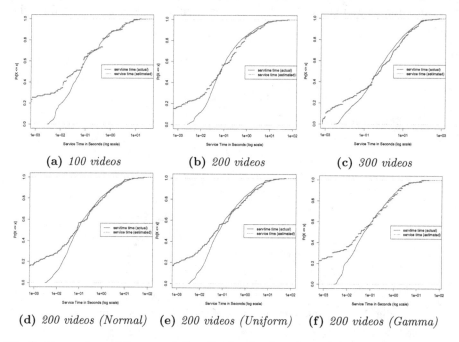

(a) *100 videos*   (b) *200 videos*   (c) *300 videos*

(d) *200 videos (Normal)*   (e) *200 videos (Uniform)*   (f) *200 videos (Gamma)*

Fig. 2. Comparison of the cumulative distribution functions of the experimentally obtained and the estimated service times when the available videos for downloading are (top: from left to right) 100, 200, and 300. The workloads are generated with a power law distribution initial video popularity. Bottom: the available videos on the server are 200, but the initial video popularity follows (from left to right) normal, uniform, and gamma distributions.

Fig. 2 displays the relationship between the CDFs of the actual service time we measure and the service time we estimate from the size of incoming requests. While the graphs show deviations in the lower end of the distributions, the middle parts and the upper ends match comparatively well. To evaluate the goodness of fits, we calculate the R^2 values. The best estimation is achieved when the initial video popularity has a gamma distribution ($R^2 = 0.8132$) whereas the worst estimation is achieved when the initial video popularity follows a uniform distribution ($R^2 = 0.648$). In general, the results suggest that the assumption that a linear relationship exists between the service time and the request size is a plausible assumption.

From the graphs it can be observed that the deviations between the CDFs of the actual and the estimated service times start to increase at values below 0.1 which comprise the lower 0.5 quantile of the values. There can be two reasons for this: (1) the lower-bound of the service time is fixed by the minimum wait time and the data rate of the network interface card. For our experiments the minimum wait time is 0.00257 second. (2) Due to the heavy-tailed nature of the video size distributions, the request size is heavy-tailed as well (we shall discuss this fact shortly). In this case, the linear model produces better results for larger values of the request size than for smaller values. Since the request size comprises a range between six (for 100 videos) and seven (for 300 videos) orders of magnitudes, but the service time only between five (for 100) and six (for 300 videos) orders of magnitudes, it may not come as a surprise to observe a larger deviation at the lower end of the CDFs.

4.2   Request Size vs. Video Size

Likewise, we examine the existence of a relationship between the statistics of the request size of a workload (client-side property) and the statistics of the video size on the server (server-side property). Similar to the previous test cases we vary the amount of available videos on the server and the distribution of the initial video popularity to generate different workloads.

As can be recalled, the video size for all the test cases obeys a Weibull distribution, but each test case results in a different value for the scaling parameter. On the other hand, the workloads generated for each test case are dissimilar with each other because of the different popularity distributions we selected. Regardless of these variation, the graphs in Fig. 3 confirm that the distributions of the size of videos on the servers exhibit strong similarity with the distributions of the request size produced by users (it should be noted that both the distributions of the video size and the request size are measured in byte).

Interestingly, for all the test cases, the CDFs of both the request and the video sizes can be estimated by Weibull distributions. Tab. 1 and Tab. 2 summarize the shapes and scales of the two random variables for all the test cases we considered.

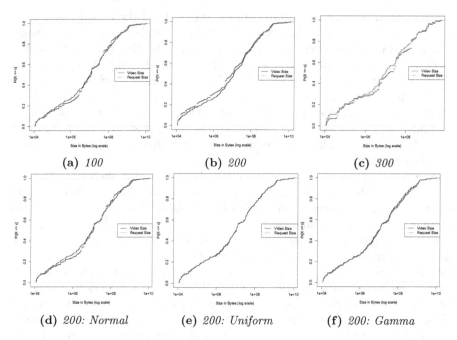

Fig. 3. Comparison between the cumulative distribution functions of the video size on the server and the request size generated by users when the available videos for downloading are (top: from left to right) 100, 200, and 300. The workloads are generated by a power-law distributed initial video popularity. Bottom: the available videos on the server are 200, but the initial video popularity follows (from left to right) normal, uniform, and gamma distributions.

Table 1. The estimated parameters and the corresponding relative error ($rel.err = \frac{actual-estimated}{actual}$) for the video and the request size distributions. The server makes 100, 200, and 300 videos available for downloading and the workloads are generated with the assumption that the initial video popularity distribution obeys a power law.

Subset	Request Size Distribution		Video Size Distribution	
	shape	scale	shape (rel. err)	scale (rel. err)
100 Videos	3.2934e-01	4.5283e+07	3.1854e-01 (.0328)	3.6322e+07 (.1979)
200 Videos	3.5754e-01	3.4089e+07	3.3792e-01 (.0549)	3.4331e+07 (.0071)
300 Videos	3.2390e-01	3.1389e+07	3.3279e-01 (.0274)	3.9909e+07 (.2714)

The above results clearly show that regardless of the distribution of view gain, the statistics of the request size can be sufficiently determined by the statistics of the size of the videos hosted by the server.

Table 2. The estimated parameters and the corresponding relative error ($rel.err = \frac{actual - estimated}{actual}$) for the video and the request size distributions. The server makes 200 videos available for downloading and the workloads are generated with the assumption that the initial video popularity distribution obeys normal, uniform, and gamma distributions.

Subset	Request Size Distribution		Video Size Distribution	
	shape	scale	shape (rel. err)	scale (rel. err)
Power law	3.576e-01	3.4724e+07	3.1854e-01 (.1092)	3.6322e+07 (.0439)
$N(1700, 300)$	3.543e-01	3.5378e+07	3.1854e-01 (.1009)	3.6322e+07 (.0267)
$U(700, 2700)$	3.350e-01	3.6285e+07	3.1854e-01 (.0491)	3.6322e+07 (.0010)
$\Gamma(.372, 2391) \cdot 10^6$	3.277e-01	3.9370e+07	3.1854e-01 (.0279)	3.6322e+07 (.0774)

5 Conclusion

In this paper we analyzed the characteristics of a video server. We generated probabilistic workload and examined the relationship between server-side statistics and workload properties. In particular, we studied whether the probability distribution function of the request size - a property which is not influenced by the server configuration - is related to the probability distribution function of the size of videos hosted by the server. We found that the distribution function of the request size resembles the distribution function of the file size despite the fact that the distribution of the workload size is technically the convolution of the distribution function of the video size and the distribution function of the video popularity. Furthermore, we examined the relationship between the statistics of the service time and the workload size. We found that a linear relationship exists between the workload size and the service time.

We thus confirmed our assertion that the performance (service time) of a video server can be sufficiently predicted by examining the statistics of the video files it hosts.

References

1. Abts, D., Marty, M.R., Wells, P.M., Klausler, P., Liu, H.: Energy proportional datacenter networks. SIGARCH Comput. Archit. News 38(3), 338–347 (2010)
2. Apparao, P., Iyer, R., Zhang, X., Newell, D., Adelmeyer, T.: Characterization & analysis of a server consolidation benchmark. In: Proceedings of the Fourth ACM SIGPLAN/SIGOPS International Conference on Virtual Execution Environments, VEE 2008, pp. 21–30. ACM, New York (2008)
3. Barford, P., Crovella, M.: Generating representative web workloads for network and server performance evaluation. ACM SIGMETRICS Performance Evaluation ... 26(1), 151–160 (1998)
4. Beloglazov, A., Buyya, R.: Adaptive threshold-based approach for energy-efficient consolidation of virtual machines in cloud data centers. In: Proceedings of the 8th International Workshop on Middleware for Grids, Clouds and e-Science - MGC 2010, pp. 1–6. ACM Press, New York (2010)

5. Cha, M., Kwak, H., Rodriguez, P., Ahn, Y.-Y., Moon, S.: Analyzing the Video Popularity Characteristics of Large-Scale User Generated Content Systems. IEEE/ACM Transactions on Networking 17(5), 1357–1370 (2009)
6. Chase, J.S., Anderson, D.C., Thakar, P.N., Vahdat, A.M., Doyle, R.P.: Managing energy and server resources in hosting centers. In: Proceedings of the Eighteenth ACM Symposium on Operating Systems Principles - SOSP 2001, p. 103. ACM Press, New York (2001)
7. Cheng, X., Dale, C., Liu, J.: Statistics and Social Network of YouTube Videos. In: 2008 16th Interntional Workshop on Quality of Service, pp. 229–238 (June 2008)
8. Dargie, W., Strunk, A., Schill Energy-aware, A.: service execution. In: 2011 IEEE 36th Conference on Local Computer Networks, pp. 1064–1071 (October 2011)
9. Delimitrou, C., Kozyrakis, C.: Quasar: Resource-Efficient and QoS-Aware Cluster Management. In: Proceedings of the Nineteenth International Conference on Architectural Support for Programming Languages and Operating Systems (ASPLOS), Salt Lake City, UT, USA (2014)
10. Dreslinski, R.G., Wieckowski, M., Blaauw, D., Sylvester, D., Mudge, T.: Near-Threshold Computing: Reclaiming Moore's Law Through Energy Efficient Integrated Circuits. Proceedings of the IEEE 98(2), 253–266 (2010)
11. Fan, X.: W.-d. Weber, L.A. Barroso. Power provisioning for a warehouse-sized computer. ACM SIGARCH Computer Architecture News 35(2), 13 (2007)
12. Gill, P., Arlitt, M., Li, Z., Mahanti, A.: Youtube traffic characterization: A view from the edge. In: Proceedings of the 7th ACM SIGCOMM conference on Internet measurement - IMC 20, San Diego, California, USA, pp. 15–28. ACM Press (2007)
13. Gummadi, K.P., Dunn, R.J., Saroiu, S., Gribble, S.D., Levy, H.M., Zahorjan, J.: Measurement, modeling, and analysis of a peer-to-peer file-sharing workload. In: Proceedings of the Nineteenth ACM Symposium on Operating Systems Principles - SOSP 2003, p. 314 (2003)
14. Kansal, A., Zhao, F., Liu, J., Kothari, N., Bhattacharya, A.A.: Virtual machine power metering and provisioning. In: Proceedings of the 1st ACM symposium on Cloud computing - SoCC 2010, p. 39 (2010)
15. Koomey, J.G.: Growth in data center electricity use 2005 to 2010. Technical report (2011)
16. Kusic, D., Kephart, J.O., Hanson, J.E., Kandasamy, N., Jiang, G.: Power and Performance Management of Virtualized Computing Environments Via Lookahead Control. In: 2008 International Conference on Autonomic Computing, June 2008, pp. 3–12 (2008)
17. Liu, H.: Host server CPU utilization in Amazon EC2 cloud (2012)
18. Mitra, S., Agrawal, M., Yadav, A., Carlsson, N., Eager, D.: A. Mahanti Characterizing Web-Based Video Sharing Workloads. ACM Transactions on the Web 5(2), 1–27 (2011)
19. Möbius, C., Dargie, W., Schill, A.: Power Consumption Estimation Models for Processors, Virtual Machines, and Servers. IEEE Transactions on Parallel and Distributed Systems, 1 (2013)
20. Nathuji, R., Schwan, K.: Vpm tokens: Virtual machine-aware power budgeting in datacenters. In: Proceedings of the 17th International Symposium on High performance Distributed Computing - HPDC 2008, p. 119. ACM Press, New York (2008)
21. Padala, P., Shin, K.G., Zhu, X., Uysal, M., Wang, Z., Singhal, S., Merchant, A., Salem, K.: Adaptive control of virtualized resources in utility computing environments. ACM SIGOPS Operating Systems Review 41(3), 289 (2007)
22. Paxson, V., Floyd, S.: Wide area traffic: The failure of Poisson modeling. IEEE/ACM Transactions on Networking 3(3), 226–244 (1995)

23. Petrucci, V., Carrera, E.V., Loques, O., Leite, J.C.B., Mossé, D.: Optimized Management of Power and Performance for Virtualized Heterogeneous Server Clusters. In: 2011 11th IEEE/ACM International Symposium on Cluster, Cloud and Grid Computing, pp. 23–32 (May 2011)
24. Raghavendra, R., Ranganathan, P., Talwar, V., Wang, Z., Zhu, X.: No "Power" Struggles: Coordinated Multi-level Power Management for the Data Center. SIGARCH Comput. Archit. News 4, 48–59 (2008)
25. Reiss, C., Tumanov, A., Ganger, G.R., Katz, R.H., Kozuch, M.A.: Heterogeneity and dynamicity of clouds at scale: Google trace analysis. In: Proceedings of the Third ACM Symposium on Cloud Computing - SoCC 2012, pp. 1–13. ACM Press, New York (2012)
26. Sotomayor, B., Montero, R.S., Llorente, I.M., Foster, I.: Virtual infrastructure management in private and hybrid clouds. IEEE Internet Computing 13(5), 14–22 (2009)
27. Strunk, A., Dargie, W.: Does Live Migration of Virtual Machines cost Energy? In: 2013 IEEE 27th International Conference on Advanced Information Networking and Applications (AINA), Barcelona, Spain, pp. 514–521 (2013)
28. Tang, W., Fu, Y., Cherkasova, L., Vahdat, A.: Internet Systems. Long-term Streaming Media Server Workload Analysis and Modeling. Technical report, HP Laboratories (2003)
29. Veeraraghavan, K., Chen, P.M., Flinn, J., Narayanasamy, S.: Detecting and surviving data races using complementary schedules. In: Proceedings of the Twenty-Third ACM Symposium on Operating Systems Principles, SOSP 2011, pp. 369–384. ACM, New York (2011)
30. Veloso, E., Almeida, V., Meira, W., Bestavros, A.: A hierarchical characterization of a live streaming media workload. IEEE/ACM Transactions on Networking 14(1), 133–146 (2006)
31. Verma, A., Dasgupta, G., Nayak, T.K., De Ravi Kothari, P.: Server workload analysis for power minimization using consolidation. In: Proceedings of the 2009 Conference on USENIX Annual Technical Conference. USENIX Association (2009)
32. Willinger, W., Taqqu, M.S., Sherman, R., Wilson, D.V.: Self-similarity through high-variability: Statistical analysis of Ethernet LAN traffic at the source level. IEEE/ACM Transactions on Networking 5(1), 71–86 (1997)
33. Wood, T., Tarasuk-Levin, G., Shenoy, P., Desnoyers, P., Cecchet, E., Corner, M.D.: Memory buddies: Exploiting page sharing for smart colocation in virtualized data centers. ACM SIGOPS Operating Systems Review 43(3), 27 (2009)
34. Zink, M., Suh, K., Gu, Y., Kurose, J.: Characteristics of YouTube network traffic at a campus network Measurements, models, and implications. Computer Networks 53(4), 501–514 (2009)

Throughput Maximization with Multiclass Workloads and Resource Constraints

Davide Cerotti[1], Marco Gribaudo[1], Ingolf Krüger[2], Pietro Piazzolla[1],
Filippo Seracini[2], and Giuseppe Serazzi[1]

[1] Dip. di Elettronica e Informazione, Politecnico di Milano,
via Ponzio 34/5, 20133 Milano, Italy
`name.surname@polimi.it`
[2] Department of Computer Science and Engineering
University of California San Diego, La Jolla, CA 92023-0404, USA
`{fseracini,ikrueger}@ucsd.edu`

Abstract. In this paper we study the impact of different types of constraints on the maximum throughput that a system can handle. In particular, we focus on constraints limiting the use of resources and/or the allowed response time. The problem is made even more difficult by the pronounced diversity in resource requirements of the different applications in execution, i.e., by the multiclass characteristic of the workloads. The proposed approach allows to determine the maximum load of the different classes, while still satisfying the considered performance objectives. An experimental validation of the described technique through the study of a realistic e-commerce application is presented.

1 Introduction

Over the last few years, the growth of available physical resources was a very evident phenomenon thanks to the widespread diffusion of cloud computing. Concurrently, the capacity requirements of the new applications has also increased significantly. Modern computing infrastructures are characterized by a huge amount of resources with heterogeneous capacities (e.g. [14,23]) that are shared among several applications with very different requirements. Such features have made the allocation of resources a very critical problem because the capacity required to sustain the flow of requests may not be always available. The performance of the servers remains a crucial component of many computing infrastructures. In order to address this problem in the case of shared systems, different types of constraints are imposed to the resources deployed to the various applications.

In this paper we study the effects of a variety of resource and time based constraints on a performance objective function. The constraints at the resource level are based on the utilization, and on the maximum number of jobs in the system. As time based constraints we consider thresholds on residence time and on the mean system response time. The performance objective function to be maximized is the system throughput.

B. Sericola, M. Telek, and G. Horváth (Eds.): ASMTA 2014, LNCS 8499, pp. 238–252, 2014.

In this paper, peculiar properties of open multiclass queuing networks subject to different types of constraints are investigated and their applicability to some practical problems is proposed. In particular, the problem solved is the following: find the maximum throughput per class of requests that the system can sustain while satisfying the given time and/or resource based constraints.

Even if multiclass open queuing networks are well established mathematical models, the specific way in which they are used in this paper constitute a novelty. This application of known theory can provide new interesting insights and be useful to solve stream-line research problems about the allocation of resources in contexts such as cloud computing and multi-tier architectures.

The structure of the paper is the following. Section 2 analyzes the related works, and Section 3 presents a brief overview of some basic results of open queuing networks that will be used throughout the paper. Sections 4 and 5 address the identification of the maximum throughput without and with constraints. Section 6 applies the results to a realistic system, and Section 7 concludes the paper.

2 Related Work

A common problem in data center management is resource allocation and provisioning in the presence of loads that can vary frequently with Internet applications. Resource over-provisioning leads to low average server utilization and high recurring utility costs. On the other hand, under-provisioning translates in a potential shortage of computing resources. Both strategies may cause serious economic losses. Provisioning decisions are usually taken by either hardware, platform or application providers, even if in many cases the responsibility of provisioning is demanded to end users (see e.g [23,25,24]).

Several techniques have been recently introduced to deal with the identification of the proper set of resources. Autonomic data centers, referred sometimes to as self-tuning, self-adaptive or self-aware systems (e.g.[11,28]) try to adapt the allocated resources to the fluctuations of requests in order to meet agreed operational objectives. In [20,21] the authors take into consideration the response time only, typically defined as the aggregated value across all the request classes. Our solution differs from these approaches as it deals also with multi-class workloads. Moreover, our approach enables the data center resource management to identify the workload mix that maximizes both the throughput and the utilization of resources under a set of constraints, not only response time. Other bottleneck identification techniques for queueing networks are considered in [7]. Optimization of a cost function has been addressed in many different ways: using combinatorial search algorithms in [5,9,10], linear programming in [2], game theory in [26], while in other cases the maximization of some utility functions like response time (e.g. [22]) or power consumption (e.g. [3]) is sought. An heuristic approach is discussed in [1].

The problem of scalability has been approached by dynamically allocating and deallocating resources. Admission control schemes have been devised in order to guarantee given objectives. Usually, performance goals are achieved by rejecting some types of requests during peak periods [8,10] or by maximizing of provider's revenue [13,15]. Different approaches are also based on policies that control the arrival rates of the classes of applications in order to saturate simultaneously multiple resources to maximize a given metric as in [27,12].

3 Background

In this section, we briefly review the basic notations that will be used in the following. We consider a workload consisting of C classes of requests and a system with M resources that operate at a fixed rate. Requests cannot change class during their execution. Let D_{mc} be the global service demand of a class c job on station m (with $1 \leq c \leq C$ and $1 \leq m \leq M$). The service demands of the system are described by the following $M \times C$ demand matrix:

$$\mathbf{D} = \begin{vmatrix} D_{11} & \cdots & D_{1C} \\ \vdots & & \vdots \\ D_{M1} & \cdots & D_{MC} \end{vmatrix} \tag{1}$$

Class c requests enter the system at a Poisson rate λ_c. We collect the workload intensities of all the classes in a vector $\boldsymbol{\lambda} = (\lambda_1, \dots, \lambda_C)$. The *overall arrival rate*, i.e., the *global load* of the system, is given by $\Lambda = \sum_c \lambda_c$. Let $\boldsymbol{\beta} = (\beta_1, \dots, \beta_C)$ be the *population mix* vector, where β_c is the fraction of arriving requests that belong to class c. The following relations between Λ, $\boldsymbol{\lambda}$, $\boldsymbol{\beta}$ exists:

$$\boldsymbol{\lambda} = \Lambda\boldsymbol{\beta}, \quad \lambda_c = \Lambda\beta_c, \quad \sum_c \lambda_c = \Lambda,$$

$$\boldsymbol{\beta} = \frac{\boldsymbol{\lambda}}{\Lambda}, \quad \beta_c = \frac{\lambda_c}{\Lambda}, \quad \sum_c \beta_c = 1 \tag{2}$$

We define the *population mix scaled demand* for resource m D_m as: $D_m(\boldsymbol{\beta})$

$$D_m(\boldsymbol{\beta}) = \sum_c \beta_c D_{mc} \tag{3}$$

$D_m(\boldsymbol{\beta})$ represents the mean service demand generated on resource m by a given population mix $\boldsymbol{\beta}$.

We use the subscript 'mc' to denote an index computed for class c at resource m (U_{mc}, R_{mc} and Q_{mc}). We also denote aggregated metrics related to class c by the index $\star c$: $U_{\star c}$, $R_{\star c}$ and $Q_{\star c}$. The metrics at the resource level are denoted by the index of the resource: R_m and Q_m.

4 Maximization of Unconstrained Systems

With a multiclass workload, the maximum throughput that a system can handle is a function of the fraction of the jobs of the different classes in concurrent execution, i.e., is *mix dependent*. Indeed, for a given workload intensity there are mixes in correspondence to which the system perform better than with other, i.e., that provide the maximum throughput and the minimum response time. We are considering open systems, and it is known that as the arrival rate of customers of the various classes increases, the number of customers in the system, and thus the response time, tends to grow *without bounds*, (i.e. the system *saturate*). In particular, with a given population mix β, the system is not in saturation if the utilization of each resource is strictly less than 1, that is: $\Lambda D_m(\beta) < 1$, $\forall m :$ $1 \leq m \leq M$. This ensures that the system is stable, and it can be used to determine the maximum arrival rates that it can handle. For a given population mix β, the maximum possible arrival rate $\hat{\Lambda}(\beta)$ can be determined by inverting the stability condition:

$$\hat{\Lambda}(\beta) = \frac{1}{\max_m\{D_m(\beta)\}}. \tag{4}$$

It can then be interesting to determine the population mix β for which the system can experience the maximum throughput $\hat{\Lambda}(\beta)$. Since the utilization of the resources are linear functions, the population mix β^* corresponding to the maximum throughput can be obtained solving the following Linear Programming Problem (LPP):

$$
\begin{aligned}
&\textbf{Variables:} && \lambda_c, && 1 \leq c \leq C \\
&\textbf{Objective:} \quad \text{maximize} && \sum_c \lambda_c \\
&\textbf{Constraints:} \sum_d \lambda_d D_{md} \leq 1, && 1 \leq m \leq M \\
&&& \lambda_c \geq 0, && 1 \leq c \leq C
\end{aligned}
\tag{5}
$$

The interpretation of the LPP of Eq. 5 is the following: the objective function corresponds to the total arrival rate, expressed as the sum of the arrival rates of the individual classes ($\Lambda = \sum_c \lambda_c$). Constraints ensure that arrival rates are positive ($\lambda_c \geq 0$), and that the utilization of each resource is less than 1 ($\sum_c \lambda_d D_{md} \leq 1$). If we call λ_c^*, $1 \leq c \leq C$ the optimizer, then the maximum allowed throughput corresponds to the value assumed by the objective function $\Lambda^* = \sum_c \lambda_c^*$. The optimal population mix can then be computed as $\beta^* = \{\beta_c^*\}$ with $\beta_c^* = \lambda_c^*/\Lambda^*$.

The set of the population mixes that can achieve the maximum throughput corresponds to the solution of the linear program. If the solution of the LPP of Eq.5 is a single point, then the maximum throughput can be obtained only for single population mix β^*. If the solution of the LPP in Eq.5 is a segment or a convex polyhedron, then there exists a set of population mixes corresponding to the maximum overall throughput Λ^*.

We now see the application of the technique through the description of an e-commerce system consisting of four resources (*Management*, *CMS System*, *Inventory* and *Shipping*) and three classes of customers (*Intranet*, *On-line Purchase*, *In Store*). The service demands of the three classes are shown in Table 1. The *Management* resource is the storage server for the financial, transactions and customers data. The different contents of the website (mainly products catalog) are managed by the *CMS system*, while the administration of the catalog and the warehouse inventory involve the *Inventory* server. A dedicated resource is provided to handle the *Shipping* of purchased items. The back-end operations, including tasks related to the organization of the web site, the products catalog update, and the customers data update, are executed by requests of the *Intranet* class. *On-Line Purchase* represents the process of buying one or more products through the web in a single transaction, while *In Store purchase* transactions take place off-line, and usually are started by a customer entering a physical store. During their execution the requests of each class visit all the four servers.

Table 1. Service demands **D** of the three classes of requests at the three resources of the e-commerce model

	D	Class 1 (*ms.*) Intranet	Class 2 (*ms.*) On-Line Purchase	Class 3 (*ms.*) In Store
Resource 1	**Management**	80	65	60
Resource 2	**CMS system**	30	130	80
Resource 3	**Inventory**	45	30	135
Resource 4	**Shipping**	65	115	45

Fig.1 shows the maximum throughput $\hat{\Lambda}(\beta)$ for all the possible mixes $\beta = |\beta_1, \beta_2, (1 - \beta_1 - \beta_2)|$ of the requests of the three classes. The mixes of requests of each surface of the polyhedron shown correspond to a bottleneck on a different resource (indicated in the figure by its own label). The edges at the intersection of two surfaces are the *common saturation sectors*, and represent the mixes whose execution generates more than one bottleneck, i.e., that concurrently saturate two resources [4]. It is clear from the figure that there exists a point, corresponding to the population mix β^*, for which the throughput is maximized (i.e. the top of the pyramid-like polyhedron). In this point, more than two resources saturate concurrently.

Solving the LPP of Eq. 5 to determine the maximum system throughput with respect to the mixes β, we obtain:

$$\beta^* = |0.4512\ 0.2307,\ 0.3181| \ ,\ \Lambda^* = 14.25\ jobs/s \tag{6}$$

It is interesting to point out that the maximum system throughput is obtained with $\beta_1 = 0.4512$, that is, when 45% of the requests in execution are of the *Intranet* class. This is not surprising, since the global service demand of the *Intranet* requests, i.e., the sum of its service demands over all the resource, is smaller with respect to the ones of the other two classes.

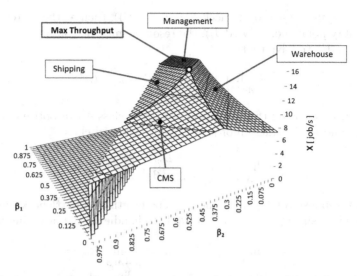

Fig. 1. Maximum system throughput vs the population mixes β in a three class, four resources system

5   Performance Constraints

We then consider the case in which constraints are imposed to some performance metrics of the system. Typical constraints are, for example, thresholds to the maximum mean response time, to the utilization, or upper bounds for the queue lengths of a resource. Constraints may reduce the maximum allowed arrival rate Λ^* to values that are smaller than $\hat{\Lambda}(\beta)$ given by Eq.4. They can also change the population mixes β^* for which the maximum throughput Λ^* can be achieved. To address this issue, we divide the performance constraints in two categories: *system-level* and *resource-level*. *System-level* constraints concern either the whole system or a specific class c. For all possible β, as the total arrival rate Λ approaches $\hat{\Lambda}(\beta)$, both the queue length and the response time tend to infinity:

$$\lim_{\Lambda \to \hat{\Lambda}(\beta)} R_{*c} = \infty, \quad \lim_{\Lambda \to \hat{\Lambda}(\beta)} R_S = \infty,$$
$$\lim_{\Lambda \to \hat{\Lambda}(\beta)} Q_{*c} = \infty, \quad \lim_{\Lambda \to \hat{\Lambda}(\beta)} Q_S = \infty. \tag{7}$$

Resource-level constraints refer to a station m, either for a specific class c or for the aggregate of all the classes. In this case the utilization, the mean queue length and the response time can have very different behaviors depending on the considered population mix β. Let us call $d(\beta) = \{m : 1 \leq m \leq M \wedge D_m(\beta) = \max_l\{D_l(\beta)\}$, the set of resources that are bottleneck with the population mix β, and let us call $\hat{u}_m$, $\hat{b}_m$, $\hat{q}_m$, ($\hat{u}_{mc}$, $\hat{b}_{mc}$, $\hat{q}_{mc}$) the maximum utilization, mean response time and mean queue length that can be obtained for a resource m (respectively for class c requests at resource m). Let us focus on resource utilization first. If a resource m is a bottleneck (i.e. $m \in \boldsymbol{d}(\boldsymbol{\beta})$), then as the system reaches its instability point (i.e. $\Lambda \to \hat{\Lambda}(\boldsymbol{\beta})$), the resource must be completely saturated. In other words, we have that:

$$\hat{u}_m = \lim_{\Lambda \to \hat{\Lambda}(\boldsymbol{\beta})} U_m = 1, \qquad \forall m \in \boldsymbol{d}(\boldsymbol{\beta}). \tag{8}$$

However, if we focus on the utilization of a given class c at a bottleneck station m, it will tend to the fraction of jobs of the considered class:

$$\hat{u}_{mc} = \lim_{\Lambda \to \hat{\Lambda}(\boldsymbol{\beta})} U_{mc} = \frac{\beta_c D_{mc}}{D_m(\boldsymbol{\beta})}, \qquad \forall_r \in \boldsymbol{d}(\boldsymbol{\beta}). \tag{9}$$

The mean response time and the mean queue length at a bottleneck resource m tend to infinity since the system is not able to handle all the incoming requests. That is:

$$\hat{b}_{mc} = \lim_{\Lambda \to \hat{\Lambda}(\boldsymbol{\beta})} R_{mc} = \infty, \; \hat{b}_m = \lim_{\Lambda \to \hat{\Lambda}(\boldsymbol{\beta})} R_m = \infty,$$
$$\hat{q}_{mc} = \lim_{\Lambda \to \hat{\Lambda}(\boldsymbol{\beta})} Q_{mc} = \infty, \; \hat{q}_m = \lim_{\Lambda \to \hat{\Lambda}(\boldsymbol{\beta})} Q_m = \infty, \tag{10}$$
$$\forall m \in \boldsymbol{d}(\boldsymbol{\beta}).$$

Instead if we consider a resource k that is not a bottleneck (i.e, $k \notin \boldsymbol{d}(\boldsymbol{\beta})$), the limiting value of the considered performance index is finite, and it can be computed with standard queueing network formulas using an arrival rate $\hat{\Lambda}(\boldsymbol{\beta})$. Let us call such limiting values:

$$\hat{u}_{kc} = \lim_{\Lambda \to \hat{\Lambda}(\boldsymbol{\beta})} U_{kc} = U_{kc}(\hat{\Lambda}(\boldsymbol{\beta})), \quad \hat{u}_k = \lim_{\Lambda \to \hat{\Lambda}(\boldsymbol{\beta})} U_k = U_k(\hat{\Lambda}(\boldsymbol{\beta})),$$
$$\hat{b}_{kc} = \lim_{\Lambda \to \hat{\Lambda}(\boldsymbol{\beta})} R_{kc} = R_{kc}(\hat{\Lambda}(\boldsymbol{\beta})), \quad \hat{b}_k = \lim_{\Lambda \to \hat{\Lambda}(\boldsymbol{\beta})} R_k = R_k(\hat{\Lambda}(\boldsymbol{\beta})),$$
$$\hat{q}_{kc} = \lim_{\Lambda \to \hat{\Lambda}(\boldsymbol{\beta})} Q_{kc} = Q_{kc}(\hat{\Lambda}(\boldsymbol{\beta})), \quad \hat{q}_k = \lim_{\Lambda \to \hat{\Lambda}(\boldsymbol{\beta})} Q_k = Q_k(\hat{\Lambda}(\boldsymbol{\beta})), \tag{11}$$
$$\forall_k \notin \boldsymbol{d}(\boldsymbol{\beta}).$$

5.1   Resource-Level Constraints

Constraints on resources (either per class, or aggregate) can be computed with closed form expressions. Let us denote with u_{mc}, b_{mc}, q_{mc}, the maximum utilization, the response time, and the mean queue length allowed for a class c job at resource m, and with u_m, b_m and q_m the corresponding thresholds at resource m regardless of the class.

Constraints are automatically satisfied if they are based on thresholds that are greater than the maximum values determined in Eq.11: that is if either $u_{mc} > \hat{u}_{mc}$, $b_{mc} > \hat{b}_{mc}$, $q_{mc} > \hat{q}_{mc}$, $u_m > \hat{u}_m$, $b_m > \hat{b}_m$ or $q_m > \hat{q}_m$. Let us consider a non-trivial case, and let us initially focus on the response time of a class c job at resource m. The constraint is not trivial if $b_{mc} < \hat{b}_{mc}$. In this case we have that:

$$\frac{D_{mc}}{1 - \Lambda D_m(\boldsymbol{\beta})} < b_{mc}. \tag{12}$$

Since we consider that the system must be stable, we know that $U_m < 1$ $\forall r$, which implies that $1 - \Lambda D_m(\boldsymbol{\beta}) > 0$ (since we have that $U_m = \Lambda D_m(\boldsymbol{\beta})$). We can then multiply both sides of the equation with the denominator obtaining:

$$D_{mc} < b_{mc}\left(1 - \Lambda D_m(\boldsymbol{\beta})\right). \tag{13}$$

We can thus derive Λ from Eq. 13:

$$\Lambda < \Lambda^* = \frac{b_{mc} - D_{mc}}{b_{mc}D_m(\boldsymbol{\beta})}, \tag{14}$$

that expresses the maximum total arrival rate Λ^* allowed for the population mix $\boldsymbol{\beta}$ to ensure that the mean response time is $R_{mc} < b_{mc}$.

The expressions for the maximum arrival rate that satisfy the other type of resource level constraints can be derived with similar computations, and they are shown in Table 2.

Table 2. Maximum throughput to assure resource-level constraints as function of the control parameters

	Utilization	Response time	Mean queue length
Class c, res. m	$\Lambda < \dfrac{u_{mc}}{\beta_c D_{mc}}$	$\Lambda < \dfrac{b_{mc} - D_{mc}}{b_{mc}D_m(\boldsymbol{\beta})}$	$\Lambda < \dfrac{q_{mc}}{q_{mc}D_m(\boldsymbol{\beta}) + \beta_c D_{mc}}$
Res. m	$\Lambda < \dfrac{u_m}{D_m(\boldsymbol{\beta})}$	$\Lambda < \dfrac{b_m - D_m(\boldsymbol{\beta})}{b_m D_m(\boldsymbol{\beta})}$	$\Lambda < \dfrac{q_m}{D_m(\boldsymbol{\beta})(q_m + 1)}$

5.2 System-Level Constraints

The maximum throughput for system-level constraints does not have a closed form expression, and it must be computed numerically. However, it is easy to show that the considered indices are monotone with respect to Λ, and have a closed form expression for their first derivative. These features allow the use of efficient numerical solution techniques to compute the maximum throughput with an iterative procedure. In particular, the Newton-Raphson method [18] is able to converge to the solution in very few iterations.

The technique requires an initial guess x_0 for the unknown that has to be computed. This initial solution should be greater than the actual value, in order to keep the guess decreasing (and thus always included in the stability region of the model). Let us focus on limiting the mean number of jobs in the system below a threshold q_s. From Eq.7 we know that $\lim\limits_{\Lambda \to \hat{\Lambda}(\boldsymbol{\beta})} Q_S = \infty$, which means that it is always possible to find an initial guess greater than the threshold in an interval close to x_0. We can then set $x_0 = \hat{\Lambda}(\boldsymbol{\beta}) - \delta$, where $\delta > 0$ is a small number such that $Q_S(x_0) > q_s$ (i.e the corresponding performance index computed in x_0 is greater than the required threshold). This parameter δ can be efficiently computed with an exponential scaling step. The procedure to compute the maximum throughput satisfying $Q_S < q_S$, is the following:

$$
\begin{aligned}
&\delta = 1; \\
&\textbf{do } \{ \\
&\qquad \delta = \delta \cdot \delta_0; \\
&\qquad x = \hat{\Lambda}(\beta) \cdot (1 - \delta); \\
&\} \textbf{ while } (Q_S(x) < q_S); \\
&\textbf{while } \left(\left| \frac{Q_S(x) - q_S}{q_S} \right| > \epsilon \right); \\
&\qquad x = x - \frac{Q_S(x) - q_S}{Q'_S(x)}; \\
&\} \\
&\Lambda = x;
\end{aligned}
\tag{15}
$$

where ϵ is a term that represents the relative precision of the solution, δ_0 is a constant $0 < \delta_0 < 1$ corresponding to the exponential scaling factor and $Q'_S(x)$ is the first derivative of the mean number of requests in the system (as defined in Table 3). In the experiments presented in this paper we used $\delta_0 = 0.1$ and $\epsilon = 10^{-6}$.

Table 3. First derivative of system-level indices

	Response time	Mean queue length
Class c	$\dfrac{\partial R_{*c}}{\partial \Lambda} = \displaystyle\sum_m \dfrac{D_{mc} D_m(\beta)}{(1 - \Lambda D_m(\beta))^2}$	$\dfrac{\partial Q_{*c}}{\partial \Lambda} = \beta_c \displaystyle\sum_m \dfrac{D_{mc}}{(1 - \Lambda D_m(\beta))^2}$
System	$\dfrac{\partial R_S}{\partial \Lambda} = \displaystyle\sum_m \dfrac{D_m(\beta)^2}{(1 - \Lambda D_m(\beta))^2}$	$\dfrac{\partial Q_S}{\partial \Lambda} = \displaystyle\sum_m \dfrac{D_m(\beta)}{(1 - \Lambda D_m(\beta))^2}$

5.3 Constrained Maximum Throughput

Also for constrained systems, we can determine the population mix for which the maximum arrival rate could be achieved without violating the given performance constraints. This can be done by solving an Optimization Problem (OP), similar to the one proposed in Section 4, where extra constraints are added to account for the desired performance requirements:

$$
\begin{aligned}
&\textbf{Variables:} && \lambda_c, && 1 \le c \le C \\
&\textbf{Objective:} && \text{maximize } \sum_c \lambda_c \\
&\textbf{Constraints:} \sum_d \lambda_d D_{md} \le 1, && 1 \le m \le M \\
& && \lambda_c \ge 0, && 1 \le c \le C \\
& && \textit{Performance constraints}_1 \\
& && \quad\vdots \\
& && \textit{Performance constraints}_n
\end{aligned}
\tag{16}
$$

Constraint expressions are given in Table 4. For five of the performance indexes shown in Table 4, (the one with the white background), the corresponding

Table 4. Constraints. Equations in gray cells are of non-linear constraints

	Utilization	Response time	Mean queue length
Class c, res. m	$\lambda_c D_{mc} \leq u_{mc}$	$b_{mc}\left(1 - \sum_d \lambda_d D_{md}\right) \geq D_{mc}$	$q_{mc}\left(1 - \sum_d \lambda_d D_{md}\right) \geq$ $\lambda_c D_{mc}$
Res. m	$\sum_d \lambda_d D_{md} \leq u_m$	$b_{mc}\left(1 - \sum_d \lambda_d D_{md}\right)\sum_d \lambda_d \geq$ $\sum_d \lambda_d D_{md}$	$q_m\left(1 - \sum_d \lambda_d D_{md}\right) \geq$ $\sum_d \lambda_d D_{md}$
Class c	//	$\sum_m \dfrac{D_{mc}}{1 - \sum_d \lambda_d D_{md}} \leq b_c$	$\lambda_c \sum_m \dfrac{D_{mc}}{1 - \sum_d \lambda_d D_{md}} \leq q_c$
System	//	$\sum_m \dfrac{\sum_d \lambda_d D_{md}}{1 - \sum_d \lambda_d D_{md}} \leq$ $b_S\left(\sum_d \lambda_d\right)$	$\sum_m \dfrac{\sum_d \lambda_d D_{md}}{1 - \sum_d \lambda_d D_{md}} \leq q_S$

constraints are linear in λ_c. This means that when only such constraints are present, Eq. 16 is a LPP which can be efficiently and accurately solved using the simplex algorithm. In presence of the other constraints (the one with gray background in Table 4), non linear optimization techniques should be employed. However, since all the functions are convex, the problem can still be solved using simple techniques such as Successive Quadratic Programming (SQP) [16].

As an example, we can use the result to investigate the effect of combining several different constraints on the total throughput of the system described in Section 4. In particular, we set the following requirements:

S1: The utilizations of the (**S1a**) *Inventory* and (**S1b**) *Shipping* resources should be less than 70%.

S2: The requests of the *Intranet* class should be executed with a mean transfer time less than 650 *ms*.

S3: We mush have $\beta_3 = 0.2$.

The maximum throughput with the given set of constraints corresponds to the solution of the following OP:

Variables: $\lambda_c, \ 1 \leq c \leq C$

Objective: maximize $\sum_c \lambda_c$

Constraints:

$$
\begin{aligned}
&1) \quad u_{r_3} - \sum_d \lambda_d D_{3d} \geq 0 \qquad 2) \ u_{r_4} - \sum_d \lambda_d D_{4d} \geq 0 \\
&3) \ b_1 - \sum_m \frac{D_{m1}}{1 - \sum_d \lambda_d D_{md}} \geq 0 \qquad 4) \ \lambda_3 - \beta_3\left(\sum_d \lambda_d\right) \geq 0 \\
&5) \qquad 1 - \sum_d \lambda_d D_{md} \geq 0, \qquad\qquad 1 \leq m \leq M \\
&6) \qquad\qquad \lambda_c \geq 0, \qquad\qquad\qquad 1 \leq c \leq C
\end{aligned}
\tag{17}
$$

The constraints in Eq. 17 are derived from those in Table 4 and expressed as inequalities greater than or equal to 0 to conform with OP conventions. Constant parameters u_{r_3} and u_{r_4} of the *1)* and *2)* constraints (which corresponds to constraints **S1a** and **S1b**) are both set to 0.7. In constraint *3)* (corresponding to **S2**) the Response time requirement of class 1 is set to $b_1 = 650$ ms, while constraint *4)* corresponds to constraint **S3**. It is interesting to note that among all the constraints in Eq.17, *3)* (**S2**) is the only one non-linear. If we exclude it from the OP, the Eq. 17 becomes a LPP which can be solved very efficiently with the simplex algorithm. In this case, the maximum Λ is 9.1864 job/s. with $\beta_1 = 0.38857$, $\beta_2 = 0.41143$ and $\beta_3 = 0.20006$. If we consider the complete set of constraints, we can use the SQP non-linear optimization technique, to find a maximum $\Lambda = 8.7508$ jobs/s. with $\beta_1 = 0.35245$, $\beta_2 = 0.44755$ and $\beta_3 = 0.2$. Results were computed using GNU Octave [17] on a standard laptop PC in few seconds.

5.4 Computational Complexity

The proposed performance bounds can be computed very efficiently. The population mix scaled demands $D_m(\boldsymbol{\beta})$ must be computed for all the resources $1 \leq m \leq M$: this has time complexity $O(M \cdot C)$ and storage complexity $O(M)$. Value $D_m(\boldsymbol{\beta})$ can then be used to compute the population mix maximum throughput $\hat{\Lambda}(\boldsymbol{\beta})$ with complexity $O(M)$.

Knowing the population mix scaled demand, resource-level constraints can be obtained in $O(1)$ time since they do not include any iterative procedure and can be computed with closed form expressions.

System-level constraints are more complex since they must be computed iteratively. However, thanks to the Newton-Raphson, usually less than ten iterations are enough to reach a solution within the desired precision. During each iteration, both the value of the performance index and of its derivative must be computed: the complexity of both operations is $O(M)$ since these expressions iterate over all the system resources.

The most time-consuming operation is thus the computation of $D_m(\boldsymbol{\beta})$, which however could be cached and reused if several constraints have to be computed (i.e., a set of J constraints must be satisfied at the same time). The final computational complexity is $O(M \cdot \max\{C, J\})$, which gives the possibility to explore very large parameter spaces in term of different $\boldsymbol{\beta}$, even when considering a large number of classes, resources and performance objectives.

The convergence to the optimal solution of the LPP to determine either the unconstrained maximum throughput, or to consider linear constraints is also not an issue. The LPPs have always C variables and $M + \#LK + C$ (LP) constraints, where $\#LK$ is the total number of linear constraints. If the problem includes also $\#LK$ non-linear constraints, then the OP has a total of $M + \#LK + \#NLK + C$ (OP) constraints, which can still be handled by todays commodity hardware, provided that the total number of classes or resources is not extremely high.

6   Experimental Results

To show the applicability of the proposed technique in a real scenario, we applied it to study the maximum throughput of the RUBiS [6] benchmark application. RUBiS is a prototype of an auction site that mimics eBay, which is available in three different technologies: Java servlets, PHP, and Enterprise Java Bean. We used the servlet version of the benchmark, which is organized as a three-tier architecture using standard HTML, Java Servlet, and SQL technology. RUBiS comes with Apache server as the web server, JBoss as application server and MySQL as database. Each tier was deployed on a different physical machine equipped with a single core Intel Xeon processor running at 2.66 GHz with Ubuntu 12.04 LTS; the client emulator and the load balancer were deployed on servers running Microsoft Windows Server 2008 R2. A dedicated gigabit LAN provided the network functionality.

Since servlets' execution times are mostly related with the queries sent to the database, and with the amount of data returned, we focused our study on two servlets with fixed queries: *ViewBidHistory* and *PutComment*. Even though RUBiS is composed of a larger set of servlets, we selected these two as representatives of two types of requests: the application server intensive and the database intensive type, respectively. We consider this not to be an over-simplification; on the contrary, it is quite common in other related works as well, where techniques such as K-means clustering are commonly used to group requests into fewer clusters with similar profiles e.g. [19]. The technique proposed in Sec. 4 and Sec. 5 requires the determination of the demands of the considered classes at the resources that compose the system. In order to estimate such parameters, we studied the system by performing a set of test workloads. In particular, we loaded our client emulator with $\Lambda \in \{150, 200, 250, 350, 400\}$ job/s and $\beta_1 \in \{0, 0.2, 0.4, 0.6, 0.8, 1\}$. Not all the configurations were stable: in particular we experienced a large number of requests being dropped for $\Lambda = 350$ job/s and $\beta_1 = 0$, and for $\Lambda = 400$ job/s and $\beta_1 \leq 0.6$. We discarded the configurations where we experienced drops in the system. We used the JBoss' and database's mean service times to determine the demands that better describes our system through a simple fitting procedure: we minimized the difference between the service time expected by the model and the one measured on the real system. Figure 2a shows both the model and the measures: as it can be seen, we have very small errors for the JBoss component, while the DB experiences a larger deviation. Table 5a shows the estimated service times.

By applying the values from Table 5a to Eq. 5, we determined that the systems is capable of offering a maximum throughput $\Lambda^* = 443.5$ jobs/s for a population mix of $\beta^* = |0.0558, 0.9442|$. In particular, in Fig. 2b we can see that the system should be stable for all the possible population mixes for $\Lambda < 308$ jobs/s, while it will present instabilities for some β for $\Lambda > 308$. These results were confirmed by measurements, where we found the system being unstable at $\Lambda = 350$ job/s for $\beta_{ViewBidHistory} = 1$, and at $\Lambda = 400$ job/s for $\beta_{ViewBidHistory} \leq 0.4$.

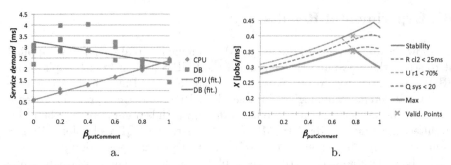

Fig. 2. (a.) Estimated vs. measured mean service demands and (b.) Maximum throughput of the considered RUBiS application

Table 5. (a.) Estimated demands of the considered RUBiS servlets (in msec) and (b.) Performance indices of the RUBiS application at β^*

D	PutComment	viewBidHistory	Λ [job/sec]	$R_{ViewBidHistory}$ [ms.]	U_{JBoss}	Q_{sys}
JBoss	0.5855	2.3532	350	11.0	55%	11.9
DB	3.2463	2.1959	400	55.0	74%	373.6

a.   b.

We then adds three constraints to the system:

S2: The requests of the *ViewBidHistory* class should be executed with a mean transfer time less than 25 ms.
S2: The utilizations of the *JBoss* resources should be less than 70%.
S3: The mean number of jobs in the system should be not greater than 20.

Figure 2b shows the effect of the constraints to the maximum throughput for various values of the class mix β. Using the optimization procedure described in Sec. 5.3, we can determine that the maximum throughput is $\Lambda^* = 358.34$ jobs/s. for a population mix of $\beta^* = |0.2262, 0.7738|$. To check such requirements, we measured the system at $\beta^* = |0.2262, 0.7738|$ for $\Lambda_{low} = 350$ job/s. and $\Lambda_{high} = 400$ job/s. The system was stable for both arrival rates. The results are shown in Table 5b: as it can be seen, all the constraints are met at Λ_{low}, but they are all violated for Λ_{high}, even if the system is still stable.

7   Conclusions

In this paper we described a technique to identify the maximum throughput that a system can provide, given an SLA for each class of applications. Limiting values of several performance metrics were considered, i.e., response times, utilizations, and queue lengths, and the maximum throughput for each class was computed. We demonstrated that the predicted values can be obtained in a real environment through experiments executed on a commonly adopted benchmark that simulates an e-commerce web site.

The future work will be concentrated on policies for the performance control of each node of a large set of interconnected systems in order to maximize the throughput of the global network.

References

1. Anselmi, J., Casale, G.: Heavy-traffic revenue maximization in parallel multiclass queues. Perform. Eval. 70(10), 806–821 (2013), http://dx.doi.org/10.1016/j.peva.2013.08.008
2. Anselmi, J., Cremonesi, P., Amaldi, E.: On the consolidation of data-centers with performance constraints. In: Mirandola, R., Gorton, I., Hofmeister, C. (eds.) QoSA 2009. LNCS, vol. 5581, pp. 163–176. Springer, Heidelberg (2009)
3. Ayoub, R., Ogras, U., Gorbatov, E., Jin, Y., Kam, T., Diefenbaugh, P., Rosing, T.: Os-level power minimization under tight performance constraints in general purpose systems. In: 2011 International Symposium on Low Power Electronics and Design (ISLPED), pp. 321–326 (2011)
4. Balbo, G., Serazzi, G.: Asymptotic analysis of multiclass closed queueing networks: Multiple bottlenecks. Performance Evaluation 30(3), 115–152 (1997)
5. Barford, P., Crovella, M.: Generating representative web workloads for network and server performance evaluation. SIGMETRICS Perform. Eval. Rev. 26, 151–160 (1998), http://doi.acm.org/10.1145/277858.277897
6. Benchmark, R.: http://rubis.ow2.org/
7. Casale, G., Serazzi, G.: Bottlenecks identification in multiclass queueing networks using convex polytopes. In: Proc. of IEEE MASCOTS Symposium, pp. 223–230. IEEE Press (2004)
8. Cherkasova, L., Phaal, P.: Session-based admission control: A mechanism for peak load management of commercial web sites. IEEE Trans. Comput. 51, 669–685 (2002), http://dx.doi.org/10.1109/TC.2002.1009151
9. Eager, D.L., Sevcik, K.C.: Bound hierarchies for multiple-class queuing networks. J. ACM 33, 179–206 (1986), http://doi.acm.org/10.1145/4904.4992
10. Elnikety, S., Tracey, J., Nahum, E., Zwaenepoel, W.: A method for transparent admission control and request scheduling in e-commerce web sites. In: Proceedings of the 13th WWW, pp. 276–286 (2004)
11. Leitner, P., Michlmayr, A., Rosenberg, F., Dustdar, S.: Monitoring, prediction and prevention of sla violations in composite services. In: ICWS 2010, pp. 369–376 (July 2010)
12. Litoiu, M.: A performance analysis method for autonomic computing systems. ACM Trans. Auton. Adapt. Syst. 2 (March 2007), http://doi.acm.org/10.1145/1216895.1216898
13. Liu, Z., Squillante, M.S., Wolf, J.L.: On maximizing service-level-agreement profits. In: Proceedings of EC 2001, pp. 213–223. ACM, New York (2001), http://doi.acm.org/10.1145/501158.501185
14. Mars, J., Tang, L.: Whare-map: Heterogeneity in "homogeneous" warehouse-scale computers. In: ISCA 2013, ACM, New York (2013), http://doi.acm.org/10.1145/2485922.2485975

15. Menascé, D.A., Almeida, V.A.F., Fonseca, R., Mendes, M.A.: Business-oriented resource management policies for e-commerce servers. Perform. Eval. 42, 223–239 (2000),
    http://dx.doi.org/10.1016/S0166-5316(00)00034-1
16. Nocedal, J., Wright, S.J.: Numerical optimization. Springer, New York (2006),
    http://site.ebrary.com/id/10228772
17. Octave, G.: http://www.gnu.org/software/octave/
18. Press, W.H., Teukolsky, S.A., Vetterling, W.T., Flannery, B.P.: Numerical recipes in C: The art of scientific computing, 2nd edn. Cambridge University Press, New York (1992)
19. Singh, R., Sharma, U., Cecchet, E., Shenoy, P.: Autonomic mix-aware provisioning for non-stationary data center workloads. In: ICAC 2010, pp. 21–30. ACM, New York (2010),
    http://doi.acm.org/10.1145/1809049.1809053
20. Urgaonkar, B., Chandra, A.: Dynamic provisioning of multi-tier internet applications. In: Proceedings of the Second International Conference on Automatic Computing, USA, pp. 217–228. IEEE Computer Society, Washington, DC (2005),
    http://dx.doi.org/10.1109/ICAC.2005.27
21. Villela, D., Pradhan, P., Rubenstein, D.: Provisioning servers in the application tier for e-commerce systems. ACM Trans. Internet Technol. 7(1) (February 2007),
    http://doi.acm.org/10.1145/1189740.1189747
22. Walsh, W.E., Tesauro, G., Kephart, J.O., Das, R.: Utility functions in autonomic systems. In: Proceedings of the International Conference on Autonomic Computing, pp. 70–77 (2004)
23. website, T.A.E, http://aws.amazon.com/ec2/#instance
24. Website, T.I.S.C, http://www.ibm.com/cloud-computing/us/en/
25. Website, T.M.A, http://www.microsoft.com/windowsazure/
26. Xu, X., Yu, H., Cong, X.: A qos-constrained resurce allocation game in federated cloud. In: IMIS 2013, pp. 268–275 (2013)
27. Xue, J.W.J., Chester, A.P., He, L., Jarvis, S.A.: Model-driven server allocation in distributed enterprise systems. In: ABIS 2009 (2009)
28. Zhou, X., Ippoliti, D.: Resource allocation optimization for quantitative service differentiation on server clusters. Journal of Parallel and Distributed Computing 68(9), 1250–1262 (2008),
    http://www.sciencedirect.com/science/article/pii/S074373150800107X

Checkpointing in Failure Recovery
in Computing and Data Transmission

Søren Asmussen[1], Lester Lipsky[2], and Stephen Thompson[2]

[1] Dept. of Mathematics, Aarhus University
Ny Munkegade, 8000 Aarhus C, Denmark
[2] Dept. of Computer Science and Engineering, University of Connecticut
Storrs, CT 06269-2155, USA
asmus@imf.au.dk, lester@engr.uconn.edu, stevet@alexzilla.com

Abstract. A task with ideal execution time T such as the execution of a computer program or the transmission of a file on a data link may fail. A number of protocols for failure recovery have been suggested and analyzed, in particular RESUME, REPLACE and RESTART. We consider here RESTART with particular emphasis on *checkpointing* where the task is split into K subtasks by inserting $K-1$ checkpoints. If a failure occurs between two checkpoints, the task needs to be restarted from the last checkpoint only. Various models are considered: the task may have a fixed ($T \equiv t$) or a random length, and the spacing of checkpoints may be equidistant, non-equidistant or random. The emphasis here is on tail asymptotics for the total task time X in the same vein as the study of Asmussen *et al.* [5] on simple RESTART. For a fixed task length ($T \equiv t$) and certain types of failure mechanism, for example Poisson, the conclusion of the study is clear and not unexpected: the essential thing to control for minimizing large delays is making the maximal distance between checkpoints as small as possible. For random unbounded task length, it is seen that the effect of checkpointing is limited in the sense that the tail of X remains power-like as for simple RESTART ($K = 1$).

Keywords: computer reliability, coupling, data transmission, failure rate, fault-tolerant computing, geometric sum, order statistics, renewal equation, stochastic ordering, tail asymptotics, Tauberian theorem, uniform spacings.

1 Introduction

For many systems failure is rare enough that it can be ignored. For others, failure is common enough that the design choice of how to deal with it may have a significant impact on the performance of the system. Failures may occur for many reasons: external ones such as power failure, disk failure, processor failure, and internal ones due to problems with the task itself. Consider a job that ordinarily would take a time T to be executed on some system (e.g., a CPU). If at some time $U < T$ the processor fails, the job may take a *total time*

B. Sericola, M. Telek, and G. Horváth (Eds.): ASMTA 2014, LNCS 8499, pp. 253–272, 2014.

X to complete. We let F, G be the distributions of T, U and H the distribution of X which in addition to F, G depends on the failure recovery scheme.

Many papers discuss methods of failure recovery and analyze their complexity in one or more metrics, like *restartable processors* in Chlebus *et al.* [11], or *stage checkpointing* in De Prisco *et al.* [13], etc. There are many specific and distinct failure recovery schemes, but they can be grouped into three broad classes. In the *RESUME* scenario, if there is a processor failure while a job is being executed, after repair is implemented the job can continue where it left off. In the *REPLACE* situation, if a job fails, it is replaced by a different job from the same distribution. Here, no details concerning the previous job are necessary in order to continue. The analysis of the distribution function $H(x) = \mathbb{P}(X \leq x)$ when the policy is RESUME or REPLACE was carried out by Kulkarni *et al.* [23], [24].

There are many examples where the RESTART scenario is relevant. The obvious one alluded to above involves execution of a program on some computer. If the computer fails, and the intermediate results are not saved externally (e.g., by *checkpointing*), then the job must restart from the beginning. As another example, one might wish to copy a file from a remote system using some standard protocol as FTP or HTTP. The time it takes to copy a file is proportional to its length. A transmission error immediately aborts the copy and discards the partially received data, forcing the user to restart the copy from the beginning. Yet another example is call centers where connections may be broken. However, despite its relevance the RESTART policy resisted detailed analysis until the recent work of Asmussen *et at.* [5] where the tail of H was found in a variety of combinations of tail behavior of F and G. A main finding was that the tail of H is always heavy-tailed if F has unbounded support. The consequence is that delays can be very severe in the RESTART setting. The purpose of the present paper is to study a method of failure recovery mitigating this, *checkpointing* where $K - 1$ checkpoints are inserted in the task, splitting it into K subtasks. If a failure occurs between two checkpoints, the task needs to be restarted from the last checkpoint only.

We start the paper in Section 2 by a summary of the results of [5] for simple RESTART. In Section 3, we introduce the models for checkpointing to be studied later in the paper. Various possibilities are considered: the task may have a fixed ($T \equiv t$) or a random length, and the spacing of checkpoints may be equidistant, non-equidistant or random. There could be a fixed number of checkpoints, or a variable number proportional to T.

Sections 4–10 contains our main new results, a detailed analysis of the suggested checkpointing models. The emphasis is on tail asymptotics for the total task time X in the same vein as the study of [5] on simple RESTART, whereas early literature on checkpointing tends to concentrate on expected values (see Nicola [28] for a survey with an extensive list of references). The results show that sometimes checkpointing is a considerable improvement, sometimes not (in particular not for random spacings and or unbounded task lengths).

It should be noted that a feature neglected here is the cost of checkpointing, which may be considerable. If checkpointing was free, the optimal solution would obviously be to insert an infinity of checkpoints, in which case $X = T$ so that studies like ours would be irrelevant. We have also (as in [5]) ignored that repair times may be non-zero.

The asymptotic results on H are illustrated numerically in Section 8, and comparisons in the sense of stochastic ordering are given for the different models in Section 7.

Notation: constants are throughout denoted by subscripts referring to the specific models (RESTART, A, and checkpoint models A–E), $C_{R,1}, C_{C,2}, c_{A,3}, c_{D,2}$ etc., such that upper case C refers to a constant whose value is important and lower case c to one where it is less so.

The distribution of the (ideal) task time T is throughout denoted by F, and the distribution of the failure times $U_1, U_2, \ldots$ by G. For convenience, the density g of G is asssumed to exist, and F is taken either degenerate at T (say), i.e. $T \equiv t$, or with a density f. Assume $\mathbb{E}U < \infty$ and write $\mu = 1/\mathbb{E}U$.

For most applications, it would be of particular interest to assume G to be exponential, say at rate μ, and F to be either degenerate, say at t, gamma-like in the sense that

$$f(t) \sim ct^{\alpha-1}e^{-\lambda t}, \quad t \to \infty, \tag{1}$$

or of power-form in the sense that $\log f(t)/t \to -\alpha - 1$; the main example of this is a regularly varying f where $f(t) = L(t)/t^{\alpha+1}$ with L slowly varying at ∞ and $\alpha > 0$, cf. [7]. We shall therefore often specialize to one of these cases.

The proofs are sometimes sketchy and more detail can be found in a full paper available upon request from the authors.

2 Simple RESTART

Consider in this section the simple RESTART model without checkpoints. The approach of [5] is to first obtain the tail of H in the case of a deterministic $T \equiv t$ and next to mix over t with weights $f(t)$ (the density of F) for the case of a random T.

2.1 Deterministic Task Time $T \equiv t$

Consider a deterministic $T \equiv t$ and let $X_R(t)$ be the corresponding simple RESTART total time (without checkpointing), $H_R(x|t) = \mathbb{P}(X_R(t) \leq x)$. Write $U_1, U_2, \ldots$ for the failure times, assumed i.i.d. with distribution G

Let $\widehat{H}_R[a|t] = \mathbb{E}e^{aX_R(t)}$ be the m.g.f. of $H_R(\cdot|t)$ and $m_R(t) = \mathbb{E}X_R(t)$ the mean. Then:

Proposition 1. *Define* $m_1(t) = \int_0^t y\, G(dy)$. *Then*

$$m_R(t) = t + \frac{m_1(t)}{\overline{G}(t)}, \quad \widehat{H}_R[a|t] = e^{at}\frac{\overline{G}(t)}{1 - \int_0^t e^{ay}\, G(dy)}. \tag{2}$$

Proof. Similar formulas appear already in early literature. A short proof goes by noting that if $U_1 = y \leq t$, we have $X_R(t) = y + \widetilde{X}_R(t)$ where $\widetilde{X}_R(t)$ is an independent copy of $X_R(t)$. Given $U_1 > t$, we simply have $X_R(t) = t$. This gives equations that are readily solved for the unknowns. □

Now consider the tail. As in [5], we can write $X_R(t) = t + S_R(t)$ where $S_R(t) = \sum_1^N U_i(t)$ is a geometric sum: $N, U_1(t), U_2(t), \ldots$ are independent such that $\mathbb{P}(U_i(t) \leq s) = G(s)/G(t)$ for $s \leq t$, $\mathbb{P}(U_i(t) \leq s) = 1$ for $s > t$, and $\mathbb{P}(N = n) = (1 - \rho)\rho^n$ with $\rho = G(t)$. From [5] we know that

$$\mathbb{P}(S_R(t) > x) \sim C_{R,1}(t)e^{-\gamma(t)x}, \tag{3}$$

where $\gamma(t)$ is the solution of

$$1 = \int_0^t e^{\gamma(t)y} G(dy) \tag{4}$$

(note for the following that $\gamma(t)$ is non-increasing in t) and

$$C_{R,1}(t) = \frac{\overline{G}(t)}{\gamma(t)m_2(t)} \quad \text{where } m_2(t) = \int_0^t y e^{\gamma(t)y} G(dy). \tag{5}$$

Since $\mathbb{P}(X_R(t) > x) = \mathbb{P}(S_R(t) > x - t)$, we therefore have

$$\overline{H}_R(x|t) = \mathbb{P}(X_R(t) > x) \sim C_{R,2}(t)e^{-\gamma(t)x} \tag{6}$$

where $C_{R,2}(t) = e^{\gamma(t)t}C_{R,1}(t)$. In summary:

Theorem 1. *Assume $T \equiv t$ and $\overline{G}(t) > 0$. Then*

$$\overline{H}(x|t) \sim D(t)e^{-\gamma(t)x}, \quad x \to \infty,$$

where $\gamma(t) > 0$ is given by (4) and $D(t) = \overline{G}(t)e^{\gamma(t)t}/\gamma(t)B(t)$ where $B(t) = \int_0^t y e^{\gamma(t)y}g(y)\, dy$.

2.2   Unbounded Task Time

In the case of an infinite support of f, Theorem 1 shows that the tail of H is heavier than $e^{-\gamma(t)x}$ for all t. Now note that $\gamma(t) \downarrow 0$ as $t \to \infty$. It therefore follows that H is heavy-tailed in the sense that $e^{\gamma x}\overline{H}(x) \to \infty$. for all $\gamma > 0$. This is remarkable by giving an example where combining two light-tailed distributions F, G can lead to heavy tails. We illustrate this via the following particularly important case:

Theorem 2. *Assume that $f(t) \sim c_F t^{\alpha-1}e^{-\lambda t}$, $t \to \infty$, and that $\overline{g}(t) = e^{-\mu t}$, i.e. failures follow a Poisson(μ) process. Then*

$$\overline{H}(x) \sim \frac{c_F \Gamma(\lambda/\mu)}{\mu^{\alpha+\lambda/\mu}} \frac{\log^{\alpha-1} x}{x^{\lambda/\mu}}.$$

Sketch of proof. Clearly,

$$\overline{H}(x) \sim \int_0^\infty \mathbb{P}(X > x \,|\, T = t) f(x) \,\mathrm{d}x.$$

From $\gamma(t) \to 0$, one expects that only large t matter so that (using Theorem 1)

$$\overline{H}(x) \sim \int_0^\infty \overline{c}_F t^{\alpha-1} \mathrm{e}^{-\lambda t} D(t) \mathrm{e}^{-\gamma(t)x} \,\mathrm{d}t.$$

Now a simple Taylor expansion gives $\gamma(t) \sim \mu \overline{G}(t) = \mu \mathrm{e}^{-\mu t}$. From this one concludes $D(t) \to 1$ and expects that

$$\overline{H}(x) \sim I(x) \quad \text{where} \quad I(x) = \int_0^\infty \overline{c}_F t^{\alpha-1} \mathrm{e}^{-\lambda t} \mathrm{e}^{-\mu \mathrm{e}^{-\mu t} x} \,\mathrm{d}t.$$

Substituting $z = \mu \mathrm{e}^{-\mu t} x$, we have

$$\mathrm{e}^{-\lambda t} = \left(\frac{z}{\mu x}\right)^{\lambda/\mu}, \quad -\mu z \,\mathrm{d}t = \mathrm{d}z, \quad t = \frac{1}{\mu}\left(\log x + \log(\mu/z)\right),$$

and so $I(x)$ is asymptotically

$$\int_0^{\mu x} c_F \frac{1}{\mu^{\alpha-1}} \left(\log x + \log(\mu/z)\right)^{\alpha-1} \left(\frac{z}{\mu x}\right)^{\lambda/\mu} \mathrm{e}^{-z} \frac{1}{\mu z} \,\mathrm{d}z$$

$$= c_F \frac{1}{\mu^{\alpha+\lambda/\mu}} \frac{1}{x^{\lambda/\mu}} \int_0^{\mu x} \left(\log x + \log(\mu/z)\right)^{\alpha-1} z^{\lambda/\mu-1} \mathrm{e}^{-z} \,\mathrm{d}z$$

which has the same asymptotics as asserted.   □

In general, the asymptotic form of $\overline{H}(x)$ becomes the simpler (and easier to derive!) the more F and G are alike. For example:

Corollary 1. *Assume f, g belong to the class of regularly varying densities of the form $L(t)/t^{1+\alpha}$ where L is slowly varying, with parameters α_F, L_F for f and α_G, L_G for g. Then $\overline{H}(x) = L_H(x)/x^{\alpha_H}$, where $\alpha_H = \alpha_F/\alpha_G$ and L_H is slowly varying with*

$$L_H(x) \sim \frac{\Gamma(\alpha_H) \alpha_G^{\alpha_H-1}}{\mu^{\alpha_H}} \frac{L_F(x^{1/\alpha_G})}{L_G^{\alpha_H}(x^{1/\alpha_G})}.$$

3   Checkpoint Models

The checkpoints are $t_0 = 0 < t_1 < \ldots < t_{K-1} < T$, with corresponding spacings $h_k = t_k - t_{k-1}$, $k = 1, \ldots, K$ (with the convention $t_K = T$).

It is assumed that system failures regenerate at a checkpoint t_k (no memory on the previous checkpoint period $[t_{k-1}, t_k)$). Obviously, the case $K = 1$ corresponds to the simple RESTART setting.

We shall study the following models (here throughout $t_0 = 0$):

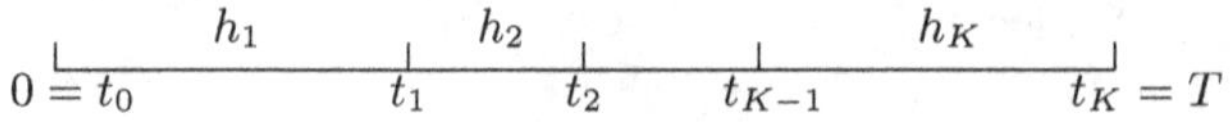

Fig. 1. The checkpoints and their spacings

A: T is deterministic, $T \equiv t$, and the checkpoints are deterministic and equidistant, $t_1 = t/K$, $t_2 = 2t/K, \ldots$, $t_{K-1} = (K-1)t/K$. Equivalently, $h_k = t/K$.

B: T is deterministic, $T \equiv t$, and the checkpoints are deterministic but not equidistant, for simplicity $h_k \neq h_\ell$ for $k \neq \ell$. In particular, there is a unique maximal checkpoint interdistance h_{k^*}.

C: T is deterministic, $T \equiv t$, and the checkpoints are random. More precisely, the set $\{t_1, \ldots, t_{K-1}\}$ is the set of outcomes of $K-1$ i.i.d. uniform r.v.'s $V_1, \ldots, V_{K-1}$ on $(0, t)$. That is, the $t_k = V_{(k)}$, $k = 1, \ldots, K-1$, are the order statistics of $K-1$ i.i.d. uniform r.v.'s on $(0, t)$.

D: T is random and the checkpoints equally spaced, $h_k \equiv h$ for $k < K$. Thus, $K = \lceil T/h \rceil$ is random (note that $h_K = T - (K-1)h$).

E: T is random and the checkpoints are given by $t_k = t'_k T$ for a deterministic set of constants $0 = t'_0 < t'_1 < \ldots < t'_{K-1} < t'_K = 1$.

Model A is the most obvious choice of checkpoints for a fixed $T \equiv t$. In transmission of a stream of data, it corresponds to breaking the stream into packets of equal size. In computing, it could be implemented by the operating system by interrupting the program execution at will. Similar remarks apply to Model D. However, in practice it is not always possible to place checkpoints everywhere, as this depends to a large extent on the structure of the task. This motivates model B. A more realistic plan would be to assume that the checkpoints are selected randomly, and Model C is the simplest case of this. Finally, Model E could again arise in data transmission where a simple case of the model is that any message is routinely split into a fixed number K packets of equal size.

Note that Models, A, B, C, E have a fixed K, but model D has variable K, proportional to T. This is important for controlling the tail of X.

We write $X_A(t), X_B(t), X_C(t)$ for the total task times in Models A, B, C where $T \equiv t$, X_D, X_E in Models D, E, and $X_R(t)$ for simple RESTART. Also, we define $H_A(\cdot|t), H_D$ etc. as the corresponding distributions. Our results will deal mainly with tail behaviour of these distributions subject to various assumptions, but we also give some discussion of comparison of the models.

It should be noted that the list of models A–E for checkpointing is by no means exhaustive. For example, Nicola *et al.* [29] consider a scheme where at a failure a fraction $0 < q < 1$ of the work performed since the previous failure is lost. This model has the fundamental difference from A–E that the total task time X given $T = t$ is bounded and is therefore not considered here.

4   Model A

We assume in this section that T is deterministic, $T \equiv t$, and the checkpoints are deterministic and equidistant, $t_1 = t/K$, $t_2 = 2t/K, \ldots, t_{K-1} = (K-1)t/K$.

Equivalently, $h_k = t/K = h_K^t$ where here and in the following $h_K^t = t/K$. Thus, the total task time $X_A(t)$ is the sum of K i.i.d. copies $X_{R,1}(h_K^t), \ldots, X_{R,K}(h_K^t)$ of RESTART total times with task time h_K^t and failure time distribution G.

Recall the definitions of $C_{R,1}(\cdot)$, $C_{R,2}(\cdot)$ from Section 2.1 and define

$$C_{A,1,k} = \frac{\overline{G}(h_K^t)^{k-1} C_{R,1}(h_K^t)}{(k-1)! m_2 (h_K^t)^{k-1}}.$$

Theorem 3. *Let* $C_{A,2} = C_{A,1,K} e^{\gamma(h_K^t)t}$. *Then*

$$\overline{H}_A(x|t) = \mathbb{P}\big(X_A(t) > x\big) \sim x^{K-1} e^{-\gamma(h_K^t)x} \cdot C_{A,2}$$

There are several proofs of Theorem 3, all fairly straightforward. Three different ones can be found in the full version of the paper. Here we just recall that $X_A(t) = X_{R,1}(h_K^t) + \cdots + X_{R,K}(h_K^t)$, so that we can appeal to Theorem 1 and the following simple lemma on convolutions of gamma-like tails:

Lemma 1. *Let* $V_1, \ldots, V_K \geq 0$ *be independent r.v.'s such that* $\mathbb{P}(V_i > x) \sim D_i x^{\alpha_i - 1} e^{-\gamma x}$ *for some* $\gamma > 0$, $\alpha_1, \ldots, \alpha_K > 0$ *and some* $D_1, \ldots, D_K > 0$. *Then*

$$\mathbb{P}(V_1 + \cdots + V_K > x) \sim D x^{\alpha - 1} e^{-\gamma x},$$

where $\alpha = \alpha_1 + \cdots + \alpha_m$, $D = D_1 \Gamma(\alpha_1) \cdots D_K \Gamma(\alpha_K) \gamma^{K-1} / \Gamma(\alpha)$.

Proof. The lemma is standard if each V_i has a gamma(α_i, γ) distribution. Indeed, then $D_i = \gamma^{\alpha_i - 1} / \Gamma(\alpha_i)$ and the sum is gamma(α, γ) with asymptotic tail $\gamma^{\alpha-1} x^{\alpha-1} e^{-\gamma x} / \Gamma(\alpha)$.

For the general case, show that the contribution to

$$\mathbb{P}(V_1 + \cdots + V_K > x) = \int \cdots \int_{\{v_1 + \cdots + v_K > x\}} \mathbb{P}(V_1 \in dv_1) \cdots \mathbb{P}(V_K \in dv_K)$$

from any region of the form $\{v_i \leq a_i\}$ is negligible, use a Riemann sum approximation for the remaining tail integral, and involve what is known for the Gamma case. We omit the details. □

5   Model B

Throughout this section, we assume $T \equiv t$ and consider the case of deterministic but not equidistant checkpoints $0 = t_0 < t_1 < \cdots < t_{K-1} < t$. The length of the kth checkpoint interval is then $h_k = t_k - t_{k-1}$, $k = 1, \ldots, K$, where $t_K = t$. Not surprising is the asymptotics of X then determined by the largest h_k:

Theorem 4. *Assume* $T \equiv t$ *and* $h_k \neq h_\ell$ *for* $k \neq \ell$, *and define* $k^* = \arg\max h_k$. *Then* $\mathbb{P}(X_B > x) \sim C_{B,1} e^{-\gamma(h_{k^*})x}$, *where*

$$C_{B,1} = C_{R,2}(h_{k^*}) \prod_{k \neq k^*} \widehat{H}_R[\gamma(h_{k^*}) \mid h_k].$$

[For $\widehat{H}_R[\cdot|\cdot]$, see (2)]. In the proof, we need the following elementary lemma:

Lemma 2. *Let $U_1, U_2 \geq 0$ be independent r.v.'s such that $\mathbb{P}(U_1 > x) \sim Ce^{-\eta x}$ and $\mathbb{P}(U_2 > x) = o(e^{-\eta x})$ for some $\eta > 0$. Then*

$$\mathbb{P}(U_1 + U_2 > x) \sim Ce^{-\eta x}\mathbb{E}e^{\eta U_2}.$$

Proof. Clearly, $\mathbb{P}(U_1 + U_2 > x, U_2 > x) = o(e^{-\eta x})$ and hence

$$\frac{\mathbb{P}(U_1 + U_2 > x)}{e^{-\eta x}} \sim \int_0^x \frac{\mathbb{P}(U_1 > x - y)}{e^{-\eta(x-y)}} e^{\eta y} \, \mathbb{P}(U_2 \in dy) \rightarrow \int_0^\infty Ce^{\eta y} \, \mathbb{P}(U_2 \in dy)$$

by dominated convergence.   □

Proof of Theorem 4. The contributions $X_{R,1}(h_1), \ldots, X_{R,K}(h_K)$ from the individual checkpoint intervals are independent, have tails which decay exponentially at rates $\gamma(h_1), \ldots, \gamma(h_k)$, and m.g.f.'s $\widehat{H}_R[\cdot|h_1], \ldots, \widehat{H}_R[\cdot|h_K]$. Now just use Lemma 2 inductively, starting by $X_{R,k^*}(h_{k^*})$ and adding one $X_{R,k}(h_k)$ with $k \neq k^*$ at a time.   □

The assumption that $h_k \neq h_\ell$ for $k \neq \ell$ is not essential: if some h_k have multiplicities > 1, it is straightforward to modify the result and proof by combining with what has been shown for Model A. The asymptotics takes the form $x^{M-1}e^{-\gamma(h^*)x}C_{B,3}$ where M is the multiplicity of h^*. We omit the details for the ease of exposition.

6   Model C

We next consider a random set-up. Recall that $g(\cdot)$ is the density of the failure distribution.

Theorem 5. *Assume $T \equiv t$ and that the checkpoints are $0, t$ and $K - 1$ i.i.d. uniform r.v.'s on $(0, t)$. That is, the checkpoints are $0 = t_0 < t_1 < \cdots < t_{K-1} < t_K = t$ where $t_1 < \cdots < t_{K-1}$ are the order statistics of $K - 1$ i.i.d. uniform r.v.'s $V_1, \ldots, V_{K-1}$ on $(0, t)$. Assume that g is left continuous at t with $g(t) > 0$ and let*

$$C_{C,1}(t) = \frac{C_{R,2}(t)\Gamma(K)}{C_{C,2}(t)^{K-1}t^{K-1}}, \quad C_{C,2}(t) = e^{\gamma(t)t}g(t)/m_R(t).$$

Then

$$\mathbb{P}(X_C > x) \sim C_{C,1}(t)\frac{e^{-\gamma(t)x}}{x^{K-1}} \tag{7}$$

The implication is that in this setting, the effects of checkpointing for large processing times is minimal, reduction from order $e^{-\gamma(t)x}$ in the case of no checkpointing to order $e^{-\gamma(t)x}/x^{K-1}$. The heuristics behind this is clear: in the notation of Model B, h_{k^*} is now random but has t as upper endpoint of its support, so that the tail of X must decay slower than $e^{-\gamma(s)x}$ for any $s < t$. On the other hand, the decay can not be faster than exponential at rate $\gamma(t)$. What may not be obvious is the modification of the exponential decay by the factor $1/x^{K-1}$ (and of course the form of $C_{C,1}$!).

The uniformity assumption in Theorem 5 is maybe the first possible model for random checkpoints that comes to mind, but of course, there are other possibilities depending on the context. Thus, Theorem 5 should largely be seen as a first example on random checkpoints.

Proof of Theorem 5. The heuristics above indicate that the form of the distribution of h_{k^*} close to t must play a key role. It is a classical result in order statistics, going all the way back to Fisher [16] (see also Feller [15] and [12]), that this distribution can be found in closed form: for $t = 1$, the tail is

$$(K-1)(1-y)^{K-2} - \binom{K-1}{2}(1-2y)^{K-2} + \cdots + (-1)^{i-1}\binom{K-1}{i}(1-iy)^{K-2}$$

where the final index i is the last with $1 - iy > 0$. This result is not trivial at all, but we are only concerned with the upper tail (y close to 1), and here a simple direct argument is available. Indeed, since the distribution of the set $V_1, \ldots, V_{K-1}$ is exchangeable, one gets for $y > t - t/K$ and a general t the density of h_{k^*} as

$$\mathbb{P}(h_{k^*} \in dy) = (K-1)\mathbb{P}(V_1 \in dy, k^* = 1)$$
$$= (K-1)\mathbb{P}(V_1 \in dy, V_2 > y, \ldots, V_{K-1} > y) = \frac{K-1}{t^{K-1}}(t-y)^{K-2}.$$

In Corollary 3.1 of [5], a uniformity property of the Cramér-Lundberg asymptotics is noted (this relies on uniform renewal theorems, cf. Kartashov [22], [21] and Wang & Woodroofe [31]. Together with some easy extension of the arguments of Section 5 this can be seen to imply that Theorem 4 holds uniformly in h_{k^*}. It follows as in the proof of Theorem 4 that

$$\int_{t/K}^{t} \mathbb{P}(h_{k^*} \in dy)\mathbb{P}(X_C > x \mid h_{k^*} = y)\, dy = \int_{t/K}^{t} \mathbb{P}(h_{k^*} \in dy)c_{C,1}(y)e^{-\gamma(y)x}(1+o(1)),$$

where the o(1) term is uniform in y and

$$c_{C,1}(y) = C_{R,2}(y)\mathbb{E}\Big[\prod_{k \neq k^*} \widehat{H}_R[\gamma(y) \mid h_k] \Big| h_{k^*} = y\Big].$$

Here $c_{C,1}(y) \to C_{R,2}(t)$ as $y \uparrow t$ because $h_k \leq t - y$ when $h_{k^*} = y$ and $\widehat{H}_R[s \mid a] \to 1$ uniformly in $a \leq a_0$ as $s \downarrow 0$.

It is shown in [5] by a simple Taylor expansion that

$$\gamma(y) - \gamma \sim (t-y)C_{C,2} \quad \text{as } y \uparrow t. \tag{8}$$

Choose $t_1 < t$ with

$$\gamma(t_1) > \gamma(t), \quad c_{C,1}(y) \leq (1+\epsilon)C_{R,2}(t), \gamma(y) \geq (1-\epsilon)(t-y)C_{C,2}$$

for $t_1 < y \leq t$. We then get

$$\mathbb{P}(X_C > x) = \int_{t_1}^{t} \mathbb{P}(h_{k^*} \in dy) c_{C,1}(y) e^{-\gamma(y)x} (1 + o(1)) \, dy + O(e^{-\gamma(t_1)x}),$$

where the integral can be bounded above by

$$(1 + o(1))(1 + \epsilon) C_{R,2}(t)(K - 1) \int_{t_1}^{t} \frac{(t - y)^{K-2}}{t^{K-1}} e^{-(1-\epsilon)C_{C,2}(t-y)x} \, dy$$

$$= (1 + o(1)) \frac{(1 + \epsilon) C_{R,2}(t)(K - 1)}{[(1 - \epsilon) C_{C,2} t x]^{K-1}} e^{-\gamma(t)x} \int_{0}^{(1-\epsilon)C_{C,2}(t-t_1)x} z^{K-2} e^{-z} \, dz$$

$$= (1 + o(1)) \frac{(1 + \epsilon) C_{R,2}(t)(K - 1)\Gamma(K - 1)}{[(1 - \epsilon) C_{C,1} t x]^{K-1}} e^{-\gamma(t)x},$$

where we substituted $z = (1 - \epsilon)C_{C,2}(t - y)x$. This shows that

$$\limsup_{x \to \infty} \frac{x^{K-1}}{e^{-\gamma(t)x}} \mathbb{P}(X_C > x) \leq \frac{1 + \epsilon}{(1 - \epsilon)^{K-1}} C_{C,1}.$$

Letting $\epsilon \downarrow 0$ and combining with a similar lower bound completes the proof. □

7 Orderings for Models A, B, C

In models A, B, C, we have $T \equiv t$. In this section, we study comparisons of the total RESTART time $X_R(t)$ with checkpointing total times $X_A(t), X_B(t), X_C(t)$ for the different models.

For a fixed K, one expects Model A to be preferable to Model B, which in turn as a minimum should improve RESTART. With expected values as performance measure, this means mathematically that

$$K\mathbb{E}X_R(t/K) \leq \sum_{k=0}^{K} \mathbb{E}X_R(h_k) \leq \mathbb{E}X_R(t) \tag{9}$$

when $h_0 + \cdots + h_K = t$. However, depending on t and the failure distribution G, this is not always the case:

Example 1. Let $t = 2$, $K = 2$ and let $g(u) > 0$ be arbitrary on $(0, 1]$, $g(u) = 0$ on $(1, 2]$. Then $\overline{G}(1) = \overline{G}(2)$ and so by (2)

$$2\mathbb{E}X_R(1) = 2\left(1 + \frac{1}{\overline{G}(1)} \int_{0}^{1} yg(y) \, dy\right) = 2 + \frac{2}{\overline{G}(2)} \int_{0}^{2} yg(y) \, dy$$

$$= \mathbb{E}X_R(2) + \frac{1}{\overline{G}(2)} \int_{0}^{2} yg(y) \, dy > \mathbb{E}X_R(2).$$

By a slight perturbation, the counterexample is easily modified to a $g(u)$ that is strictly positive. □

Remark 1. The intuition behind Example 1 is of course that checkpointing at 1 forces taking the high risk of early failures in $[0,1]$ twice because the failure mechanism regenerates at each checkpoint. Instead, simple RESTART benefits from the low risk in the task period $(1,2]$ of the execution of the task.

Intuitively, one expects checkpointing at $t_1 = 1/2$ (Model B) to be preferable to $t = 1$ (Model A), and the analysis below shows that indeed this is the case for example if $g(u) = e^{-u}$ for $0 < u \leq 2$.

Continuing with analysing the case $K = 2$, let $\varphi(s) = \mathbb{E}X_R(s)$. Equation (9) can then be rewritten

$$2\varphi(t/2) \leq \varphi(s) + \varphi(t-s) \leq \varphi(t) \tag{10}$$

for $0 < s < t$. Since $\varphi(s) + \varphi(t-s)$ is symmetric around $s = 1/2$, the function must have either a local minimum (when $\varphi''(t/2) \geq 0$) or a local maximum (when $\varphi''(t/2) \leq 0$) at $t/2$, and Example 1 shows that $\varphi''(t/2) \leq 0$ may indeed arise.

The observation that equidistant checkpointing is not always optimal is not new. For example, it appears already in Tantawi & Ruschitzka [30].   □

A simple sufficient condition avoding the above pathologies is the following result formulated in the more general framework of the stochastic ordering $\preceq_{st}$. It covers, e.g., an exponential G, i.e. Poisson.

Theorem 6. *Assume that the failure rate $\mu(t) = g(t)/\overline{G}(t)$ of G is non-decreasing. Then $X_A(t) \preceq_{st} X_B(t) \preceq_{st} X_R(t)$. In particular, equation (9) on ordering of means holds. Further, $X_A(t) \preceq_{st} X_C(t) \preceq_{st} X_R(t)$.*

Recall (e.g. Müller & Stoyan [27]) that $X \preceq_{st} Y$ can, for example, be defined by the requirement that X, Y can be defined on a common probability space such that $X \leq Y$ a.s.). It implies, of course, ordering of means, but also ordering of tails. That the Models A, B ,C are tail ordered follows already by comparisons of Theorems 3, 4, 5 since $\gamma(t)$ is non-increasing in t.

The key step in the proof is the following coupling lemma. Here we consider two RESTART total times $X'_R(t), X''_R(t)$ with the same t, but failure time distributions G', G'' such that the corresponding failure rates satisfy $\mu'(s) \leq \mu''(s)$ for all $s \leq t$. Let M', M'' be the number of failures (restarts) and U'_i, U''_j the failure times, $i = 1, \ldots, M', j = 1, \ldots, M''$.

Lemma 3. *The G' and G'' systems can be defined on a common probability space in such a way that $M' \leq M''$ and*

$$\{U'_i\}_{i \leq M'} \subseteq \{U''_j\}_{j \leq M''}. \tag{11}$$

Proof. Assume given the G''-system. At the first failure time U''_1, we then flip a coin coming up heads w.p. $\mu'(U''_1)/\mu''(U''_1)$. If a head comes up, we take $U'_1 = U''_1$. Otherwise, the G'' system will restart, and after a number of restarts, the elapsed time since the last restart will reach U''_1 for the first time. If a further failure occurs before t, say at U''_{k_1}, a new coin flip is performed w.p. $\mu'(U''_{k_1})/\mu''(U''_{k_1})$

for heads. If a head comes up, we take $U_1' = U_{k_1}''$. Otherwise, the G'' system will restart, and after a number of restarts, the elapsed time since the last restart will reach U_{k_1}'' for the first time, and the procedure is repeated. The construction ensures that the first failure of the G' system occurs at rate $\mu'(\cdot)$, as should be. An illustration is in Fig. 2. The ● mark G''-failure times at which a coin flip is performed unsuccesfully, the * the succesful one (if any), and the ○ the ones at which no coin flipping at all is performed.

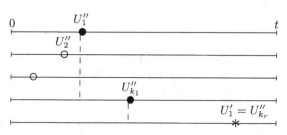

Fig. 2. Construction of U_1'

We face two possibilities: there are no successful coin flips before the G'' system has completed its task. In this case, the conclusion of the lemma is obvious (the l.h.s. of (11) is just the empty set and $M' = 0$). The second is that of a successful coin flip before completion as on Fig. 2. Then $M' \geq 1$, U_1' corresponds to precisely one U_{k_r}'' and so U_1' belongs to the set of the r.h.s. of (11). Further, after the successful coin flip both systems restart, and the whole construction can be repeated to either construct U_2' as a member of the r.h.s. of (11) or to conclude that the G' system have reached completion, that $M' = 1 \leq M''$, and that the conclusion of the lemma holds. Continuing in this manner until completion completes the proof. □

Proof of $X_B(t) \preceq_{st} X_R(t)$. Let $K = 2$, $t = u + v$, $h_1 = u$, $h_2 = v$. Then $X_B(t) \stackrel{\mathcal{D}}{=} X_R(u) + X_R(v)$ and

$$X_R(t) \stackrel{\mathcal{D}}{=} X_R(u) + v + \sum_{i=1}^{M(u,v)} \left[U_i^{\#}(u,v) + X_{R,i}(u) \right] \qquad (12)$$

where $M(u,v)$ is the number of restarts in simple RESTART with t replaced by v, $\mu(\cdot)$ changed to $\mu(\cdot + u)$, $U_1^{\#}(u,v)$, $U_2^{\#}(u,v), \ldots$ the corresponding failure times, $X_{R,1}(u,v)$, $X_{R,2}(u,v), \ldots$ replicates of the corresponding RESTART total time, and obvious independence assumptions apply. Indeed, the part u of the total task time must first be completed which gives the term $X_R(u)$. At each failure after that, we must again reach u which explains the presence of the $X_{R,i}(u)$ in the sum, and finally v is the contribution from the final successful passage from u to t.

Now since $\mu(\cdot)$ is non-decreasing, Lemma 3 implies $M(0,v) \preceq_{\mathrm{st}} M(u,v)$ and that we may assume $M(0,v) \leq M(u,v)$ and

$$\left\{U_1^{\#}(0,v)\right\}_{i \leq M(0,v)} \subseteq \left\{U_1^{\#}(u,v)\right\}_{i \leq M(u,v)}.$$

It follows that

$$X_{\mathrm{R}}(t) \succeq_{\mathrm{st}} X_{\mathrm{R}}(u) + v + \sum_{i=1}^{M(0,v)} U_i^{\#}(0,v) \overset{\mathcal{D}}{=} X_{\mathrm{R}}(u) + X_{\mathrm{R}}(v) \overset{\mathcal{D}}{=} X_{\mathrm{B}}(t).$$

This proves $X_{\mathrm{B}}(t) \preceq_{\mathrm{st}} X_{\mathrm{R}}(t)$ for $K = 2$. For $K = 3$, we then get

$$X_{\mathrm{B}}(t) \overset{\mathcal{D}}{=} X_{\mathrm{R}}(h_1) + X_{\mathrm{R}}(h_2) + X_{\mathrm{R}}(h_3) \preceq_{\mathrm{st}} X_{\mathrm{R}}(h_1 + h_2) + X_{\mathrm{R}}(h_3)$$
$$\preceq_{\mathrm{st}} X_{\mathrm{R}}(h_1 + h_2 + h_3) = X_{\mathrm{R}}(t).$$

Continuing in this manner yields $X_{\mathrm{B}}(t) \preceq_{\mathrm{st}} X_{\mathrm{R}}(t)$ for all K.   □

Proof of $X_{\mathrm{A}}(t) \preceq_{\mathrm{st}} X_{\mathrm{B}}(t)$. Let first $K = 2$, $h_1 = s < h_2 = t - s$ (thus $s < t/2 < t - s/2$). Then as in (12),

$$X_{\mathrm{R},1}(t/2) \overset{\mathcal{D}}{=} X_{\mathrm{R}}(s) + t/2 - s + \sum_{i=1}^{M(s,t/2-s)} \left\{U_i^{\#}(s,t/2-s) + X_{\mathrm{R},i}(t/2-s)\right\},$$

$$X_{\mathrm{R}}(t-s) \overset{\mathcal{D}}{=} X_{\mathrm{R},2}(t/2) + t/2 - s + \sum_{i=1}^{M(t/2,t/2-s)} \left\{U_i^{\#}(t/2,t/2-s) + X_{\mathrm{R},i}(t/2-s)\right\}$$

By Lemma 3, we may assume $M(s, t/2-s) \leq M(t/2, t/2-s)$ and, reordering if necessary, that the $U_i^{\#}(s, t/2-s)$ and $X_{\mathrm{R},i}(t/2-s)$ coincide for $i \leq M(s, t/2-s)$. It follows that

$$X_{\mathrm{B}}(t) \overset{\mathcal{D}}{=} X_{\mathrm{R}}(s) + X_{\mathrm{R},1}(t/2) + t/2 - s$$
$$+ \sum_{i=1}^{M(t/2,t/2-s)} \left\{U_i^{\#}(t/2,t/2-s) + X_{\mathrm{R},i}(t/2-s)\right\}$$

$$\geq X_{\mathrm{R}}(s) + X_{\mathrm{R},1}(t/2) + t/2 - s + \sum_{i=1}^{M(s,t/2-s)} \left\{U_i^{\#}(s,t/2-s) + X_{\mathrm{R},i}(t/2-s)\right\}$$

$$= X_{\mathrm{R},1}(t/2) + X_{\mathrm{R},2}(t/2) \overset{\mathcal{D}}{=} X_{\mathrm{A}}(t),$$

completing the proof for $K = 2$.

For $K > 2$, one could believe that an extension of the argument or a simple induction argument as in the proof of $X_{\mathrm{B}}(t) \preceq_{\mathrm{st}} X_{\mathrm{R}}(t)$ would apply, but we were not able to proceed along these lines. We first illustrate our alternative proof for $K = 3$. We may assume $h_1 \leq h_2 \leq h_3$ and define $h_{0,i} = h_i$ for $i = 1, 2, 3$ and $\Delta_0 = h_3 - h_1$. Using the result for $K = 2$ twice, we obtain

$$X_{\mathrm{B}}(t) = X_{\mathrm{R}}(h_1) + X_{\mathrm{R}}(h_2) + X_{\mathrm{R}}(h_3)$$
$$\succeq_{\mathrm{st}} X_{\mathrm{R}}\big((h_1 + h_2)/2\big) + X_{\mathrm{R}}\big((h_1 + h_2)/2\big) + X_{\mathrm{R}}(h_3)$$
$$\succeq_{\mathrm{st}} X_{\mathrm{R}}\big((h_1 + h_2)/2\big) + X_{\mathrm{R}}\big((h_1 + h_2)/4 + h_3/2\big) + X_{\mathrm{R}}\big((h_1 + h_2)/4 + h_3/2\big),$$

with the understanding that all r.v.'s are independent. We therefore obtain a set

$$h_{1,1} = (h_1 + h_2)/2, \quad h_{1,2} = h_{1,3} = (h_1 + h_2)/4 + h_3/2$$

with $h_{1,1} \le h_{1,2} \le h_{1,3}$ and satisfying

$$X_B(t) \preceq_{st} X_R(h_{1,1}) + X_R(h_{1,2}) + X_R(h_{1,3}).$$
$$\Delta_1 = h_{1,3} - h_{1,1} \le 3h_3/4 + h_1/4 - h_1 = 3\Delta_0/4.$$

See Fig. 3.

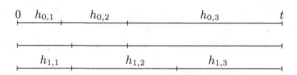

Fig. 3. $K = 3$: first recursive step

It follows by induction that there exists a set $h_{n,1} \le h_{n,2} \le h_{n,3}$ satisfying

$$t = h_{n,1} + h_{n,2} + h_{n,3}, \tag{13}$$
$$X_B(t) \preceq_{st} X_R(h_{n,1}) + X_R(h_{n,2}) + X_R(h_{n,3}). \tag{14}$$
$$\Delta_n = h_{n,3} - h_{n,1} \le \left(\frac{3}{4}\right)^n \Delta_0. \tag{15}$$

It follows from (15) and 13 that $h_{n,i} \to t/3$, $i = 1, 2, 3$. Hence (14) combined with easy continuity properties gives

$$X_B(t) \preceq_{st} X_R(t/3) + X_R(t/3) + X_R(t/3) \overset{\mathcal{D}}{=} X_A(t).$$

For a general $K > 2$, the same construction is easily seen to provide a set $h_{n,1} \le \cdots \le h_{n,K}$ satisfying $h_{0,i} = h_i$ and

$$t = h_{n,1} + \cdots \le h_{n,K},$$
$$X_B(t) \preceq_{st} X_R(h_{n,1}) + \cdots + X_R(h_{n,K}).$$
$$\Delta_n = h_{n,K} - h_{n,1} \le (1 - 2^{K-1})^n \Delta_0,$$

where $\Delta_0 = h_K - h_1$. The rest of the argument is then as for $K = 3$.   □

Proof of $X_A(t) \preceq_{st} X_C(t) \preceq_{st} X_R(t)$. Let $X_C(t|v_1, \ldots, v_{K-1})$ have the conditional distribution of $X_C(t)$ given $V_1 = v_1, \ldots, V_{K-1} = v_{K-1}$. This is a special case of Model B, so

$$X_A(t) \preceq_{st} X_C(t|v_1, \ldots, v_{K-1}) \preceq_{st} X_R(t).$$

Now just note that $\preceq_{st}$ is closed under mixtures.   □

8   Numerical Examples for Models A, B, C

In the numerical examples, we took $T \equiv t = 1$ and assumed exponential failures, $g(u) = \mu e^{-\mu u}$. We considered the following scenarios: (a) Model A, $K = 1$; (b) Model A, $K = 2$; (c) Model A, $K = 4$; (d) Model A, $K = 16$; (e) Model B, $K = 2$, $t_1 = 0.4$; (f) Model B, $K = 2$, $t_1 = 0.2$; (g) Model B, $K = 4$, $t_1 = 0.3$, $t_2 = 0.5$, $t_3 = 0.85$; (h) Model C, $K = 2$; (i) Model C, $K = 4$; (j) Model C, $K = 16$. Thus, (a) is the simple RESTART setting, the benchmark with which the gain of checkpointing in the other scenarios is to be compared. For each scenario, we considered exponential(μ) failure distributions with μ chosen such that $e^\mu - 1$ took the values $10, 5, 1, 0.5, 0.1, 0.01, 0.001$ (meaning $\mu = 2.398, 1.792, 0.693, 0.405, 0.0953$, resp. 0.00995). Note that (cf. [5]) $e^\mu - 1$ has a simple interpretation as $\mathbb{E}N$ where N is the number of restarts. As performance measures, we computed first the expected total time $\mathbb{E}X$. In scenarios (a)–(g), this follows in a straigthforward way by expressing the total time as sums of independent RESTART times and using formula (2) for $\mathbb{E}X_R(t)$. In scenarios (h), (i), (j), $\mathbb{E}X_C(t)$ is in principle explicitly computable using known formulas for uniform spacings, but since these are complicated sums, we used simulation for the ease of computation.

Table 1. Expected values of the total time

$\mathbb{E}N$	10	5	1	0.5	0.1	0.01	0.001
(a)	4.17	2.79	1.44	1.23	1.05	1.01	1.00
(b)	1.93	1.62	1.20	1.11	1.02	1.00	1.00
(c)	1.37	1.26	1.09	1.05	1.01	1.00	1.00
(d)	1.08	1.06	1.02	1.01	1.00	1.00	1.00
(e)	2.01	1.66	1.21	1.11	1.03	1.00	1.00
(f)	2.68	2.02	1.28	1.15	1.03	1.00	1.00
(g)	1.42	1.30	1.10	1.06	1.01	1.00	1.00
(h)	2.65	2.00	1.28	1.15	1.03	1.00	1.00
(i)	1.77	1.50	1.16	1.09	1.02	1.00	1.00
(j)	1.16	1.12	1.04	1.02	1.01	1.00	1.00

It is seen that $\mathbb{E}X$ is very close to $t = 1$ even in the simple RESTART setting if $\mathbb{E}N$ is moderate or small so that the numbers for $\mathbb{E}X$ only show a notable advantage of checkpointing if $\mathbb{E}N$ is 10 or 5 or larger. For a comparison of the effects of unevenly spaced checkpoints, compare entry (b) with (e) and (f), and (c) with (g).

In contrast, one should see an effect of checkpointing also for small $\mathbb{E}N$ when considering the tail or equivalently the quantiles. We considered the and the $1 - \alpha$ quantile $q_{1-\alpha}$ for $\alpha = 0.05, 0.01, 0.001, 0.0001$ (that is, the exact value of $q_{1-\alpha}$ is the solution of $\mathbb{P}(X > q_{1-\alpha}) = \alpha$). For each combination of scenario, μ and quantile, the first number (in Roman) given is computed by crude Monte Carlo (which is sufficient for the present purposes; note, however, that more sophisticated algorithms are available, cf. Asmussen [3]). The second number (in Italics) is the one provided by our approximations. The numbers are given in the following four tables, corresponding in lexicograhical order to $\alpha = 0.05, 0.01, 0.001, 0.0001$.

As we see it, some main conclusions to be drawn are: (a) checkpointing is less efficient in reducing the probability of long processing times than one maybe would have expected; (b) our approximations are reasonably precise.

EN	10	5	1	0.5	0.1	0.01	0.001
(a)	11.1	6.90	2.84	2.16	1.47	1.00	1.00
	11.0	6.90	2.85	2.18	1.49	1.20	1.10
(b)	3.57	2.83	1.78	1.50	1.24	1.00	1.00
	3.60	2.87	1.84	1.60	1.30	1.15	1.09
(c)	1.94	1.73	1.36	1.24	1.11	1.00	1.00
	2.13	1.90	1.51	1.39	1.23	1.13	1.09
(d)	1.19	1.15	1.08	1.06	1.03	1.00	1.00
	1.47	1.41	1.29	1.24	1.17	1.12	1.09
(e)	3.92	3.03	1.84	1.57	1.23	1.00	1.00
	3.93	3.03	1.84	1.58	1.26	1.11	1.05
(f)	6.34	4.42	2.24	1.78	1.27	1.00	1.00
	6.34	4.41	2.23	1.81	1.36	1.15	1.07
(g)	2.16	1.87	1.41	1.30	1.12	1.00	1.00
	2.22	1.93	1.46	1.33	1.16	1.07	1.03
(h)	6.54	4.43	2.20	1.79	1.27	1.00	1.00
(i)	3.46	2.68	1.68	1.48	1.15	1.00	1.00
(j)	1.48	1.37	1.19	1.14	1.03	1.00	1.00

EN	10	5	1	0.5	0.1	0.01	0.001
(a)	16.4	10.2	4.01	2.97	1.93	1.00	1.00
	16.5	10.2	4.01	2.98	1.92	1.45	1.28
(b)	4.72	3.68	2.20	1.87	1.46	1.00	1.00
	4.74	3.71	2.26	1.92	1.49	1.26	1.17
(c)	2.34	2.03	1.54	1.41	1.23	1.00	1.00
	2.46	2.17	1.66	1.52	1.31	1.18	1.13
(d)	1.26	1.21	1.12	1.10	1.06	1.00	1.00
	1.51	1.45	1.31	1.27	1.19	1.13	1.10
(e)	5.39	4.07	2.33	1.95	1.52	1.00	1.00
	5.38	4.07	2.33	1.94	1.48	1.24	1.15
(f)	9.33	6.36	3.02	2.38	1.72	1.00	1.00
	9.30	6.35	3.01	2.38	1.67	1.34	1.21
(g)	2.66	2.27	1.64	1.48	1.28	1.00	1.00
	2.70	2.31	1.67	1.50	1.27	1.14	1.09
(h)	10.6	6.90	3.06	2.40	1.71	1.00	1.00
(i)	5.54	3.40	2.21	1.84	1.45	1.00	1.00
(j)	1.77	1.61	1.32	1.25	1.14	1.00	1.00

EN	10	5	1	0.5	0.1	0.01	0.001
(a)	24.1	14.8	5.64	4.13	2.57	1.91	1.00
	24.4	15.0	5.67	4.13	2.53	1.80	1.53
(b)	6.59	4.84	2.78	2.31	1.74	1.45	1.00
	-	4.87	2.84	2.36	1.76	1.43	1.29
(c)	2.80	2.43	1.78	1.61	1.35	1.23	1.00
	2.91	2.54	1.88	1.70	1.43	1.26	1.18
(d)	1.32	1.28	1.18	1.14	1.08	1.06	1.00
	1.56	1.49	1.35	1.30	1.21	1.14	1.11
(e)	7.50	5.54	3.03	2.47	1.78	1.50	1.00
	7.48	5.56	3.03	2.47	1.80	1.44	1.30
(f)	13.4	9.18	4.17	3.20	2.13	1.70	1.00
	13.5	9.13	4.14	3.19	2.13	1.61	1.41
(g)	3.36	2.83	1.97	1.73	1.41	1.28	1.00
	3.38	2.84	1.98	1.75	1.43	1.25	1.17
(h)	16.8	10.8	4.42	3.35	2.14	1.69	1.00
(i)	9.61	6.51	3.14	2.49	1.75	1.44	1.00
(j)	2.36	2.04	1.56	1.43	1.26	1.13	1.00

EN	10	5	1	0.5	0.1	0.01	0.001
(a)	31.6	19.4	7.40	5.24	3.10	1.99	1.89
	32.3	19.7	7.33	5.28	3.14	2.15	1.78
(b)	8.48	6.02	3.34	2.77	1.97	1.50	1.45
	-	-	3.40	2.79	2.02	1.59	1.41
(c)	3.17	2.79	2.01	1.79	1.47	1.25	1.22
	3.34	2.89	2.10	1.87	1.54	1.34	1.24
(d)	1.39	1.35	1.22	1.18	1.11	1.06	1.06
	1.62	1.54	1.38	1.33	1.23	1.16	1.12
(e)	9.36	7.04	3.78	2.99	2.10	1.59	1.50
	9.57	7.05	3.73	3.00	2.11	1.63	1.44
(f)	17.1	11.8	5.25	4.07	2.54	1.79	1.71
	17.6	11.9	5.27	4.00	2.58	1.88	1.61
(g)	4.07	3.44	2.28	1.99	1.58	1.34	1.28
	4.06	3.38	2.29	1.99	1.59	1.35	1.25
(h)	23.3	14.8	5.93	4.38	2.67	1.91	1.68
(i)	13.9	9.43	4.22	3.26	2.20	1.68	1.43
(j)	3.20	2.63	1.86	1.65	1.39	1.25	1.13

Remark 2. Giving confidence limits for quantiles is possible, but tedious, cf. [4] III.4a, III.5.3. We have not implemented this. To get an idea of the order of the uncertainly, we instead repeated a few of the simulation runs 5 time. The results are given in Table 2 and indicate that the simulation results are reasonably precise.

Table 2. Indications of uncertainty on quantiles

(a), $EN = 10$, $\alpha = 0.05$	11.06	11.05	11.03	11.06	11.05
(a), $EN = 0.001$, $\alpha = 0.05$	1.00	1.00	1.00	1.00	1.00
(a), $EN = 10$, $\alpha = 0.0001$	31.56	31.55	31.71	31.60	32.04
(a), $EN = 0.001$, $\alpha = 0.0001$	1.894	1.912	1.896	1.903	1.895

9 Model D

Next consider a random T with infinite support and equally spaced checkpoints at $0, h, 2h, \ldots, h\lfloor T/h \rfloor$, with the same regeneration assumption as above.

The key representation is the independent sum

$$X_{\mathrm{D}} = S_N + X_{\mathrm{R}}(T - Nh) \quad \text{where} \quad S_N = X_{\mathrm{R},1}(h) + \cdots + X_{\mathrm{R},N}(h), \quad (16)$$

where $N = \lfloor T/h \rfloor$. The easy case when deriving the asymptotics for this expression is the heavy-tailed one [recall the expression (2) for $m_{\mathrm{R}}(t) = \mathbb{E}X_{\mathrm{R}}(t)$]:

Theorem 7. *Assume that the distribution of T is regularly varying, $\mathbb{P}(T > t) = L(t)/x^\alpha$ with $\alpha > 0$ and $L(\cdot)$ slowly varying. Then*

$$\mathbb{P}(X_{\mathrm{D}} > x) \sim \mathbb{P}\big(T > xh/m_{\mathrm{R}}(h)\big) \sim \frac{m_{\mathrm{R}}(h)^\alpha L(x)}{h^\alpha x^\alpha}. \quad (17)$$

Proof. Clearly, $\mathbb{P}(N \geq n) = \mathbb{P}(T \geq nh)$. Therefore the term S_N is a light-tailed random walk sampled at a regularly varying time, and therefore by Asmussen, Klüppelberg & Sigman [6] $\mathbb{P}(S_N > x) \sim \mathbb{P}(N > x/m_{\mathrm{R}}(t))$ which is the same asymptotics as is claimed in (17) (the last identity follows from L being slowly varying). In particular, S_N is subexponential and since $X_{\mathrm{R}}(T - Nh)$ is light-tailed and independent of S_N, this implies that $S_N + X_{\mathrm{R}}(T - Nh)$ has the same tail asymptotics as S_N. This completes the proof. □

In the light-tailed case, we will consider the example of a gamma-like T:

Theorem 8. *Assume that the density $f(t)$ of T satisfies $f(t) \sim ct^\alpha e^{-\lambda t}$ as $t \to \infty$ where $-\infty < \alpha < \infty$. Then $\mathbb{P}(X_{\mathrm{D}} > x) \sim C_{\mathrm{D},1}e^{-\gamma_2 x}$ where $\gamma_2 = \gamma_2(h)$ is the solution of $\widehat{H}_{\mathrm{R}}[\gamma_2] = e^{\lambda h}$ and $C_{\mathrm{D},1}$ a constant.*

[$C_{\mathrm{D},1}$ can be evaluated explicitly by collecting expressions given below; for simplicity, we omit the details]. Note that since $\widehat{H}_{\mathrm{R}}[s|h]$ has a singularity at $s = \gamma(h)$, one has $\gamma_2(h) < \gamma(h)$, as was to be expected since the tail of each individual $X_{\mathrm{R},k}(h)$ decays exponentially at rate $\gamma(h)$.

Proof. It is easy to see that $\mathbb{P}(N \geq n) = \mathbb{P}(T \geq nh) \sim c_1 n^\alpha \rho^n$ where $c_1 = ch^\alpha/\lambda$, $\rho = e^{-\lambda h}$, and that

$$\lim_{n \to \infty} \mathbb{P}\big(T - Nh \leq y \,\big|\, N \geq n\big) = \frac{1 - e^{-\lambda y}}{1 - e^{-\lambda h}}, \quad 0 \leq y \leq h$$

(that is, the asymptotic distribution is exponential truncated to $(0, h)$). By Lemma 1, we therefore are able to evaluate the asymptotic tail of.

$$\mathbb{P}\big(S_N + X_R(T - Nh) > x \,\big|\, N = n\big)\,,.$$

The proof is completed by an extension to random sums with $\mathbb{P}(N = n) \sim cn^{\alpha-1}\rho^n$ of the Cramér-Lundberg asymptotics for geometric sums (the Cramér-Lundberg theory is the case $\alpha = 1$) given in the concluding remarks of Embrechts et al. [14].

10   Model E

We assume in this section that T is random with infinite support and that the checkpoints are given by $t_k = t'_k T$ for a deterministic set of constants $0 = t'_0 < t'_1 < \ldots < t'_{K-1} < 1$.

The most natural case is probably that of equally spaced checkpoints, $t'_k = k/K$. As in [5], one can show that the contribution to $\mathbb{P}(X(T) > x)$ from the event $\{T \leq t^*\}$ is negligible for any $t^* < \infty$. Replacing the root $\gamma(s)$ by its aymptote $\mu\overline{G}(s)$ for large s and $m(s)$ by its asymptote $m(\infty)$, similar estimates as in [5] combined with Theorem 3, then yields the following extension of Lemma 1.1 of [5] (recall that $h^t_K = t/K$):

Lemma 4. *Assume* $t_k = kT/K$. *Let* $\mu = 1/\mathbb{E}U$ *and define*

$$I_\pm(x, \epsilon) = \int_{t^*}^{\infty} \frac{\overline{G}(h^t_K)^{K-1}}{(K-1)!m(\infty)^{K-1}} \cdot \exp\big\{-\mu\overline{G}(h^t_K)x(1 \pm \epsilon)\big\} f(t)\, dt$$

Then for each $\epsilon > 0$,

$$1 - \epsilon \leq \liminf_{x \to \infty} \frac{\overline{H}(x)}{I_+(x, \epsilon)} \leq \limsup_{x \to \infty} \frac{\overline{H}(x)}{I_-(x, \epsilon)} \leq 1 + \epsilon.$$

From this we get the following exact asymptotics:

Theorem 9. *Assume* $t_k = kT/K$ *and that*

$$f(t) = g(h^t_K)\overline{G}(h^t_K)^{\beta-1}L_0\big(\overline{G}(h^t_K)\big)$$
$$= g\big(t/(K+1)\big)\overline{G}\big(t/(K+1)\big)^{\beta-1}L_0\big(\overline{G}(t/(K+1))\big) \tag{18}$$

for some β *and some function* $L_0(s)$ *that is slowly varying at* $s = 0$. *Then*

$$\overline{H}(x) \sim \frac{\Gamma(\beta + K - 1)}{(K-1)!m(\infty)^{K-1}} \frac{L_0(1/x)}{x^{\beta+K-1}\mu\beta + K - 1} \tag{19}$$

As in [5], the assumption (18) covers the case where F and G are not too different, in particular when they are both gamma-like or both regularly varying.

Proof of Theorem 9. Substituting $s = \overline{G}(t/(K+1))$ and using (18), we get

$$I_\pm(x, \epsilon) = \frac{1}{(K-1)!m(\infty)^{K-1}} \int\limits_0^{\overline{G}(t^*/(K+1))} \psi(s)\,\mathrm{d}s$$

where

$$\psi(s) = \frac{(K+1)\overline{G}(s)^{K-1}}{(K-1)!m(\infty)^{K-1}} \exp\left\{-\mu s x(1 \pm \epsilon)\right\} s^{\beta+K-2} L_0(s).$$

Then by Karamata's Tauberian theorem ([7, Theorems 1.5.11 and 1.7.1]),

$$\begin{aligned}
I_\pm &\sim \frac{\Gamma(\beta+K-1)}{(K-1)!m(\infty)^{K-1}} \frac{L_0\left(1/(x\mu(1 \pm \epsilon))\right)}{x^{\beta+K-1}\mu^{\beta+K-1}(1 \pm \epsilon)^\beta} \\
&\sim \frac{\Gamma(\beta+K-1)}{(K-1)!m(\infty)^{K-1}} \frac{L_0(1/x)}{x^{\beta+K-1}\mu\beta + K - 1(1 \pm \epsilon)^{\beta+K-1}}.
\end{aligned}$$

Let $\epsilon \downarrow 0$. □

References

1. Andersen, L.N., Asmussen, S.: Parallel computing, failure recovery and extreme values. J. Statist. Theory. Pract. 2, 279–292 (2008)
2. Asmussen, S.: Applied Probability and Queues, 2nd edn. Springer (2003)
3. Asmussen, S.: Importance sampling for failure probabilities in computing and data transmission. Journal Applied Probability 46, 768–790 (2009)
4. Asmussen, S., Glynn, P.W.: Stochastic Simulation: Algorithms and Analysis. Springer (2007)
5. Asmussen, S., Fiorini, P., Lipsky, L., Rolski, T., Sheahan, R.: On the distribution of total task times for tasks that must restart from the beginning if failure occurs. Math. Oper. Res. 33, 932–944 (2008)
6. Asmussen, S., Klüppelberg, C., Sigman, K.: Sampling at a subexponential time, with queueing applications. Stoch. Proc. Appl. 79, 265–286 (1999)
7. Bingham, N.H., Goldie, C.M., Teugels, J.L.: Regular Variation. Cambridge University Press (1987)
8. Bobbio, A., Trivedi, K.: Computation of the distribution of the completion time when the work requirement is a PH random variable. Stochastic Models 6, 133–150 (1990)
9. Castillo, X., Siewiorek, D.P.: A performance-reliability model for computing systems. In: Proc FTCS-10, Silver Spring, MD, pp. 187–192. IEEE Computer Soc. (1980)
10. Chimento Jr., P.F., Trivedi, K.S.: The completion time of programs on processors subject to failure and repair. IEEE Trans. on Computers 42(1) (1993)
11. Chlebus, B.S., De Prisco, R., Shvartsman, A.A.: Performing tasks on synchronous restartable message-passing processors. Distributed Computing 14, 49–64 (2001)
12. David, H.A.: Order Statistics. Wiley (1970)

13. DePrisco, R., Mayer, A., Yung, M.: Time-optimal message-efficient work performance in the presence of faults. In: Proc. 13th ACM PODC, pp. 161–172 (1994)
14. Embrechts, P., Maejima, M., Teugels, J.L.: Asymptotic behaviour of compound distributions. Astin Bulletin 15(1) (1985)
15. Feller, W.: An Introduction to Probability Theory and its Applications II, 2nd edn. Wiley (1971)
16. Fisher, R.A.: Tests of significance in harmonic analysis. Proc. Roy. Soc. A 125, 54–59
17. Gut, A.: Stopped Random Walks. Springer (1988)
18. Jelenković, P., Tan, J.: Can retransmissions of superexponential documents cause subexponential delays? In: Proceedings of IEEE INFOCMO 2007, Anchorage, pp. 892–900 (2007)
19. Jelenković, P., Tan, J.: Dynamic packet fragmentation for wireless channels with failures. In: Proc. MobiHoc 2008, Hong Kong, May 26-30 (2008)
20. Jelenković, P., Tan, J.: Characterizing heavy-tailed distributions induced by retransmissions. Adv. Appl. Probab. 45(1) (2013)
21. Kartashov, N.V.: A uniform asymptotic renewal theorem. Th. Probab. Appl. 25, 589–592 (1980)
22. Kartashov, N.V.: Equivalence of uniform renewal theorems and their criteria. Teor. Veoryuatnost. i Mat. Statist. 27, 51–60 (1982) (in Russian)
23. Kulkarni, V., Nicola, V., Trivedi, K.: On modeling the performance and reliability of multimode systems. The Journal of Systems and Software 6, 175–183 (1986)
24. Kulkarni, V., Nicola, V., Trivedi, K.: The completion time of a job on a multimode system. Adv. Appl. Probab. 19, 932–954 (1987)
25. Lipsky, L.: Queueing Theory. A Linear Algebraic Approach, 2nd edn. Springer (2008)
26. Lipsky, L., Doran, D., Gokhale, S.: Checkpointing for the RESTART problem in Markov networks. J. Appl. Probab. 48A, 195–207 (2011)
27. Müller, A., Stoyan, D.: Comparison Methods for Stochastic Models and Risks. Wiley (2002)
28. Nicola, V.F.: Checkpointing and the modeling of program execution time. In: Lyu, M.R. (ed.) Software Fault Tolerance, ch. 7, pp. 167–188 (1995)
29. Nicola, V.F., Martini, R., Chimento, P.F.: The completion time of a job in a failure environment and partial loss of work. In: Proceedings of the 2nd International Conference on Mathematical Methods in Reliabiliy (MMR 2000), Bordeaux, pp. 813–816 (2000)
30. Tantawi, A.N., Rutschitzka, M.: Performance analysis of checkpointing strategies. ACM Trans. Comp. Syst. 2, 123–144 (1984)
31. Wang, M., Woodroofe, M.: A uniform renewal theorem. Sequential Anal. 15, 21–36 (1996)

Author Index